# Diagnosis, Epidemiology, and Vaccine Development of Swine Viral Diseases

# Diagnosis, Epidemiology, and Vaccine Development of Swine Viral Diseases

Guest Editors

**Wentao Li**
**Yifei Lang**

MDPI

Basel • Beijing • Wuhan • Barcelona • Belgrade • Novi Sad • Cluj • Manchester

*Guest Editors*

Wentao Li
College of Veterinary Medicine
Huazhong Agricultural University
Wuhan
China

Yifei Lang
College of Veterinary Medicine
Sichuan Agricultural University
Chengdu
China

*Editorial Office*
MDPI AG
Grosspeteranlage 5
4052 Basel, Switzerland

This is a reprint of the Special Issue, published open access by the journal *Veterinary Sciences* (ISSN 2306-7381), freely accessible at: https://www.mdpi.com/journal/vetsci/special_issues/RP60G56243.

For citation purposes, cite each article independently as indicated on the article page online and as indicated below:

Lastname, A.A.; Lastname, B.B. Article Title. *Journal Name* **Year**, *Volume Number*, Page Range.

**ISBN 978-3-7258-5917-7 (Hbk)**
**ISBN 978-3-7258-5918-4 (PDF)**
**https://doi.org/10.3390/books978-3-7258-5918-4**

# Contents

veterinary sciences

MDPI

Article

# RNase Hybridization-Assisted Amplification (RHAM) Technology: A High-Sensitivity, Field-Deployable Alternative to Quantitative Polymerase Chain Reaction for the Rapid Detection of African Swine Fever Virus

Sakchai Ruenphet [1,2], Nitipon Srionrod [3], Teera Nipakornpun [4] and Supphathat Wutthiwitthayaphong [1,*]

1 Animal Biotechnology, Mahanakorn University of Technology, 140 Cheum-Sampan Rd., Nong Chock, Bangkok 10530, Thailand; sakchai@mut.ac.th
2 Immunology and Virology Department, Mahanakorn University of Technology, 140 Cheum-Sampan Rd., Nong Chock, Bangkok 10530, Thailand
3 Clinic for Swine, Mahanakorn University of Technology, 140 Cheum-Sampan Rd., Nong Chock, Bangkok 10530, Thailand; nitipon@mut.ac.th
4 Pacific Biotech, 42 Moo 4, Petchaboon-Chalianglub Rd., Napa, Muang, Petchaboon 67000, Thailand; teera@brianet.com
* Correspondence: wsupphathat@mut.ac.th; Tel.: +66-95-456-0888

**Simple Summary:** African swine fever is a devastating and costly pig disease with no vaccine, making fast and accurate testing essential to stop its spread. Current rapid on-farm tests are often unreliable and miss infected animals, while accurate lab tests are too slow for effective outbreak control. This study aimed to evaluate a new, portable testing technology, RNase hybridization-assisted amplification, to determine if it could provide a better solution for on-site diagnosis. We compared its performance against a common test strip and the best available laboratory method using real samples from pig farms in Thailand. Our results showed the new test was highly effective, correctly identifying nearly 95% of infected pigs, which is almost as accurate as the laboratory test, while delivering results in under 35 min. In contrast, the standard test strip missed over 40% of infected pigs. We conclude that this new technology is a powerful and reliable tool for on-farm diagnosis. Its adoption will be valuable to society by empowering veterinarians and authorities to quickly detect the disease, prevent its spread, and protect the livelihoods of farmers and the global pork supply.

**Abstract:** African Swine Fever (ASF) control is severely limited by a diagnostic gap, as laboratory-based quantitative polymerase chain reaction (qPCR) is highly sensitive but slow, whereas field-deployable immunochromatographic assays (ICAs) are rapid but unreliable. To address this limitation, this study evaluated a novel, rapid isothermal assay, RNase hybridization-assisted amplification (RHAM), as a high-sensitivity, point-of-need diagnostic solution. This study compared the performance of RHAM and a conventional p72-based ICA against the qPCR reference standard using 106 diverse clinical field samples, including oral swabs, blood, serum, and organs, collected from suspected ASF cases in Thailand. The ICA exhibited markedly low diagnostic performance, achieving only 56.76% sensitivity and showing moderate agreement ($\kappa = 0.421$) with qPCR, highlighting the need for a more reliable alternative. In contrast, the RHAM assay achieved 94.59% sensitivity and 96.88% specificity, providing results rapidly within 35 min. This statistically superior performance (McNemar's test, $p < 0.0001$) demonstrated almost perfect agreement ($\kappa = 0.891$) with the qPCR reference standard, missing only four samples with very high Ct values (>30). In conclusion, RHAM is a powerful, accurate, and field-deployable diagnostic tool that effectively bridges the diagnostic gap, offering qPCR-like sensitivity for the rapid containment of ASF outbreaks.

**Keywords:** African swine fever virus; immunochromatographic assay; quantitative polymerase chain reaction; RNase hybridization-assisted amplification

## 1. Introduction

African swine fever (ASF) is a devastating hemorrhagic disease capable of causing mortality rates approaching 100% in domestic pigs, posing a catastrophic threat to the global swine industry [1,2]. Since its transcontinental expansion began accelerating in 2018, the virus has rapidly spread across Asia, Europe, and. The Americans, resulting in the loss of hundreds of millions of pigs and causing severe economic destabilization, disproportionately affecting small-scale farmers and threatening global food security [3–6]. The disease is caused by the African swine fever virus (ASFV), and its complex epidemiology, combined with the absence of a commercially available effective vaccine, makes control exceptionally challenging [3–5]. Consequently, management strategies rely heavily on strict biosecurity, culling of infected herds, and robust surveillance to prevent further spread [7]. Recognizing its global significance, the World Organization for Animal Health (WOAH) lists ASF as a notifiable disease [8].

Effective control of ASF is contingent on surveillance programs that ensure early, rapid, and accurate detection of the virus [9]. According to international guidelines from WOAH and Food and Agriculture Organization (FAO), a definitive. ASF diagnosis must be based on laboratory confirmation [10]. Quantitative polymerase chain reaction (qPCR) is the designated gold-standard method of choice, valued for its high sensitivity and specificity [8]. However, qPCR requires specialized laboratory equipment and highly trained personnel, often resulting in substantial logistical delays between sampling on the farm and receiving a diagnosis [11]. This logistical bottleneck frequently hampers the timely containment efforts necessary to control an outbreak.

To address the need for faster, on-site results, rapid field-deployable tests have been developed. Among the most common are immunochromatographic assays (ICAs), or lateral flow devices, which detect viral proteins like p72 [12]. While their simplicity, portability, and speed make them useful for preliminary screening during active outbreaks in resource-limited areas, international guidelines mandate that all field-based results, particularly negatives, must be confirmed by laboratory testing (i.e., qPCR) [9]. This is because ICAs notoriously suffer from limited diagnostic sensitivity, increasing the risk of false negatives, especially in animals with low viral loads during early infection or in asymptomatic carriers [10].

These contrasting limitations—the slow turnaround of highly accurate lab tests versus the low reliability of rapid field tests—highlight an urgent need for a diagnostic tool that bridges this gap, combining the high analytical sensitivity of qPCR with the speed and portability of an ICA. Isothermal amplification technologies, such as loop-mediated isothermal amplification (LAMP), have emerged as promising alternatives, as they enable rapid nucleic acid amplification without the need for complex thermal cycling equipment [13]. However, many traditional LAMP assays still require separate nucleic acid extraction steps or nuanced result interpretation. The novel RNase hybridization-assisted amplification (RHAM) technology evaluated in this study builds upon the isothermal principle but is designed to improve field usability by integrating sample lysis, amplification, and a simple fluorescent readout into a single, portable 'sample-to-answer' device. Therefore, this study evaluates the performance of this novel RHAM assay and a conventional ICA, comparing their diagnostic accuracy against the qPCR reference standard using a diverse set of clinical field samples from commercial farms in Thailand.

## 2. Materials and Methods

### 2.1. Ethical Approval

The animal procedures were approved by the Animal Research Ethics Committee of the Faculty of Veterinary Medicine, Mahanakorn University of Technology, Thailand (Approval code: ACUC-MUT-2024/004). All samples used in this study were obtained from field outbreaks or diagnostic submissions.

### 2.2. Sample and Sample Preparation

A total of 106 samples were collected from pigs with suspected or confirmed ASF to support diagnostic testing. The samples—comprising oral swabs, whole blood, serum, and internal organs—were obtained from commercial farms located in central and northeastern Thailand, including Nakorn Pathom, Ratchaburi, and Nakhon Ratchasima provinces. Sample selection was based on the presence of clinical signs consistent with ASF. All laboratory work was performed at the Virology and Molecular Diagnostic Center, Faculty of Veterinary Medicine, Mahanakorn University of Technology, Thailand.

For organ samples were collected from dead pigs. Pooled internal organs (lung, spleen, lymph nodes) were homogenized in phosphate-buffered saline (PBS), centrifuged at $2000\times g$ for 10 min, and the supernatant was stored at −80 °C until analysis. For oral sampling, because a large sample volume was required for multiple testing methods, five swabs per pig were pooled into a conical tube containing 3 mL of PBS, vortexed, centrifuged at $2000\times g$ for 10 min, and the resulting supernatant was stored at −80 °C until analysis. Blood samples were processed to obtain both whole blood and serum, which were stored at −80 °C until analysis.

### 2.3. Immunochromatography Assay

Monoclonal antibodies (mAbs) targeting the ASFV p72 protein were generated by Pacific Biotech Co., Ltd. (VetDiag®, Muang, Petchaboon, Thailand) under a patented process. Briefly, BALB/c mice were immunized intraperitoneally with p72 protein emulsified in Freund's complete adjuvant (Sigma-Aldrich, St. Louis, MO, USA). Two booster injections were administered at two-week intervals. One week after the final booster, blood was collected to measure antibody titers, and the mouse with the highest titer received a final injection of p72 protein without adjuvant. Three days later, the spleen was harvested, and splenocytes were fused with myeloma cells using 50% polyethylene glycol (Sigma-Aldrich, St. Louis, MO, USA) at a 5:1 ratio. Hybridoma cells were cultured in Hypoxanthine-Aminopterin-Thymidine (HAT) selective medium ($10^5$ cells/mL) in 96-well plates at 37 °C with 5% $CO_2$. After 10 days, culture supernatants were screened for anti-p72 antibodies by indirect enzyme-linked immunosorbent assay (ELISA). Positive clones were subjected to limiting-dilution cloning to establish stable monoclonal cell lines.

The generated mAbs were subsequently used for ICA development. Nitrocellulose membranes were striped with goat anti-rabbit IgG (control line) and anti-p72 mAb (test line) at 1 μL/cm, then dried under <20% humidity for >4 h. Colloidal gold was prepared by boiling distilled water, adding 1% gold chloride trihydrate and 1% sodium citrate, and stirring until a red solution formed. The pH was adjusted to 7.2, and anti-p72 mAb was added dropwise and incubated for 30 min. The mixture was then blocked with 10% bovine serum albumin (BSA) and centrifuged at $8000\times g$ for 30 min. The resulting pellet was resuspended in 1% BSA in PBS and stored at 4 °C. Conjugate pads were soaked in buffer containing sucrose, sodium azide, BSA, and gold-labeled antibodies (OD 2.0 for mAb, OD 0.7 for rabbit IgG), and then dried (>4 h, <20% humidity). Finally, the membranes, conjugated pads, and sample pads were assembled into plastic housings to produce the ICA kit (Figures 1 and 2).

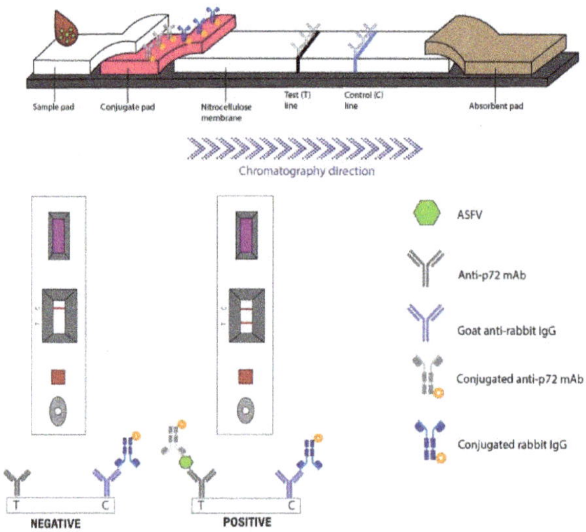


**Figure 1.** Schematic diagram of the p72 immunochromatography assay.

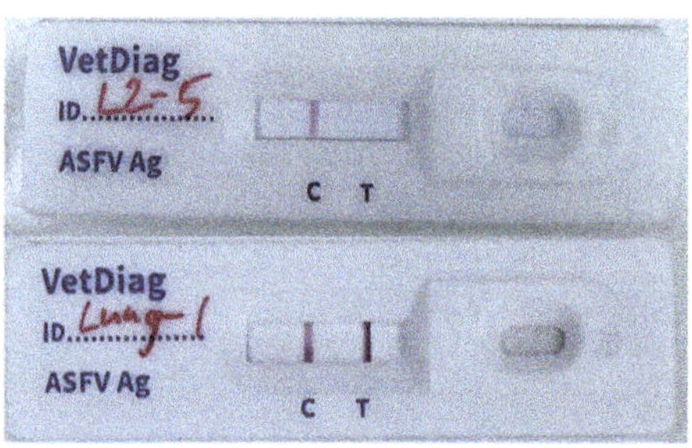


**Figure 2.** Exemplary immunochromatography assay results. Above shows a valid negative result, below shows a valid positive result.

*2.4. RNase Hybridization-Assisted Amplification (RHAM) Test*

The RHAM test kit comprises three main components: the Pluslife Integrated Nucleic Acid Testing Device, a nucleic acid releaser tube, and the Pluslife ASFV card (patented by Guangzhou Pluslife Biotech, Guangzhou, China). The procedure for ASFV detection is summarized as follows. Before testing, the device is preheated. A sample swab is then inserted into the nucleic acid releaser tube, twisted ten times while pinching the tip, and incubated at 65 °C for 5 min. The resulting lysate is poured into the Pluslife ASFV card between the injection lines. The card is placed on the Pluslife stand for 15 s, gently shaken, and then inserted into the device. Pressing the "Start Test" button initiate the amplification process. Test results are visually indicated: a red light denotes a positive ASFV result, whereas a blue light indicates a negative result. A schematic representation of the complete workflow is provided in Figures 3 and 4 showed negative and positive results.

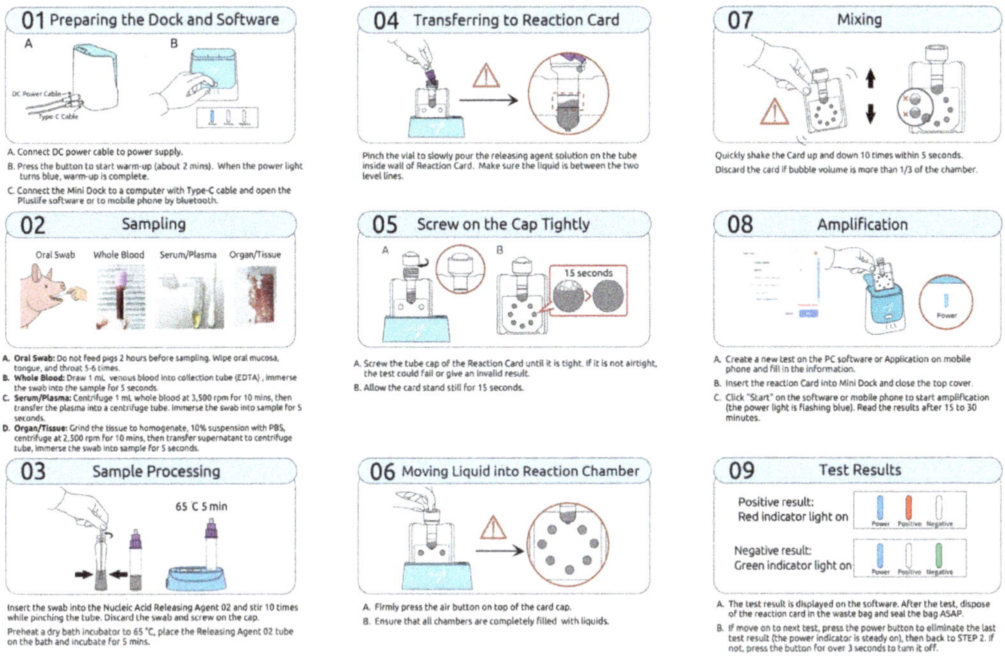


**Figure 3.** The operation process illustration of the RNase hybridization-assisted amplification (RHAM) test kit for nucleic acid detection of African swine fever virus.

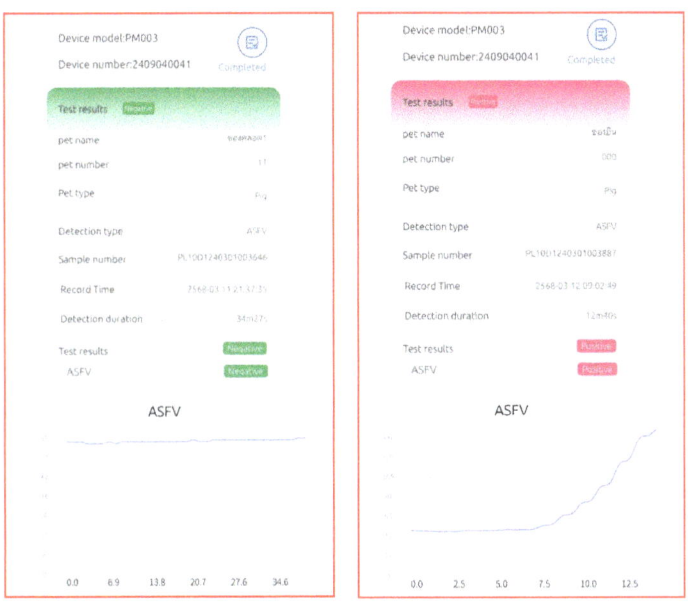


**Figure 4.** An exemplary RNase hybridization-assisted amplification (RHAM) test kit for nucleic acid detection of African swine fever virus (ASFV). The left panel shows a negative result, whereas the right panel shows a positive result. The results present fluorescence intensity curves recorded during the testing period, along with the corresponding detection duration. In ASFV-positive samples with high titers, the detection duration is shorter than in samples with low viral titers. In contrast, the detection duration for negative results is approximately 35 min.

### 2.5. Quantitative Polymerase Chain Reaction

Viral DNA was extracted from all samples using the TAN Bead® Nucleic Acid Extraction Kit (Taiwan Advanced Nanotech, Taoyuan, Taiwan) in conjunction with the Automated Nucleic Acid Extractor (Smart LabAssist SLA-E13200, Taiwan Advanced Nanotech). Quantitative detection of ASFV DNA was carried out using the Virotype® ASFV 2.0 PCR Kit (Indical Bioscience, Leipzig, Germany) on a C1000 Touch Thermal Cycler (Bio-Rad, Hercules, CA, USA). The assay incorporates two internal controls: an endogenous β-actin control and an exogenous control added during DNA purification to verify both extraction efficiency and amplification integrity. The qPCR cycling conditions consisted of an initial denaturation at 95 °C for 2 min, followed by 40 cycles of 95 °C for 5 s and 60 °C for 30 s. Samples were interpreted as positive when the cycle threshold (Ct) value was <35, suspect when between 35 and 40, and negative when >40.

### 2.6. Statistical Analysis

Sensitivity, specificity, accuracy, and precision were evaluated by comparing the ICA and RHAM results with those obtained by qPCR, which served as the reference standard. Sample yielding results consistent with qPCR were categorized as true positives or true negatives, whereas discrepancies were classified as false positives or false negatives. The counts of each classification were then used to calculate diagnostic performance metrics according to standard evaluation methods previously described [14–16].

Sensitivity, specificity, accuracy, and precision for each diagnostic modality were calculated using the following formulas:

$$Sensitivity = \frac{No.\ of\ true\ positive}{No.\ of\ true\ positive + No.\ of\ false\ negative} \times 100$$

$$Specificity = \frac{No.\ of\ true\ negative}{No.\ of\ true\ negative + No.\ of\ false\ positive} \times 100$$

$$Accuracy = \frac{(No.\ of\ true\ positive) + (No.\ of\ true\ negative)}{(No.\ of\ true\ positive) + (No.\ of\ true\ negative) + (No.\ of\ false\ positive) + (No.\ of\ false\ negative)} \times 100$$

$$Precision = \frac{No.\ of\ true\ positive}{No.\ of\ true\ positive + No.\ of\ false\ positive} \times 100$$

For all diagnostic methods evaluated, 95% confidence intervals (CIs) of sensitivity, specificity, accuracy, and precision were calculated using the Wilson score interval method for binomial proportions. This method was selected for its superior accuracy and reliability in estimating CIs, particularly for proportions derived from binary diagnostic outcomes (e.g., positive or negative results). To assess the agreement between each assay (ICA and RHAM) and the qPCR reference standard, Cohen's kappa coefficient (κ) was calculated. Kappa values were interpreted according to the following thresholds: <0.20 (slight), 0.21–0.40 (fair), 0.41–0.60 (moderate), 0.61–0.80 (substantial), and 0.81–1.00 (almost perfect agreement). Furthermore, to directly compare the diagnostic performance—specifically sensitivity and specificity—of the ICA and RHAM assays using the same paired samples, McNemar's test was employed. A $p$-value of <0.05 was considered statistically significant. All statistical analyses were conducted using GraphPad Prism version 9.0 (GraphPad Software, San Diego, CA, USA).

## 3. Results

A total of 106 field samples collected from different swine sources were tested for ASFV to evaluate the performance of an ICA and a novel RHAM assay against the reference qPCR method.

### 3.1. Distribution of Viral Load Across Sample Types

Among the 106 samples tested, 74 were positive for ASFV nucleic acid using the reference qPCR assay. Based on qPCR results, the viral load, inferred from the Ct values, varied significantly among different sample types. Tissue or organ samples exhibited the highest viral load, with a mean Ct of 19.04 ± 3.05. Whole blood samples showed slightly lower viral loads, with mean Ct of 21.63 ± 5.24. Serum and oral swab samples showed markedly lower viral loads, with mean Ct values of 25.37 ± 4.54 and 29.05 ± 2.41, respectively (Figure 5).

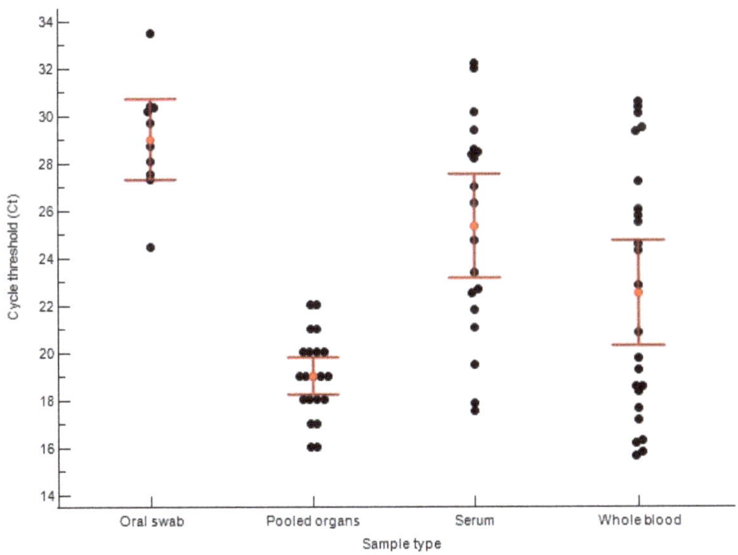


**Figure 5.** Scatter dot plot showing individual cycle threshold (Ct) values, means and standard deviation for qPCR-positive samples from oral swab (29.05 ± 2.41), pooled internal organs (19.04 ± 3.05), serum (25.37 ± 4.54) and whole blood (21.63 ± 5.24) ($n = 74$).

### 3.2. Performance of the Immunochromatography Assay

When compared with qPCR, the ICA correctly identified 42 of the 74 positive samples, resulting in 32 false-negative results. Additionally, the assays correctly identified 31 of 32 qPCR-negative samples, with one false positive (Figure 6). Further analysis of the false-negatives revealed that most corresponded to high qPCR Ct values (>27), suggesting an association between low viral load and detection failure (Table 1).

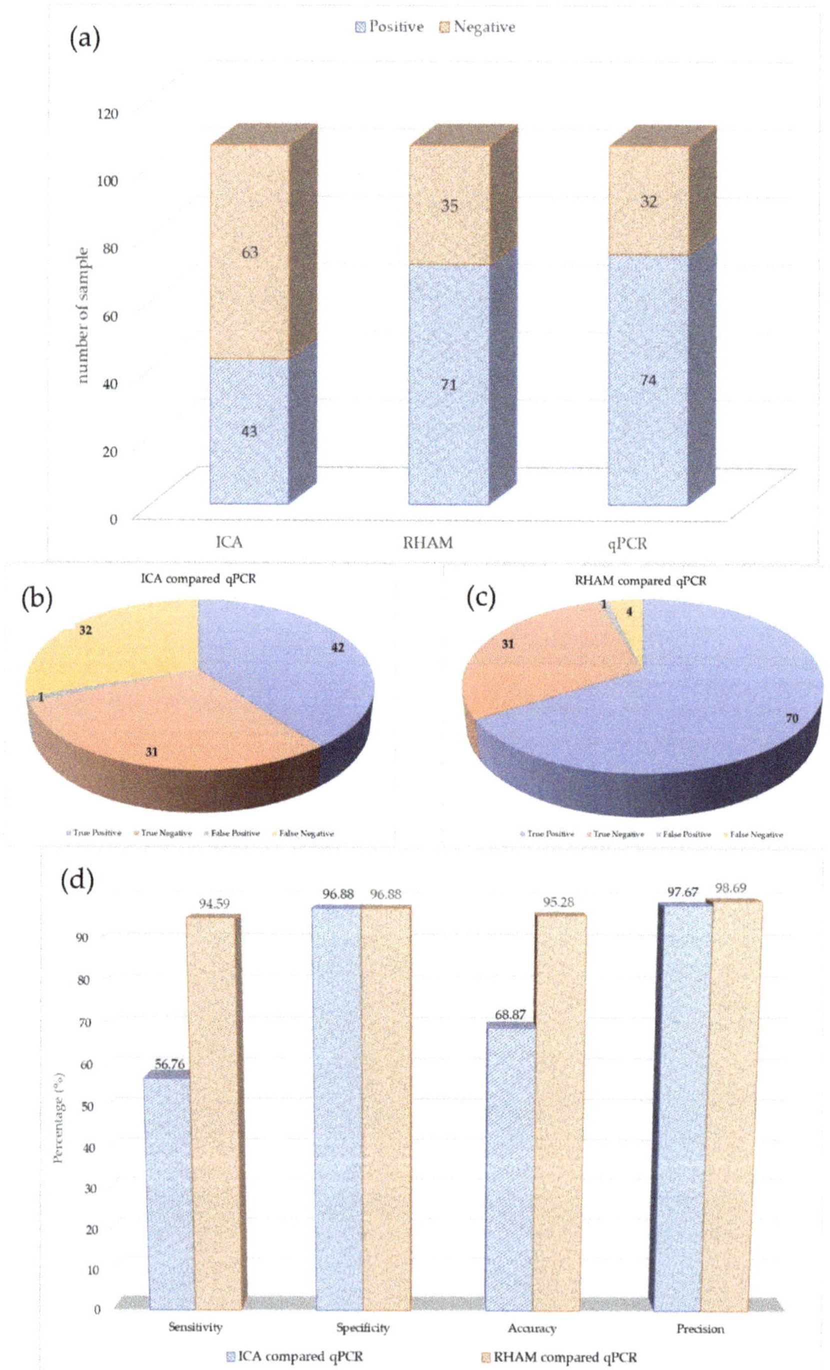


**Figure 6.** Performance graphs of ASFV detection using the immunochromatographic assay (ICA) and the RNase hybridization-assisted amplification (RHAM) ASFV test kit, compared with quantitative polymerase chain reaction (qPCR). (**a**) The ratio of positive and negative results obtained by each assay. (**b**,**c**) Comparison of the performance of the ICA and RHAM ASFV test kit with qPCR. (**d**) The analytical sensitivity, specificity, accuracy and precision of ICA and RHAM.

**Table 1.** Comparative results of p72 antigen detection using immunochromatographic assay (ICA) and genomic detection using quantitative polymerase chain reaction (qPCR) for ASFV.

| No. | ICA | qPCR | | No. | ICA | qPCR | | No. | ICA | qPCR | |
|---|---|---|---|---|---|---|---|---|---|---|---|
| | Result | Result | Ct | | Result | Result | Ct | | Result | Result | Ct |
| 1 | positive | positive | 24.51 | 37 | positive | positive | 23.59 | 73 | negative | negative | - |
| 2 | negative | positive | 30.21 | 38 | positive | positive | 24.25 | 74 | negative | negative | - |
| 3 | negative | positive | 28.77 | 39 | negative | positive | 28.13 | 75 | negative | negative | - |
| 4 | negative | positive | 27.55 | 40 | negative | positive | 31.43 | 76 | negative | negative | - |
| 5 | negative | positive | 27.33 | 41 | negative | positive | 35.01 | 77 | negative | negative | - |
| 6 | negative | positive | 29.75 | 42 | negative | negative | - | 78 | negative | negative | - |
| 7 | negative | positive | 28.11 | 43 | negative | negative | - | 79 | negative | negative | - |
| 8 | negative | positive | 30.37 | 44 | negative | negative | - | 80 | negative | negative | - |
| 9 | negative | positive | 33.51 | 45 | positive | positive | 19.81 | 81 | negative | negative | - |
| 10 | negative | positive | 30.41 | 46 | positive | positive | 18.61 | 82 | negative | negative | - |
| 11 | negative | negative | - | 47 | positive | positive | 24.32 | 83 | positive | positive | 24.74 |
| 12 | negative | negative | - | 48 | positive | positive | 25.53 | 84 | positive | positive | 23.42 |
| 13 | positive | positive | 17.26 | 49 | positive | positive | 17.18 | 85 | negative | positive | 28.51 |
| 14 | positive | positive | 20.52 | 50 | negative | negative | - | 86 | positive | positive | 22.54 |
| 15 | positive | positive | 23.98 | 51 | positive | positive | 16.24 | 87 | negative | positive | 29.39 |
| 16 | negative | positive | 28.02 | 52 | positive | positive | 19.3 | 88 | negative | positive | 28.37 |
| 17 | positive | negative | - | 53 | positive | positive | 18.59 | 89 | positive | positive | 26.31 |
| 18 | negative | negative | - | 54 | positive | positive | 16.35 | 90 | negative | positive | 30.17 |
| 19 | negative | negative | - | 55 | positive | positive | 18.37 | 91 | negative | positive | 28.58 |
| 20 | negative | negative | - | 56 | positive | positive | 15.85 | 92 | negative | positive | 28.19 |
| 21 | positive | positive | 18.03 | 57 | positive | positive | 15.68 | 93 | negative | positive | 32.21 |
| 22 | positive | positive | 19.06 | 58 | positive | positive | 17.66 | 94 | negative | negative | - |
| 23 | positive | positive | 22.40 | 59 | positive | positive | 26.08 | 95 | negative | negative | - |
| 24 | positive | positive | 26.51 | 60 | positive | positive | 20.85 | 96 | negative | negative | - |
| 25 | negative | positive | 30.14 | 61 | negative | negative | - | 97 | negative | negative | - |
| 26 | negative | positive | 33.77 | 62 | negative | negative | - | 98 | negative | negative | -s |
| 27 | negative | negative | - | 63 | positive | positive | 25.79 | 99 | positive | positive | 21.08 |
| 28 | negative | negative | - | 64 | positive | positive | 24.61 | 100 | positive | positive | 19.49 |
| 29 | positive | positive | 17.29 | 65 | negative | positive | 30.38 | 101 | positive | positive | 17.58 |
| 30 | positive | positive | 18.34 | 66 | positive | positive | 22.85 | 102 | positive | positive | 31.98 |
| 31 | positive | positive | 21.04 | 67 | negative | positive | 30.59 | 103 | positive | positive | 17.91 |
| 32 | positive | positive | 24.28 | 68 | negative | positive | 29.35 | 104 | positive | positive | 22.68 |
| 33 | negative | positive | 28.27 | 69 | negative | positive | 27.22 | 105 | negative | positive | 27.02 |
| 34 | negative | positive | 31.86 | 70 | negative | positive | 30.09 | 106 | positive | positive | 21.83 |
| 35 | negative | negative | - | 71 | negative | positive | 29.50 | | | | |
| 36 | negative | negative | - | 72 | negative | negative | - | | | | |

Ct: cycle threshold; positive: detected antigen or nucleic acid; negative: undetected antigen or nucleic acid.

Based on these results, the ICA had a diagnostic sensitivity of 56.76% (95% CI: 45.48–68.04%) and a specificity of 96.88% (95% CI: 90.85–100%). The overall accuracy and precision of the ICA were 68.87% (95% CI: 60.06–77.68%) and 97.67% (95% CI: 93.18–100%), respectively, summarizing its test performance (Figure 6).

### 3.3. *Performance of the RNase Hybridization-Assisted Amplification Test*

Table 2 showed the results of ASFV genomic detection using the RHAM ASFV test kit in comparison with qPCR. The RHAM test not only provided positive or negative results but also reported detection duration, which correlated with the Ct values obtained by qPCR across all 106 samples. Negative results using RHAM were obtained within approximately 35 min. The 70 positive samples had a mean detection time of 15.85 min (range: 10.37–30.21 min).

**Table 2.** Results of genomic detection of ASFV using the RNase hybridization-assisted amplification (RHAM) ASFV test kit compared with quantitative polymerase chain reaction (qPCR).

| No. | RHAM | | qPCR | | No. | RHAM | | qPCR | | No. | RHAM | | qPCR | |
|---|---|---|---|---|---|---|---|---|---|---|---|---|---|---|
| | Result | Min | Result | Ct | | Result | Min | Result | Ct | | Result | Min | Result | Ct |
| 1 | positive | 13.37 | positive | 24.51 | 37 | positive | 13.34 | positive | 23.59 | 73 | negative | 35.00 | negative | - |
| 2 | positive | 13.35 | positive | 30.21 | 38 | positive | 13.34 | positive | 24.25 | 74 | negative | 35.00 | negative | - |
| 3 | positive | 13.42 | positive | 28.77 | 39 | positive | 14.34 | positive | 28.13 | 75 | negative | 35.00 | negative | - |
| 4 | positive | 13.35 | positive | 27.55 | 40 | positive | 16.33 | positive | 31.43 | 76 | negative | 35.00 | negative | - |
| 5 | positive | 13.37 | positive | 27.33 | 41 | positive | 22.29 | positive | 35.01 | 77 | negative | 35.00 | negative | - |
| 6 | positive | 14.34 | positive | 29.75 | 42 | negative | 35.00 | negative | - | 78 | negative | 35.00 | negative | - |
| 7 | positive | 13.35 | positive | 28.11 | 43 | negative | 35.00 | negative | - | 79 | negative | 35.00 | negative | - |
| 8 | positive | 20.40 | positive | 30.37 | 44 | negative | 35.00 | negative | - | 80 | negative | 35.00 | negative | - |
| 9 | negative | 35.00 | positive | 33.51 | 45 | positive | 10.37 | positive | 19.81 | 81 | negative | 35.00 | negative | - |
| 10 | negative | 35.00 | positive | 30.41 | 46 | positive | 11.36 | positive | 18.61 | 82 | negative | 35.00 | negative | - |
| 11 | negative | 35.00 | negative | - | 47 | positive | 13.35 | positive | 24.32 | 83 | positive | 26.37 | positive | 24.74 |
| 12 | negative | 35.00 | negative | - | 48 | positive | 12.36 | positive | 25.53 | 84 | positive | 17.31 | positive | 23.42 |
| 13 | positive | 11.45 | positive | 17.26 | 49 | positive | 16.31 | positive | 17.18 | 85 | positive | 15.33 | positive | 28.51 |
| 14 | positive | 19.30 | positive | 20.52 | 50 | negative | 35.00 | negative | - | 86 | positive | 22.32 | positive | 22.54 |
| 15 | positive | 15.33 | positive | 23.98 | 51 | positive | 12.35 | positive | 16.24 | 87 | positive | 17.32 | positive | 29.39 |
| 16 | positive | 15.33 | positive | 28.02 | 52 | positive | 20.33 | positive | 19.3 | 88 | positive | 18.31 | positive | 28.37 |
| 17 | positive | 18.33 | negative | - | 53 | positive | 14.35 | positive | 18.59 | 89 | positive | 15.33 | positive | 26.31 |
| 18 | negative | 35.00 | negative | - | 54 | positive | 11.42 | positive | 16.35 | 90 | positive | 20.28 | positive | 30.17 |
| 19 | negative | 35.00 | negative | - | 55 | positive | 15.33 | positive | 18.37 | 91 | positive | 15.33 | positive | 28.58 |
| 20 | negative | 35.00 | negative | - | 56 | positive | 12.38 | positive | 15.85 | 92 | positive | 15.33 | positive | 28.19 |
| 21 | positive | 15.14 | positive | 18.03 | 57 | positive | 12.36 | positive | 15.68 | 93 | positive | 18.30 | positive | 32.21 |
| 22 | positive | 11.37 | positive | 19.06 | 58 | positive | 16.41 | positive | 17.66 | 94 | negative | 35.00 | negative | - |
| 23 | positive | 11.39 | positive | 22.4 | 59 | positive | 24.25 | positive | 26.08 | 95 | negative | 35.00 | negative | - |
| 24 | positive | 12.36 | positive | 26.51 | 60 | positive | 20.33 | positive | 20.85 | 96 | negative | 35.00 | negative | - |
| 25 | positive | 12.36 | positive | 30.14 | 61 | negative | 35.00 | negative | - | 97 | negative | 35.00 | negative | - |
| 26 | negative | 35.00 | positive | 33.77 | 62 | negative | 35.00 | negative | - | 98 | negative | 35.00 | negative | - |
| 27 | negative | 35.00 | negative | - | 63 | positive | 16.32 | positive | 25.79 | 99 | positive | 14.34 | positive | 21.08 |
| 28 | negative | 35.00 | negative | - | 64 | positive | 19.34 | positive | 24.61 | 100 | positive | 12.36 | positive | 19.49 |
| 29 | positive | 14.35 | positive | 17.29 | 65 | negative | 35.00 | positive | 30.38 | 101 | positive | 11.37 | positive | 17.58 |
| 30 | positive | 18.31 | positive | 18.34 | 66 | positive | 17.36 | positive | 22.85 | 102 | positive | 15.33 | positive | 31.98 |
| 31 | positive | 13.36 | positive | 21.04 | 67 | positive | 20.29 | positive | 30.59 | 103 | positive | 13.35 | positive | 17.91 |
| 32 | positive | 13.36 | positive | 24.28 | 68 | positive | 30.21 | positive | 29.35 | 104 | positive | 13.35 | positive | 22.68 |
| 33 | positive | 15.34 | positive | 28.27 | 69 | positive | 17.40 | positive | 27.22 | 105 | positive | 16.33 | positive | 27.02 |
| 34 | positive | 21.29 | positive | 31.86 | 70 | positive | 20.32 | positive | 30.09 | 106 | positive | 13.35 | positive | 21.83 |
| 35 | negative | 35.00 | negative | - | 71 | positive | 19.30 | positive | 29.50 | | | | | |
| 36 | negative | 35.00 | negative | - | 72 | negative | 35.00 | negative | - | | | | | |

Ct: cycle threshold; positive: detected nucleic acid; negative: undetected nucleic acid.

The RHAM assay showed a strong overall concordance with the qPCR results. It correctly identified 70 of the 74 positive samples, with only four false-negative results. These false negatives occurred in samples with very high qPCR Ct values (33.51, 30.41, 33.77, and 30.38), indicating low viral loads (Sample 9, 10, 26 and 65). The assay produced one false-positive result (Sample 17) out of the 32 negative samples (Table 2).

Overall, this performance corresponds to a diagnostic sensitivity of 94.59% (95% CI: 89.42–99.76%) and a specificity of 96.88% (95% CI: 90.85–100%) for the RHAM assay. The overall accuracy and precision of the RHAM assay were 95.28% (95% CI: 91.24–99.32%), and 98.59% (95% CI: 95.85–100%), respectively (Figure 6).

### 3.4. Statistic Comparison of Assays

The level of agreement between each assay (ICA and RHAM) and the qPCR reference standard was assessed based on comparative statistical analysis. The RHAM assay demonstrated almost perfect agreement ($\kappa = 0.891$), whereas the ICA showed only moderate agreement ($\kappa = 0.421$).

McNemar's test was used to directly compare the diagnostic performance of the two assays using paired samples. For diagnostic sensitivity, analysis of the 74 qPCR-positive samples indicated that RHAM's superior performance was statistically significant ($p < 0.0001$). Specifically, RHAM correctly identified 28 positive samples that the ICA failed to detect. In contrast, analysis of the 32 qPCR-negative samples revealed no statistically significant difference in diagnostic specificity between the two assays ($p = 1.0$).

## 4. Discussion

In the absence of an effective ASFV vaccine, rapid and reliable diagnostics are essential to safeguard global pig populations [17]. It must be emphasized that the foundation of ASF surveillance and control, according to international guidelines from WOAH and FAO, relies on laboratory-based confirmation, with qPCR serving as the indispensable gold standard [9,10]. Field-deployable or penside assays, including those evaluated in this study, are not substitutes for laboratory diagnosis. Instead, they serve as critical screening tools intended strictly for rapid, preliminary investigation during active outbreaks, and their results (both positive and negative) must always be confirmed by laboratory methods. Differentiating ASF from other swine diseases is challenging because its clinical signs and post-mortem lesions often resemble those of classical swine fever (CSF) and highly pathogenic porcine reproductive and respiratory syndrome (HP-PRRS). These diseases frequently present with similar hemorrhagic lymph node lesions [18–22], which further complicates field diagnosis and underscores the need for highly accurate, specific molecular tools [23–27].

This study highlights a major limitation of the commonly used ICA, with a diagnostic sensitivity of only 56.76%. Detailed sample analysis indicates that ICA performance is strongly influenced by viral load (genome presence). The 32 false-negative samples were largely associated with high qPCR Ct values, most exceeding Ct 27. This suggests that the ICA fails to detect early infections, asymptomatic carriers, or samples with moderate to low viral antigen levels. These findings align with previous reports showing variable and often low sensitivity of ICA for ASFV antigen detection, particularly under field conditions where viral loads fluctuate [11]. Although ICA demonstrated 96.88% specificity, its poor sensitivity poses an epidemiological risk by potentially allowing undetected spread of the virus. Therefore, ICA use should be restricted to rapid confirmation of clinically symptomatic cases during outbreaks, where viral titers are typically high, rather than for definitive surveillance or control programs.

The RHAM assay demonstrated excellent diagnostic performance, with 94.6% sensitivity and 96.88% specificity, closely matching qPCR performance. The four false negatives occurred in samples with very high Ct values (>30), near the detection limit of most assays including qPCR, indicating that RHAM's performance is nearly equivalent to the reference standard. Notably, a single false-positive result from the RHAM test (sample 17) was also detected as positive by the ICA, despite testing negative by qPCR. This discrepancy could stem from potential sample cross-contamination during collection or processing. Alternatively, both RHAM and ICA may have detected a true positive missed by the qPCR protocol used—a phenomenon occasionally reported when comparing different molecular assays [28]. Further investigation of such discordant samples is warranted to clarify these discrepancies. However, the original sample was exhausted after testing by ICA, RHAM and qPCR; therefore, it was unavailable for repeat qPCR or sequencing confirmation.

Despite these promising results, this study represents only a preliminary evaluation and thus has certain limitations. The validation was based on 106 clinical samples collected in Thailand, primarily from suspected cases, rather than including a comprehensive set of confirmed negative controls. Moreover, the genetic variability of ASFV, particularly the emergence of distinct genotypes (e.g., Genotype I and II), may influence the performance of molecular assays. This study did not assess the RHAM assay's efficacy across different ASFV genotypes. Therefore, larger-scale validation studies using more diverse sample sets—

including confirmed negative populations and multiple viral genotypes—are necessary to establish the assay's diagnostic reliability in varied epidemiological settings.

Furthermore, this study underscores the critical importance of sample selection in ASFV diagnostics, which reflects the virus's pathogenesis. The data in Table 1 show that the highest viral loads (lowest mean Ct values) were detected in pooled internal organs (lung, spleen, lymph node) (Ct 19.04), followed by whole blood (Ct 21.63). This aligns with the known pathophysiology of ASFV, which is characterized by an early viremia (high viral loads in blood) followed by extensive viral accumulation and replication in lymphoid tissues (such as the spleen and lymph nodes) and other secondary target organs [19]. Conversely, pooled oral swabs yielded the highest mean Ct value (29.05), indicating lower viral shedding through this route. However, this result must be interpreted with caution, as the samples were pooled (five swabs per sample), which may have diluted the viral concentration compared to an individual swab. This finding emphasizes that non-invasive sampling via oral swabs, while advantageous for screening, necessitates the use of highly sensitive molecular assays such as RHAM or qPCR to minimize the risk of false negatives.

The RHAM ASFV test kit employs isothermal amplification coupled with enzyme digestion probe technology to detect ASFV-specific sequences, providing a rapid and accurate diagnostic alternative. This assay integrates LAMP with an RNase HII reporter system for real-time signal visualization within a single reaction [29–32]. It targets the conserved B646L (p72) gene—also utilized in qPCR—which enhances analytical comparability and detection reliability [12,23,33]. Using conventional LAMP primers, Bst DNA polymerase exponentially amplifies the target sequence during isothermal incubation. A fluorescent ribonucleotide-containing probe, labeled with a 5′-fluorophore and 3′-quencher, hybridizes to the amplicon sequence. RNase HII then cleaves the ribonucleotide, releasing the probe and producing a detectable fluorescence signal. The Pluslife Integrated Nucleic Acid Testing Device records this fluorescence signal within 10–35 min, facilitating rapid ASFV detection in a single step. The assay incorporates an internal control to verify sample integrity and amplification efficiency, thereby minimizing false negatives and enhancing reliability. Collectively, this approach provides a versatile, efficient, and timely diagnostic platform that supports ASFV surveillance and rapid outbreak response.

Our results, based on diverse sample types including oral swab, whole blood, serum, tissue, and internal organs, demonstrated that the RHAM ASFV test kit achieved high sensitivity, specificity, accuracy, and precision when compared to qPCR. Despite this strong overall performance, four false-negative results were observed—two from oral swabs, one from serum, and one from whole blood—indicating potential limitations in assay sensitivity. These false-negative samples tested positive results in qPCR, with Ct value exceeding 30 (33.51, 30.41. 33.77, and 30.38), consistent with moderate to low viral genome levels approaching the detection limit of the RHAM assay. It is important to clarify that Ct values reflect the quantity of viral genome present, not necessarily the presence of infectious virus or active replications; typically, only Ct values above 35 are considered indicative of very ASFV genome levels. These findings suggest that the false negatives likely resulted from viral genome concentrations that fall below the assay's detection threshold. Furthermore, the relatively small sample volume used in RHAM testing—approximately 150 µl collected via cotton swabbing—may have contributed to the occurrence of false negatives. In addition, qPCR involves genomic extraction prior to amplification, yielding a higher concentration of purified viral nucleic acid than RHAM and potentially explaining the discrepancy. Collectively, these factors, including differences in sample volume and nucleic acid concentration, likely account for the observed false negatives, as RHAM utilizes roughly half the sample volume used in qPCR, which may influence both sensitivity and specificity.

Currently, qPCR serves the primary molecular diagnostic method for detecting ASF [23,33], a disease with significant implications for the swine industry. Despite its sensitivity, qPCR's dependence on specialized equipment and infrastructure limits accessibility and delays result turnaround. Consequently, samples must be transported to centralized laboratories, leading to delays and additional logistical challenges. Notably, qPCR requires nucleic acid extraction and amplification, taking at least 2.0 hr, whereas the RHAM assay provides results within 10–35 min, offering a significant advantage in speed and efficiency. In our study, negative results using RHAM were obtained within 35 min. The 70 positive samples showed a mean testing time of 15.85 min (range: 10.37–30.21 min), underscoring the assay's rapidity and convenience compared with qPCR. Importantly, the time required for a positive RHAM result correlated with the qPCR Ct value, reflecting the relationship between testing time and viral load. Samples with lower Ct values—indicative of higher virus titers—produced shorter RHAM testing times, demonstrating its ability to detect high viral loads rapidly and underscoring its diagnostic efficacy. Overall, these findings indicate that RHAM functions as a semi-quantitative assay, enabling rapid and efficient ASFV detection while providing insight into viral load levels.

Several studies [34–38] have demonstrated that various isothermal methods, such as LAMP, recombinase polymerase amplification (RPA), and cross-priming amplification (CPA), enables sensitive and reliable detection of ASFV, underscoring its versatility in molecular diagnostics. Building upon this isothermal principle, the RHAM ASFV test kit functions as an on-site nucleic acid detection tool that combines LAMP-based amplification with a qPCR-like reporting system, enabling accurate and efficient ASFV detection. This integration of RHAM technology positions it within this growing landscape of field-deployable molecular tools, aiming to minimize false positives and enhances the overall reliability of test results. Additionally, the kit does not require refrigeration and provides results within 35 min, making it highly practical for rapid, field-based diagnostics. Collectively, these features establish the RHAM ASFV test kit as an accurate, rapid, and practical diagnostic platform, marking a meaningful advancement in molecular tools for animal disease surveillance and control.

Finally, the role of the RHAM assay should be considered within the broader context of a comprehensive field surveillance strategy. According to recommendations from the FAO and WOAH, effective disease control relies on the integration of multiple diagnostic approaches. On-site molecular assays, such as RHAM, are particularly effective for detecting viral genomes during the acute phase of infection. However, under field conditions—especially when assessing herd status or identifying survivors—molecular assays such as RHAM should be validated by qPCR and complemented by serological testing. This occurs because the sensitivity of viral genome detection methods (such as RHAM or qPCR) decreases after the acute phase, as animals recover and clear viremia, while antibodies remain detectable [39,40]. Therefore, combining molecular and serological diagnostics is essential for identifying both acute infections and survivors, thus providing a comprehensive epidemiological overview.

## 5. Conclusions

Effective African Swine Fever control is critically hampered by the diagnostic gap between slow, laboratory-bound qPCR and fast but unreliable field-based ICAs. This study evaluated the novel RHAM technology as a high-sensitivity, point-of-need solution. Based on 106 clinical field samples, RHAM demonstrated outstanding performance, achieving 94.59% diagnostic sensitivity and 96.88% specificity, showing "almost perfect agreement" ($\kappa = 0.891$) with the qPCR reference standard. This was statistically superior ($p < 0.0001$) to the conventional ICA, which showed poor sensitivity (56.76%) and failed to detect over 40%

of positive cases. RHAM provides these results rapidly (10–35 min) by combining isothermal amplification with a specific RNase HII probe mechanism. Although this preliminary study requires further validation across different genotypes and geographical regions, we conclude that RHAM is a powerful, accurate, and field-deployable tool, representing a significant advancement for the rapid detection and timely containment of ASF outbreaks.

**Author Contributions:** Conceptualization, S.R., N.S., T.N. and S.W.; Formal analysis, S.R., N.S. and S.W.; Funding acquisition, S.R., T.N. and S.W.; Investigation, S.R., N.S. and S.W.; Methodology, S.R., N.S. and S.W.; Project administration, S.R. and S.W.; Supervision, S.R.; Writing—original draft, S.R., N.S., T.N. and S.W.; Writing—review and editing, S.R., N.S., T.N. and S.W. All authors have read and agreed to the published version of the manuscript.

**Funding:** The present study received partial financial backing from Mahanakorn University of Technology (Thailand) and Bangkok R.I.A. Group (Thailand), through the allocation of funds provided by the Veterinary Research Grant under Contract No. Virus-001-2024.

**Institutional Review Board Statement:** The guidelines concerning the appropriate care and utilization of animals received approval from the Animal Research Ethics Committee of the Faculty of Veterinary Medicine at Mahanakorn University of Technology, Thailand. To substantiate their validation, these guidelines were identified with the specific approval code ACUC-MUT-2024/004.

**Informed Consent Statement:** Signed informed consent was provided by the owner for all procedures related to the case.

**Data Availability Statement:** The original contributions presented in this study are included in the article material. Further inquiries can be directed to the corresponding author.

**Acknowledgments:** The authors gratefully acknowledge the cooperation of all pig farmers who participated in this study. The authors also thank the sample preparation assistants, with special recognition to Wisut Prasitsuwan.

**Conflicts of Interest:** This study was funded by Bangkok R.I.A. Group Co., Ltd. (Bangkok, Thailand). Teera Nipakornpun were employed by the Pacific Biotech. The funder had no role in study design, data collection, analysis, interpretation, manuscript preparation, or the decision to submit this article for publication, thereby ensuring the independence and integrity of the research. The authors declare no competing interests.

## References

1. Costard, S.; Wieland, B.; De Glanville, W.; Jori, F.; Rowlands, R.; Vosloo, W.; Roger, F.; Pfeiffer, D.U.; Dixon, L.K. African Swine fever: How can global spread be prevented? *Philos. Trans. R. Soc. B Biol. Sci.* **2009**, *364*, 2683–2696. [CrossRef] [PubMed]
2. Sauter-Louis, C.; Conraths, F.J.; Probst, C.; Blohm, U.; Schulz, K.; Sehl, J.; Fischer, M.; Forth, J.H.; Zani, L.; Depner, K.; et al. African Swine fever in wild boar in Europe—A review. *Viruses* **2021**, *13*, 1717. [CrossRef] [PubMed]
3. Ramirez-Medina, E.; O'Donnell, V.; Silva, E.; Esoinoza, N.; Velazquez-Salinas, L.; Moran, K.; Daite, D.A.; Barrette, R.; FAburay, B.; Holland, R.; et al. Experimental infection of domestic pigs with an African swine fever virus field strain isolated in 2021 from the Dominican Republic. *Viruses* **2022**, *14*, 1090. [CrossRef] [PubMed]
4. Taesuji, M.; Rattanamas, K.; Kulthonggate, U.; Mamom, T.; Ruenphet, S. Sensitivity and specificity for African horse sickness antibodies detection using monovalent and polyvalent vaccine antigen-based dot blotting. *Vet. World* **2022**, *15*, 2760–2763. [CrossRef]
5. Rattanamas, K.; Taesuji, M.; Kulthonggate, U.; Jantafong, T.; Mamom, T.; Ruenphet, S. Sensitivity of RNA viral nucleic acid-based detection of avian influenza virus, Newcastle disease virus, and African horse sickness virus on flinders technology associates card using conventional reverse-transcription polymerase chain reaction. *Vet. World* **2022**, *15*, 2754–2759. [CrossRef]
6. Nguyen-Thi, T.; Pham-Thi-Ngoc, L.; Nguyen-Ngoc, Q.; Dang-Xuan, S.; Lee, H.S.; Nguyen-Viet, H.; Padungtod, P.; Nguyen-Thu, T.; Nguyen-Thi, T.; Tran-Cong, T.; et al. An assessment of the economic impacts of the 2019 African swine fever outbreaks in Vietnam. *Front. Vet. Sci.* **2021**, *8*, 686038. [CrossRef]
7. Ata, E.B.; Li, Z.J.; Shi, C.W.; Yang, G.L.; Yang, W.T.; Wang, C.F. African Swine Fever Virus: A Raised Global Upsurge and a Continuous Threaten to Pig Husbandry. *Microb. Pathog.* **2022**, *167*, 105561. [CrossRef]

8. World Organisation for Animal Health (WOAH). Chapter 15.1. Infection with African Swine Fever Virus. In *Terrestrial Animal Health Code*; World Organisation for Animal Health (WOAH): Paris, France, 2016; Available online: https://www.woah.org/fileadmin/Home/eng/Health_standards/tahc/2016/en_chapitre_asf.htm (accessed on 5 November 2025).
9. Beltrán-Alcrudo, D.; Arias, M.; Gallardo, C.; Kramer, S.; Penrith, M.L. African Swine Fever: Detection and Diagnosis—A Manual for Veterinarians. In *FAO Animal Production and Health Manual No. 19*; Food and Agriculture Organization of the United Nations (FAO): Rome, Italy, 2017; 88p.
10. Dixon, L.K.; Stahl, K.; Jori, F.; Vial, L.; Pfeiffer, D.U. African Swine Fever Epidemiology and Control. *Annu. Rev. Anim. Biosci.* **2020**, *8*, 221–246. [CrossRef]
11. Fernández-Pinero, J.; Gallardo, C.; Elizalde, M.; Robles, A.; Gomez, C.; Bishop, R.; Heath, L.; Couacy-Hymann, E.; Fasina, F.O.; Pelayo, V.; et al. Molecular diagnosis of African swine fever by a new real-time PCR using universal probe library. *Transbound. Emerg. Dis.* **2013**, *60*, 48–58. [CrossRef]
12. Gallardo, C.; Soler, A.; Nurmoja, I.; Cano-Gomez, C.; Cvetkova, S.; Frant, M.; Wozniakowski, G.; Simon, A.; Perez, C.; Nieto, R.; et al. Dynamics of African swine fever virus (ASFV) infection in domestic pigs infected with virulent, moderate virulent and attenuated genotype II ASFV European isolates. *Transbound. Emerg. Dis.* **2021**, *68*, 2826–2841. [CrossRef]
13. Wang, S.; Shen, H.; Lin, Q.; Huang, J.; Zhang, C.; Liu, Z.; Sun, M.; Zhang, J.; Liao, M.; Li, Y.; et al. Development of a. cleaved probe-based loop-mediated isothermal amplification assay for rapid detection of African swine fever virus. *Front. Cell. Infect. Microbiol.* **2022**, *12*, 884430. [CrossRef]
14. Prasitsuwan, W.; Suwannachote, T.; Sumalai, T.; Lertpatarakomol, R.; Trairatapiwan, T.; Ruenphart, S. A comparison of diagnostic methods for canine Ehrlichiosis: Microscopy and RNases hybridization-assisted amplification technology compared with the quantitative polymerase chain reaction. *Vet. World* **2025**, *18*, 1214–1223. [CrossRef] [PubMed]
15. Punyadarsaniya, D.; Taesuji, M.; Rattanamas, K.; Ruenphet, S. Establishment of an In-House Indirect Enzyme-Linked Immunosorbent Assay to Detect Antibodies Against African Horse Sickness Based on Monovalent and Polyvalent Live Attenuated Vaccines During the First Outbreak in Thailand. *Animals* **2025**, *15*, 1433. [CrossRef] [PubMed]
16. Suwannachote, T.; Prasitsuwan, W.; Sumalai, T.; Ruenphet, S. A Comparison of Diagnostic Methods for Feline Leukemia Virus and Feline Immunodeficiency Virus: Immunochromatographic Assay and RNases Hybridization-Assisted Amplification Test Kit Compared to Reverse Transcription Quantitative Polymerase Chain Reaction. *Animals* **2025**, *15*, 1484. [CrossRef] [PubMed]
17. Tran, X.H.; Phuong, L.T.; Huy, N.Q.; Thuy, D.T.; Nguyen, V.D.; Quang, P.H.; Ngon, Q.V.; Rai, A.; Gay, C.G.; Gladue, D.P.; et al. Evaluation of the Safety Profile of the ASFV Vaccine Candidate ASFV-G-ΔI177L. *Viruses* **2022**, *14*, 896. [CrossRef]
18. Gallardo, C.; Soler, A.; Nieto, R.; Cano, C.; Pelayo, V.; Sanchez, M.A.; Pridotkas, G.; Fernandez-Pinero, J.; Briones, V.; Arias, M. Experimental infection of domestic pigs with African swine fever virus Lithuania 2014 genotype II field isolate. *Transbound. Emerg. Dis.* **2017**, *64*, 300–304. [CrossRef]
19. Salguero, F.J.; Frossard, J.P.; Rebel, J.M.; Stadejek, T.; Morgan, S.B.; Graham, S.P.; Steinbach, F. Host-pathogen interactions during porcine reproductive and respiratory syndrome virus 1 infection of piglets. *Virus Res.* **2015**, *202*, 135–143. [CrossRef]
20. Morgan, S.B.; Frossard, J.P.; Pallares, F.J.; Gough, J.; Stadejek, T.; Graham, S.P.; Steinbach, F.; Drew, T.W.; Salguero, F.J. Pathology and virus distribution in the lung and lymphoid tissues of pigs experimentally inoculated with three distinct type 1 PRRS virus isolates of varying pathogenicity. *Transbound. Emerg. Dis.* **2016**, *63*, 285–295. [CrossRef]
21. Huong Giang, N.T.; Lan, N.T.; Nam, N.H.; Hirai, T.; Yamaguchi, R. Pathological characterization of an outbreak of porcine reproductive and respiratory syndrome in Northern Vietnam. *J. Comp. Pathol.* **2016**, *154*, 135–149. [CrossRef]
22. Choe, S.; Le, V.P.; Shin, J.; Kim, J.H.; Kim, K.S.; Song, S.; Cha, R.M.; Park, G.N.; Nguyen, T.L.; Hyun, B.H.; et al. Pathogenicity and genetic characterization of Vietnamese classical swine fever virus: 2014–2018. *Pathogens* **2020**, *9*, 169. [CrossRef]
23. King, D.P.; Reid, S.M.; Hutchings, G.H.; Grierson, S.S.; Wilkinson, P.J.; Dixon, L.K.; Bastos, A.D.S.; Drew, T.W. Development of a TaqMan® PCR assay with internal amplification control for the detection of African swine fever virus. *J. Virol. Methods* **2003**, *107*, 53–61. [CrossRef] [PubMed]
24. Notomi, T.; Okayama, H.; Masubuchi, H.; Yonekawa, T.; Watanabe, K.; Amino, N.; Hase, T. Loop-mediated isothermal amplification of DNA. *Nucleic Acids Res.* **2000**, *28*, e63. [CrossRef] [PubMed]
25. Rowlands, R.J.; Duarte, M.M.; Boinas, F.; Hutchings, G.; Dixon, L.K. The CD2v protein enhances African swine fever virus replication in the tick vector, *Ornithodoros erraticus*. *Virology* **2009**, *393*, 319–328. [CrossRef] [PubMed]
26. Taesuji, M.; Rattanamas, K.; Punyadarsaniya, D.; Mamom, T.; Nguyen, H.T.; Ruenphet, S. In vitro primary porcine alveolar macrophage cell toxicity and African swine fever virus inactivation using five commercially supply compound disinfectants under various condition. *J. Vet. Med. Sci.* **2021**, *83*, 1800–1804. [CrossRef]
27. Sovijit, W.; Taesuji, W.; Rattanamas, K.; Punyadarsaniya, D.; Mamom, T.; Nguyen, H.T.; Ruenphet, S. In vitro cytotoxicity and virucidal efficacy against African swine fever using two potassium hydrogen peroxymonosulfate compared to a quaternary ammonium compound under various concentrations, exposure times and temperatures. *Vet. World* **2021**, *14*, 2936–2940. [CrossRef]
28. Petrovan, V.; Buburuzan, L.; Zaulet, M. False positive results using PCR detection method for African swine fever virus in wild boars from northern Romanian hunting zones. *Turk. J. Vet. Anim. Sci.* **2015**, *39*, 3. [CrossRef]

29. Zhu, D.; Huang, J.; Hu, B.; Cao, D.; Chen, D.; Song, X.; Chen, J.; Zhou, H.; Cen, A.; Hou, T. Trial of the Pluslife SARS-CoV-2 Nucleic Acid Rapid Test Kit: Prospective Cohort Study. *JMIR Public Health Surveill.* **2023**, *9*, e48107. [CrossRef]
30. Xiao, Z.; Liu, X.; Kang, X.; Feng, Y.; Zheng, L.; Chen, C. Rapid and accurate detection of SARS-CoV-2 using the RHAM technology. *Sci. Rep.* **2023**, *13*, 22798. [CrossRef]
31. Herrmann, L.; Breuer, J.; Duc, T.N.; Thomé, N.; Ghazaani, F.; Kamhieh-Milz, S.; Pfützner, A. Comparison of the diagnostic accuracy of the Pluslife Mini Dock RHAM technology with Abbott ID Now and Cepheid GenXpert: A retrospective evaluation study. *Sci. Rep.* **2024**, *14*, 13978. [CrossRef]
32. Zhong, G.; Zhao, Z.; He, Y.; Yu, X.; Yang, T.; Liu, H.; Huang, B. Evaluation of the RHAM-based Pluslife Rapid SARS-CoV-2 assay and comparison with EasyNAT Rapid SARS-CoV-2 assay and Wondfo 2019-nCoV antigen assay. *Microbiol. Spectr.* **2025**, e03418-24. [CrossRef]
33. Zsak, L.; Borca, M.V.; Risatti, G.R.; Zsak, A.; French, R.A.; Kutish, G.F.; Neilan, J.G.; Callahan, J.D.; Nelson, W.M.; Rock, D.L. Preclinical diagnosis of African swine fever in contact-exposed swine by a real-time PCR assay. *J. Clin. Microbiol.* **2005**, *43*, 112–119. [CrossRef] [PubMed]
34. Bohorquez, J.A.; Lanka, S.; Rosell, R.; Perez-Simo, M.; Alberch, M.; Rodriguez, F.; Ganges, L.; Maddox, C.W. Efficient detection of African Swine Fever Virus using minimal equipment through a LAMP PCR method. *Front. Cell. Infect. Microbiol.* **2023**, *13*, 1114772. [CrossRef] [PubMed]
35. Mee, P.T.; Wong, S.; O'Riley, K.J.; da Conceicao, F.; da Costa Jong, J.B.; Phillips, D.E.; Rodoni, B.C.; Rawlin, G.T.; Lynch, S.E. Field Verification of an African Swine Fever Virus Loop-Mediated Isothermal Amplification (LAMP) Assay during an Outbreak in Timor-Leste. *Viruses* **2020**, *12*, 1444. [CrossRef] [PubMed]
36. Wang, Y.; Dai, J.F.; Liu, Y.S.; Yang, J.; Hou, Q.; Ou, Y.; Ding, Y.; Ma, B.; Chen, H.; Li, M.M.; et al. Development of a Potential Penside Colorimetric LAMP Assay Using Neutral Red for Detection of African Swine Fever Virus. *Front. Microbiol.* **2021**, *12*, 609821. [CrossRef]
37. Zuo, L.; Song, Z.; Zhang, Y.; Zhai, X.; Zhai, Y.; Mei, X.; Yang, X.; Wang, H. Loop-Mediated Isothemal Amplification Combined with Lateral Flow Dipstrick for On-Site Diagnosis of African Swine Fever Virus. *Virol. Sin.* **2020**, *36*, 325–328. [CrossRef]
38. Ji, C.; Zhou, L.; Chen, Y.; Fang, X.; Liu, Y.; Du, M.; Lu, X.; Li, Q.; Wang, H.; Sun, Y.; et al. Microfluidic-LAMP chip for the point-of-care detection of gene-deleted and wild-type African swine fever viruses and other four swine pathogens. *Front. Vet. Sci.* **2023**, *10*, 1116352. [CrossRef]
39. Gallardo, C.; Nieto, R.; Soler, A.; Pelayo, V.; Fernández-Pinero, J.; Markowska-Daniel, I.; Arias, M. Assessment of African swine fever diagnostic techniques as a response to the epidemic outbreaks in eastern european union countries: How to improve surveillance and control programs. *J. Clin. Microbiol.* **2015**, *53*, 2555–2565. [CrossRef]
40. Oura, C.A.L.; Edwards, L.; Batten, C.A. Virological diagnosis of African swine fever—Comparative study of available tests. *Virus Res.* **2013**, *173*, 150–158. [CrossRef]

veterinary sciences

MDPI

*Communication*

# Synergistic Enhancement of Humoral and Cellular Immunity in Attenuated Live Vaccines Using Chemical Nanoadjuvants: Evaluation of Sustained-Release Formulations

**Jianli Shi, Chang Liu, Chen Li * and Jun Li ***

Key Laboratory of Livestock and Poultry Multi-Omics of Agriculture and Reral Affairs, Shandong Research Center of Livestock and Poultry Biologicals Engineering, Institute of Animal Science and Veterinary Medicine Shandong Academy of Agricultural Sciences, Jinan 250100, China; liuchangsaas@163.com (C.L.)
* Correspondence: carolindia@126.com (C.L.); junli79@163.com (J.L.)

**Simple Summary:** A novel nano-adjuvant with sustained release properties was developed by combining polyethylene glycol (PEG) and other biochemical and molecular biology agents with CpG. The combination of PEG20000 and benzoic acid, along with Seppic white oil, significantly enhanced humoral immunity. In contrast, the combination of total white oil, PEG6000 and benzoic acid demonstrated superior efficacy in promoting cellular immunity. These findings underscore the differential roles of sustained-release agents in shaping immune outcomes and provide a strategic framework for optimizing attenuated live vaccine formulations through adjuvant synergy.

**Abstract:** Achieving a balance between enhancing both humoral and cellular immunity while ensuring sustained release for attenuated live vaccines has long been a challenge in the vaccine industry. However, the incorporation of chemical materials as nanoadjuvants has emerged as a promising strategy to address this challenge. In this study, a novel nano-adjuvant with sustained release properties was developed by combining polyethylene glycol (PEG) and other biochemical and molecular biology agents with CpG. The efficacy of this adjuvant was assessed using the widely applied CSF attenuated live vaccine. The body temperature and relative daily weight gain results demonstrated the safety of the formulations. Importantly, the combination of CpG immune enhancers and sustained-release agents led to a synergistic augmentation of both humoral and cellular immune responses. Specifically, the combination of PEG20000 and benzoic acid, along with Seppic white oil, significantly enhanced humoral immunity. In contrast, the combination of total white oil, PEG6000 and benzoic acid demonstrated superior efficacy in promoting cellular immunity. This study offers important insights into enhancing vaccine efficacy through novel adjuvant combinations, which can be applied to the development of more effective vaccines in the future.

**Keywords:** CSFV vaccines; adjuvants; CpG; sustained-release agents

## 1. Introduction

Classical swine fever (CSF), caused by CSF virus (CSFV), is one of the most devastating viral epizootic diseases of swine in many countries, listed as a Class A animal disease by the World Organization for Animal Health (WOAH) [1]. CSFV is often a coinfection with other pathogens such as porcine reproductive and respiratory syndrome virus (PRRSV), porcine circovirus type 2 (PCV2) and pseudorabies virus (PRV). There is no effective drug treatment after CSFV infection, which can only be prevented and protected by vaccine [2,3].

By mingling with the antigen components in the vaccine, adjuvants can enhance the immune response of the vaccine, especially for antigens with insufficient immunogenicity, enabling them to produce faster and better immune responses [4]. Among them, CpG motifs have the advantage of being able to stimulate endogenous immune responses more quickly, characterized by the production of multiple Interleukins and use as an immune adjuvant [5,6]. CpG has shown a good application prospect in the field of immunoenhancers, and has become a research hotspot to treat animal diseases and even human diseases in recent years [7]. CpG adjuvant (CpG 1018) has been developed as a vaccine adjuvant in the Dynavax's hepatitis B vaccine and COVID-19 vaccine for the emergency scope of application. Our groups have also confirmed that the CpG motif could enhance the immune effect of the PCV2 DNA vaccine [8].

Sustained-release agents are a class of substances that can slowly release and exert long-term effects after wrapping vaccines. The development of sustained-release agents for vaccines has become an expanding field for generating stronger vaccines effects in the last thirty years [9]. Biochemical and molecular biology preparations used as immunological adjuvants and sustained-release agents can help to extend the duration vaccine affect. It is reported that many compounds such as white oil and benzoic acid also have sustained-release effect [10,11]. In this work, CpG and five different sustained-release agents were emulsified and immunized with the CSF vaccine to evaluate the immune efficacy.

## 2. Materials and Methods

### 2.1. Animals, and Vaccine

Seventy 35-day-old commercial Landrace piglets tested negative for CSFV antibody from Shandong Zhongba Agriculture and Animal Husbandry Development Co., Ltd. were selected for vaccine and randomly divided into seven groups (A–G), ten piglets in each group, and housed separately. All experimental animals were euthanized after the experiment. The method of euthanasia for experimental animals is electrocution and exsanguination as a secondary step within 15 s of initial stunning. CSF live vaccine was bought from Luoyang Huizhong Biology (No. 210512) and used according to the protocol supplied by the manufacturer with 2 doses per pig of all groups.

### 2.2. Experimental Design

The required CpG doses and sustained-release agent were emulsified according to the number of immunized pigs. In order to evaluate the effect of storage time, the emulsified compounds were stored at 4 °C for 6 months. Group A–E was injected into 5 pigs, respectively, and the other 5 were injected with newly emulsified agent. The dose of emulsion injected was 2 mL. Group F was used as the control without sustained-release agent. Group G was used as the negative control without sustained-release agent and CpG. A summary of the experimental design is shown in Table 1. The piglets were vaccinated intramuscularly (i.m.) in the neck at the same time with vaccine.

### 2.3. Temperature Measurement

The rectal temperature of piglets in each group was measured daily on the first 7 days after all pigs vaccinated with CSF live vaccine at about 10:00 a.m. Meanwhile, the allergic reactions at the injection site, such as swelling and absorption caused by injection, were observed. The pigs should be as quiet as possible when measuring the body temperature using forehead thermometer.

**Table 1.** Experimental design.

| Groups | Sustained-Release Agents Name | Immunoenhancer/Dose | Sustained-Release Agents and CpG Ratio |
|---|---|---|---|
| A | Total white oil | 18 CpG/800 ug | 2:1 |
| B | Seppic white oil | 18 CpG/800 ug | 1:1 |
| C | PEG6000 and benzoic acid | 18 CpG/800 ug | 0.05% benzoic acid and |
| D | PEG20000 and benzoic acid | 18 CpG/800 ug | 0.5% PEG |
| E | Defatted milk powder and sucro | 18 CpG/800 ug | 1:1 |
| F | - | 18 CpG/800 ug | - |
| G | - | - | - |

*2.4. Relative Daily Weight Gain*

The change in all pigs' body weight was recorded at the time before vaccine (dpv 0) and 60 days later (dpv 60). Group average relative daily gain (the sum of IRDG/10) was calculated to evaluate the vaccine's effects.

$$\text{Individual relative daily gain (IRDG)} = (\text{dpv 60 weight} - \text{dpv 0 weight})/\text{dpv 0 weight}/60$$

$$\text{Group average relative daily gain (GARDG)} = \text{The sum of IRDG}/10$$

*2.5. Assay of Pig Blood Antibody Levels and Cell Immunity*

Blood samples were collected on 0, 7, 14, 21, 60 days after vaccination, centrifuged at 3000 g for 10 min at 4 °C, and the serum was separated and stored at −80 °C until testing. CSFV antibody levels were detected with CSFV ELISA kit (IDEXX) according to the manufacturer's directions. The OD450 (ODTEST), ODPOS (positive control), ODNEG (negative control) was detected and blocking rate was calculated.

$$\text{Blocking rate} = (\text{ODNEG} - \text{ODTEST})/\text{ODNEG} \times 100\%.$$

Vaccine cell immunity status was detected with pig inter-leukin-4 (IL-4) ELISA kit (CUSABIO) according to the manufacturer's directions. The absorbance of each well was read in a spectrophotometer at 450 nm. Serial dilutions of standard samples were used to quantify the contents, and served as a standard curve. Then IL-4 contents in the serum samples were calculated.

*2.6. Statistical Analysis*

All data was analyzed and presented using Graphpad Prism 5 software. One-way analysis of variance (ANOVA) was used to do the statistical analyses followed by Tukey's post hoc test, respectively. Differences were considered significant at $p < 0.05$.

## 3. Results

*3.1. Temperature Measurement*

The rectal temperature of all piglets on the first 7 days after vaccination was tested. The results showed that the temperature of all pigs fluctuated, but in no cases exceeded 40.5 °C for three consecutive days, indicating that the injection did not cause a fever reaction. There were no allergic reactions such as redness, swelling, heat or pain at the injection site, and no significant changes in the spirit and appetite of all pigs. This proves that the vaccine and the selected immune enhancer, emulsified with different sustained-release agents, is safe for piglets. At the same time, there was no significant difference between the emulsified infection stored for six months with the new emulsion.

### 3.2. Relative Daily Weight Gain

The $p$-value based on the GARDG of all immunoenhanced piglets was calculated and the results are shown in Figure 1. Compared to the control piglets (groups G), there was no significant difference in the immunoenhanced group ($p > 0.05$). The weight gain was not altered by vaccination. Also, the results showed that the emulsion stored for six months has no adverse effect on weight gain.

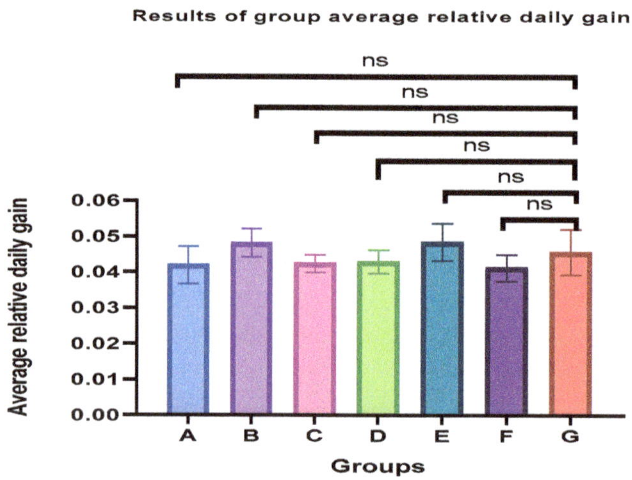


**Figure 1.** The results of group average relative daily gain. Compared to the control piglets (groups G), there was no significant difference among groups ($p > 0.05$). ns: no significant difference.

### 3.3. Serology

The blocking rate of all piglets is shown in Figure 2. It can be seen from the results that all experimental pigs can quickly produce swine fever antibodies after vaccination. On the 14th day after immunization, sustained-release agents used in group D can stimulate the rapid production of higher titers antibodies in experimental pigs. This result has important significance for emergency preventive immunity when swine fever occurs during production [12]. On the 21st day after immunization, the antibodies in the group injected with CpG and sustained-release agents were higher than those in the group injected with the vaccine alone, indicating that the CpG had shown good immune enhancement effects. Among the groups injected with emulsion, group B showed the highest antibody titer. On the 60th day after immunization, the antibody titers of all groups treated with emulsion were higher than those in group G. At the same time, there was no significant difference between the group vaccinated with the emulsion stored for 6 months and the new emulsion at the testing time point. This result showed that the emulsion stored for six months does not affect the effectiveness.

The antibody of group D is higher than other groups on the 14th day after immunization; group B showed the highest antibody titer on the 21st day after immunization. The antibody titers of groups with the immunoenhanced CpG and vaccine were higher than group G, who received only the vaccine, at 60 days post infection.

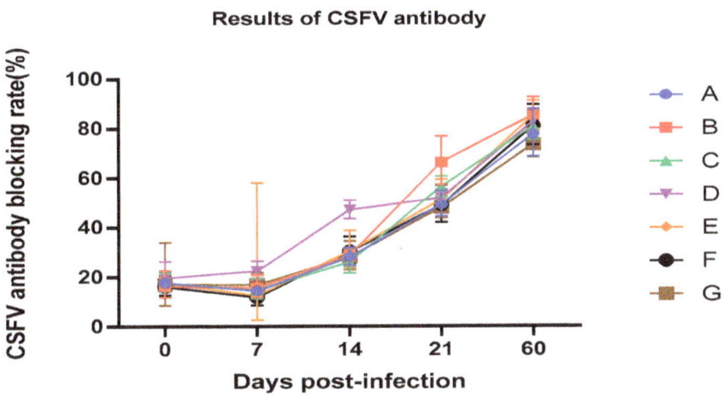


**Figure 2.** The results of classical swine fever virus (CSFV) blocking rate of all piglets detected with CSFV ELISA kit (IDEXX).

### *3.4. Changes in IL-4 Content in Piglets*

IL-4 has been called the "prototypic immunoregulatory cytokine" [13]. Like many cytokines, it can affect a variety of target cells in multiple ways. IL-4 has an important role in regulating antibody production, hematopoiesis and inflammation, and the development of effector T-cell responses. Therefore, IL-4 can be used as an important indicator of cellular immunity.

The IL-4 contents in the serum samples were calculated after serial dilutions of standard samples were used to serve as a standard curve and the results are shown in Figure 3. After vaccination, the levels of IL-4 in all pigs increased, indicating that the pigs produced a cellular immune response. On the 14th day after immunization, the IL-4 concentration reaches its peak.

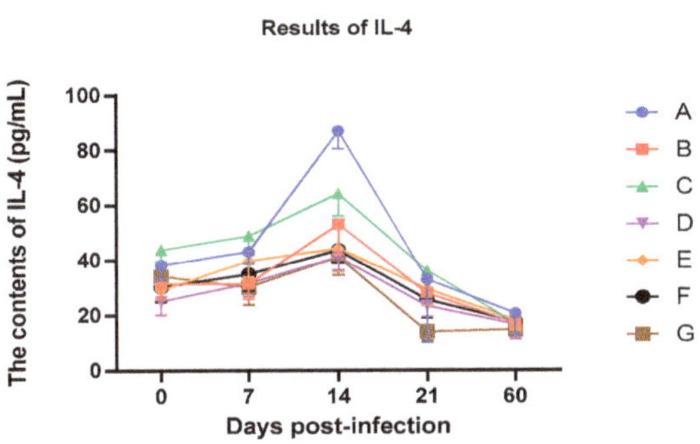


**Figure 3.** Level of interleukin-4 (IL-4) content in peripheral blood.

The groups A–F IL-4 levels were higher than those in group G. These results indicated that CpG and sustained-release agents can better stimulate the pigs to produce cellular immune responses. The A–C group, especially group A and C, showed significant effects with stronger cellular immunity. After 14 days, the IL-4 contents began to gradually decrease, but groups A–F were still higher than group G, indicating that the emulsion used still stimulated the pigs to produce cellular immune responses for a long time. On the 60th day after immunization, all groups approached normal levels. It is speculated that the reason for this being lower than before immunization was that immune and other stimulating factors can also affect the body's IL-4 levels.

On the 14th day after immunization, the IL-4 concentration reached its peak. The levels in groups A–F were higher than those in group G. After 14 days, the IL-4 contents began to gradually decrease, but groups A–F were still higher than group G.

## 4. Discussion

Immune adjuvants are now widely used in combination with various vaccines. In addition to enhancing the immune response, they also have the advantages of reducing antigen usage, efficiently activating the immune system, and prolonging the titer level. They have important value for enhancing the vaccine immune effect. Now they are an indispensable focus material in global vaccine research and development.

At present, aluminum adjuvant is still the most widely approved vaccine adjuvant used in China. Aluminum adjuvants are usually effective in increasing serum antibodies, but they can cause injection site reactions and weak ability to induce cellular immunity, which limits their application and does not have significant effects on certain vaccines. Nowadays, the direction of vaccine research is expanding and extending from preventing infectious diseases to treating chronic infectious diseases, cancer, and immune system diseases, which undoubtedly brings new challenges and opportunities for the development of new adjuvants.

CpG-ODN adjuvant can not only induce specific immune cell differentiation and proliferation, and enhance the immunogenicity of the vaccine, but also enhance antigen-specific humoral immunity in the vaccine development process [11]. However, due to the presence of nucleases, CPG is easily degraded, and sustained-release agents are necessary for its long-term effect in immune enhancement and immune regulation in the body. The sustained-release agents used in this study are commonly commercialized materials, which will not significantly increase the cost of vaccination. PEG classified as a non-immunogenic polymer has been applicated in many other products, such as cosmetics, foods and pharmaceutical preparations [12]. PEG can also coat vaccine products to maintain their stability [14].

As a kind of antibacterial and antifungal preservative, benzoic acid is widely used in foods and feeds. Recently, many studies showed that using appropriate benzoic acid can regulate enzyme activity, redox status, immunity and microbiota, which can also enhance the immunogenicity of vaccine antigens [15]. As a widely used adjuvant in inactivated vaccines, white oil can delayed the retention time of immunogens, allow them to be continuously and slowly released, and enhance the phagocytic and bactericidal abilities of macrophages [16].

Through comparative experiment, it was found that the vaccine combined with immune enhancers and sustained-release agents has no safety issues. The application effect showed that sustained-release agents PEG20000, benzoic acid and Seppic white oil have more significant effects on improving humoral immunity, and total white oil, PEG6000, benzoic acid and Seppic white oil have better effects on cellular immunity. In clinical production, there is some emergency situations, such as the occurrence of disease cases after segregation, where other healthy populations need to quickly improve their cellular and humoral immune levels. This research result can be combined with vaccines to achieve better results. Due to the different types of viruses, further experimental verification is needed to evaluate the effect of other vaccines.

## 5. Conclusions

In summary, the combination of CpG immune enhancers and sustained-release agents synergistically augmented both humoral and cellular immune responses. Specifically, PEG20000 and benzoic acid combination, and Seppic white oil significantly enhanced humoral immunity, as evidenced by elevated neutralizing antibody titers. In contrast, total white oil, PEG6000 and benzoic acid combination exhibited superior efficacy in promoting cellular immunity, with increased IL-4 levels indicating Th2-biased immune activation. These findings highlight the differential roles of sustained-release agents in tailoring immune outcomes and provide a strategic framework for optimizing CSFV vaccine formulations through adjuvant synergy.

**Author Contributions:** Methodology, C.L. (Chang Liu); Formal analysis, C.L. (Chen Li); Writing—original draft preparation, J.S.; Writing—review and editing, funding acquisition, J.L. All authors have read and agreed to the published version of the manuscript.

**Funding:** The study was partly supported by Shandong Province Pig Industry Technology System (grant no. SDAIT-08-06), The Major Scientific and Technological Innovation Project (MSTIP) (grant no. 2023CXGC010705), The Natural fund of Shandong Province (ZR2022MC011), "20 New Universities" Funding Project of Jinan City (202228112), Innovation capability improvement project of small and medium-sized technology-based enterprise (2023TSGC0734), The Agricultural Science and Technology Innovation Project of Shandong Academy of Agricultural Sciences (CXGC2025A04).

**Institutional Review Board Statement:** The whole procedure for experimental animals was performed and carried out in strict accordance with guideline (IACC20060101, 1 January 2006) of the Institutional Animal Care and Use Committee of Institute of Animal Science and Veterinary Medicine, Shandong Academy of Agricultural Sciences (project identifier: IASVM-2023-005). All animals used in this study were made to minimize suffering and euthanasia were applied for viscera.

**Informed Consent Statement:** Informed consent was obtained from all subjects involved in the study.

**Data Availability Statement:** The raw data supporting the conclusions of this article will be made available by the authors on request.

**Acknowledgments:** Thank all the authors for their contributions to this article.

**Conflicts of Interest:** The authors declare no conflicts of interest.

## References

1. Ganges, L.; Crooke, H.R.; Bohórquez, J.A.; Postel, A.; Sakoda, Y.; Becher, P.; Ruggli, N. Classical swine fever virus: The past, present and future. *Virus Res.* **2020**, *289*, 198151. [CrossRef] [PubMed]
2. Edwards, S.; Fukusho, A.; Lefèvre, P.C.; Lipowski, A.; Pejsak, Z.; Roehe, P.; Westergaard, J. Classical swine fever: The global situation. *Vet. Microbiol.* **2000**, *73*, 103–119. [CrossRef] [PubMed]
3. Saatkamp, H.W.; Berentsen, P.B.; Horst, H.S. Economic aspects of the control of classical swine fever outbreaks in the European Union. *Vet. Microbiol.* **2000**, *73*, 221–237. [CrossRef] [PubMed]
4. Kojima, Y.; Xin, K.-Q.; Ooki, T.; Hamajima, K.; Oikawa, T.; Shinoda, K.; Ozaki, T.; Hoshino, Y.; Jounai, N.; Nakazawa, M.; et al. Adjuvant effect of multi-CpG motifs on an HIV-1 DNA vaccine. *Vaccine* **2002**, *20*, 2857–2865. [CrossRef] [PubMed]
5. Krieg, A.M. CpG motifs in bacterial DNA and their immune effects. *Annu. Rev. Immunol.* **2002**, *20*, 709–760. [CrossRef] [PubMed]
6. Graham, S.P.; Haines, F.J.; Johns, H.L.; Sosan, O.; La Rocca, S.A.; Lamp, B.; Rümenapf, T.; Everett, H.E.; Crooke, H.R. Characterisation of vaccine-induced, broadly cross-reactive IFN-γ secreting T cell responses that correlate with rapid protection against classical swine fever virus. *Vaccine* **2012**, *30*, 2742–2748. [CrossRef] [PubMed]
7. Bode, C.; Zhao, G.; Steinhagen, F.; Kinjo, T.; Klinman, D.M. CpG DNA as a vaccine adjuvant. *Expert Rev. Vaccines* **2011**, *10*, 499–511. [CrossRef] [PubMed]
8. Li, J.; Shi, J.-L.; Wu, X.-Y.; Fu, F.; Yu, J.; Yuan, X.-Y.; Peng, Z.; Cong, X.-Y.; Xu, S.-J.; Sun, W.-B.; et al. Improvement of the Immunogenicity of Porcine Circovirus Type 2 DNA Vaccine by Recombinant ORF2 Gene and CpG Motifs. *Viral Immunol.* **2015**, *28*, 290–296. [CrossRef] [PubMed]
9. Aguilar, J.C.; Rodríguez, E.G. Vaccine adjuvants revisited. *Vaccine* **2007**, *25*, 3752–3762. [CrossRef] [PubMed]
10. Guo, W.; Han, R.; Xu, W.; Lu, Z.; Li, Y.; Dan, X.; Mo, Z. The protective effect of inactivated Flavobacterium columnare vaccine in grass carp (*Ctenopharyngodon idellus*). *Front. Immunol.* **2023**, *14*, 1162975. [CrossRef] [PubMed]
11. Goda, R.M.; Shohayeb, M.M. Use of Pistacia lentiscus mastic for sustained-release system of chlorocresol and benzoic acid for in vitro prevention of bacterial colonization of silicon urinary catheter. *Lett. Appl. Microbiol.* **2021**, *73*, 599–606. [CrossRef] [PubMed]

12. Tong, W.; Zheng, H.; Li, G.X.; Gao, F.; Shan, T.L.; Zhou, Y.J.; Yu, H.; Jiang, Y.F.; Yu, L.X.; Li, L.W.; et al. Recombinant pseudorabies virus expressing E2 of classical swine fever virus (CSFV) protects against both virulent pseudorabies virus and CSFV. *Antivir. Res.* **2020**, *173*, 104652. [CrossRef]
13. Melissa, A.B.; John, H. Functions of IL-4 and Control of Its Expression. *Crit Rev. Immunol.* **2017**, *37*, 181–212.
14. Zhang, D.; Liu, L.; Wang, J.; Zhang, H.; Zhang, Z.; Xing, G.; Wang, X.; Liu, M. Drug-loaded PEG-PLGA nanoparticles for cancer treatment. *Front. Pharmacol.* **2022**, *13*, 990505. [CrossRef] [PubMed]
15. Mao, X.; Yang, Q.; Chen, D.; Yu, B.; He, J. Benzoic acid used as food and feed additives can regulate gut functions. *Biomed Res. Int.* **2019**, *2019*, 5721585. [CrossRef]
16. Sjoerd, T.S.; Laura, J.K.; Matheus, H.C.; Hakan, K.; Joke, M.H.; Yvette, V.K. Immunological dynamics after subcutaneous immunization with a squalene-based oil-in-water adjuvant. *FASEB J.* **2020**, *9*, 12406–12418.

veterinary sciences

MDPI

Article

# Preparation and Characterization of Monoclonal Antibodies Against the Porcine Rotavirus VP6 Protein

Botao Sun [1], Dingyi Mao [2], Jing Chen [1], Xiaoqing Bi [1], Linke Zou [1], Jishan Bai [1], Rongchao Liu [1], Ping Hao [1], Qi Wang [1], Linhan Zhong [1], Panchi Zhang [1] and Bin Zhou [1,3,4,5,*]

1 MOE Joint International Research Laboratory of Animal Health and Food Safety, College of Veterinary Medicine, Nanjing Agricultural University, Nanjing 210095, China; 2024007018@stu.njau.edu.cn (B.S.); t2025065@stu.njau.cn (J.C.); 2024207017@stu.njau.edu.cn (X.B.); linkezou@stu.njau.edu.cn (L.Z.); baijishan@stu.njau.edu.cn (J.B.); 2022807132@stu.njau.edu.cn (R.L.); 2024207023@stu.njau.edu.cn (P.H.); 2023007092@stu.njau.edu.cn (Q.W.); zhonglinhan1115@163.com (L.Z.); 2024807252@stu.njau.edu.cn (P.Z.)

2 Jiangsu Institute of Parasitic Diseases, Wuxi 214000, China; maodingyi@jipd.com

3 Key Laboratory of Animal Bacteriology, Ministry of Agriculture and Rural Affairs, Nanjing Agricultural University, Nanjing 210095, China

4 College of Veterinary Medicine, Northeast Agricultural University, Harbin 150030, China

5 Northeast Science Observation Station for Animal Pathogen Biology, Ministry of Agriculture and Rural Affairs, Harbin 150030, China

* Correspondence: zhoubin@njau.edu.cn

**Simple Summary:** Porcine Rotavirus (PoRV), particularly genotype A, is globally prevalent and currently lacks effective antiviral therapies. Its segmented genome enables frequent genetic reassortment, posing substantial challenges for prevention and control. The development of effective vaccines is therefore imperative. VP6, a structurally conserved and highly immunogenic protein, represents an attractive target for vaccine development. In the present study, a monoclonal antibody (mAb) specifically targeting VP6 was generated, establishing a robust experimental foundation for improved PoRV diagnostics and vaccine research. This mAb demonstrated exceptional specificity against VP6, attaining a maximal detection titer of 1:3,000,000. Additionally, two previously uncharacterized B-cell epitopes were delineated, both exhibiting substantial conservation across diverse circulating PoRV strains. Collectively, these findings underscore the mAb's significant potential as a tool for the advancement of VP6-based diagnostic methodologies and therapeutic interventions against PoRV.

**Abstract:** Porcine Rotavirus (PoRV), a predominant causative agent of neonatal diarrhea in piglets, shares substantial genetic homology with human rotavirus and represents a considerable threat to both public health and the global swine industry in the absence of specific antiviral interventions. The VP6 protein, an internal capsid component, is characterized by exceptional sequence conservation and robust immunogenicity, rendering it an ideal candidate for viral genotyping and vaccine development. In the present study, the recombinant plasmid pET28a(+)-VP6 was engineered to facilitate the high-yield expression and purification of the VP6 antigen. BALB/c mice were immunized to generate monoclonal antibodies (mAbs) through hybridoma technology, and the antigenic specificity of the resulting mAbs was stringently validated. Subsequently, a panel of truncated protein constructs was designed to precisely map linear B-cell epitopes, followed by comparative conservation analysis across diverse PoRV strains. Functional validation demonstrated that all three mAbs exhibited high-affinity binding to VP6, with a peak detection titer of 1:3,000,000 and exclusive specificity toward PoRVA. These antibodies effectively recognized representative genotypes such as G3 and X1, while exhibiting no cross-reactivity with unrelated viral pathogens; however, their reactivity against other PoRV serogroups (e.g., types B and C) remains to be further elucidated. Epitope mapping identified two novel linear B-cell epitopes, $^{128}$YIKNWNLQNR$^{137}$ and $^{138}$RQRTGFVFHK$^{147}$, both displaying strong sequence conservation among circulating PoRV strains. Collectively, these findings provide a rigorous experimental framework for the functional dissection of VP6 and reinforce its potential as a valuable diagnostic and immunoprophylactic target in PoRV control strategies.

**Keywords:** Porcine Rotavirus; hybridoma technology; antibody specificity; B-cell epitopes; epitope conservation

## 1. Introduction

Rotavirus (RV) is a globally pervasive zoonotic pathogen, initially identified in diarrheic cattle in the United States in 1969 and subsequently disseminated worldwide [1]. In 1976, the International Committee on Taxonomy of Viruses designated this viral morphology as "RV" [2]. Rotavirus primarily infects a broad spectrum of mammalian hosts—including humans, bovines, ovines, canines, and swine—eliciting acute gastroenteritis. Its segmented double-stranded RNA genome facilitates frequent genetic reassortment, fostering the evolution of novel viral variants. This genomic plasticity markedly hinders the development of effective prophylactic measures and surveillance protocols, thereby amplifying both global economic losses and public health threats [3].

Currently, the Rotavirus Classification Working Group (RCWG) categorizes Rotavirus A (RVA) into GX-P[x]-Ix-Rx-Cx-Mx-Ax-Nx-Tx-Ex-Hx genotypes based on the open reading frames of its 11 genomic segments [4,5], which encode structural proteins (VP1, VP2, VP3, VP4, VP6, VP7) and non-structural proteins (NSP1–NSP6). The viral genome is encapsulated within a triple-layered capsid composed of VP2, VP6, and VP7. VP6 constitutes the intermediate layer surrounding VP2. In insect and plant cells, co-expression of VP2 and VP6 facilitates the assembly of double-layered viral particles [6]. VP6, a central structural protein, possesses high immunogenic potential and has been widely studied for its functional contributions to viral assembly and host immune modulation. Notably, recombinant virus-like particles incorporating VP2, VP6, and VP7 have demonstrated the capacity to elicit strong and protective immune responses in mouse models [7]. Moreover, VP6 is extensively employed in viral diagnostic platforms and fulfills pivotal roles in viral replication and protein biosynthesis.

Rotaviruses are taxonomically classified into ten distinct groups (A-J) based on VP6 antigenic determinants, among which RVA is the most prevalent and pathogenic [8]. In swine populations, Porcine Rotavirus A (PoRVA) constitutes a primary etiological agent of acute enteritis in suckling piglets under eight weeks of age, with morbidity rates ranging from 60% to 80% and mortality rates reaching 15–25%. This substantially compromises piglet viability and production efficiency. Furthermore, PoRVA frequently co-infects with other enteric viruses, such as Porcine Epidemic Diarrhea Virus (PEDV) and Transmissible Gastroenteritis Virus (TGEV), as well as bacterial pathogens such as *Escherichia coli* [9]. These co-infections exacerbate disease severity and transmission, resulting in significant economic losses across the swine industry [10]. Although prior research has reported the generation of monoclonal antibodies (mAbs) against VP6 for diagnostic applications and protective assessment, existing approaches are constrained by limited breadth and poor cross-reactivity across PoRVA subtypes [11]. Thus, the development of high-affinity, broadly reactive mAbs targeting VP6 is imperative to advance diagnostic capabilities and inform next-generation vaccine strategies.

Despite considerable progress in VP6-related research, there remains a pronounced deficit in the development of mAbs with diagnostic utility and therapeutic potential against

PoRV—particularly those capable of targeting conserved epitopes with high specificity and efficacy. To bridge this gap, the present study endeavors to express and purify recombinant VP6 protein using a prokaryotic expression platform, followed by the induction of antigen-specific immune responses in BALB/c mice through a combined subcutaneous and intraperitoneal immunization regimen. High-affinity, VP6-specific mAbs will be screened via hybridoma technology. In parallel, epitope mapping will be conducted to delineate evolutionarily conserved B-cell epitopes recognized by these mAbs. In conclusion, this study seeks to establish a rigorous experimental framework, with the goal of providing a robust experimental foundation and theoretical support for the development of precise diagnostic approaches for PoRV and the optimization of targeted intervention strategies.

## 2. Materials and Methods

### 2.1. Experimental Material

PoRV was isolated and characterized by our laboratory, with partial cell lysate samples from PoRV-infected cells kindly provided by senior researcher Li Bin, Institute of Veterinary Medicine, Jiangsu Academy of Agricultural Sciences., Nanjing, China. JEV was isolated and identified by our laboratory. MA104 cells and mouse myeloma SP2/0 cells were maintained in our laboratory. Specific pathogen-free (SPF) BALB/c mice were procured from Changzhou Cavens Experimental Animal Co., Ltd., Changzhou, China. and housed at the Experimental Animal Center, Nanjing Agricultural University. All immunization procedures adhered to the institutional animal care and use guidelines (Animal Welfare and Ethics Approval No. NJAU.No20220706145). The pET-28a(+)-VP6 plasmid was synthesized by the GenScript Biotech Corporation., Nanjing, China.

### 2.2. Enzymes and Related Reagents

The prokaryotic expression vector pET-28a(+) was preserved in our laboratory. Competent *Escherichia coli* DH5α (Cat. No. TSC-C01) and BL21 (Cat. No. TSC-E01) were obtained from Beijing Tsingke Biotechnology Co., Ltd., Beijing, China. Green Taq Mix (Cat. No. P131-01), a 180 kDa prestained protein marker (Cat. No. MP102-01), the BCA Protein Assay Kit (Cat. No. E112-01), and the High-Sensitivity ECL Chemiluminescence Detection Kit (Cat. No. E412-01) were purchased from Nanjing Novizan Medical Technology Co., Ltd., Nanjing, China. PrimeSTAR high-fidelity PCR polymerase (Cat. No. R040A), restriction endonucleases BamHI (Cat. No. 1010S) and XhoI (Cat. No. 1094S), DNA marker (Cat. No. 3427Q/3582Q), 10 × nucleic acid loading buffer (Cat. No. 9157), T4 DNA ligase (Cat. No. 2011A), and the DNA Gel Extraction Kit (Cat. No. 9762) were sourced from TaKaRa Bio, Inc. (Beijing, China). The 180 kDa three-color prestained protein marker (Cat. No. WJ103) and 12.5% PAGE Gel Rapid Preparation Kit (Cat. No. PG113) were purchased from Shanghai Yamei Biotechnology Co., Ltd., Shanghai, China. The plasmid extraction kit (Cat. No. D6943-01) was from Omega Bio-Tek., Nanjing, China. PoRV-positive serum was kindly provided by New Hope Liuhe Co., Ltd., Chengdu, China. HRP-conjugated goat anti-mouse IgG (Cat. No. SA00001-1), FITC-conjugated goat anti-mouse IgG (Cat. No. SA00003-1), anti-His-tag antibody (Cat. No. 66005-1-Ig), and the Mouse Monoclonal Antibody Isotyping Kit (Cat. No. PK20003) were obtained from Proteintech Group. HRP-conjugated goat anti-pig IgG (Cat. No. ab6915) was purchased from Abcam.

### 2.3. Truncated Expression of Recombinant Gene pET-28a(+)-VP6

The full-length VP6 gene sequence was obtained using RVA/Porcine-tc/KOR/174-1/2006/G8P (GenBank No. MF940552.1) as the template. Hydrophobicity analysis was conducted using TMHMM-2.0, and the three-dimensional structural model was constructed via SWISS-MODEL homology modeling. Based on antigenicity prediction and sequence conservation, the N-terminal 2–178 aa region was selected as the target fragment. Specific primers containing BamHI and XhoI restriction sites were designed and synthesized by GenScript Biotech (Table S1). PCR amplification and double digestion verification were performed for recombinant plasmid construction.

### 2.4. Expression and Purification of VP6 Recombinant Protein

The sequence-verified plasmid was extracted and subsequently transformed into *E. coli* BL21(DE3) competent cells. Upon optimizing the induction parameters, VP6 was abundantly expressed as inclusion bodies, which were subsequently subjected to partial denaturation and refolding protocols. The recombinant protein was then purified via nickel–nitrilotriacetic acid (Ni-NTA) affinity chromatography. The purity and identity of the isolated VP6 protein were validated through SDS-PAGE and Western blot (WB) analyses.

### 2.5. Immunization of Mice

Four female BALB/c mice (6–8 weeks) were immunized with emulsified VP6 protein according to the protocol in Table S2 [12]. One mouse received PBS emulsified with Freund's complete adjuvant as a control. After the third immunization, blood was collected via orbital sampling to determine antibody titers for booster immunization and cell fusion preparation.

### 2.6. Cell Fusion and Subcloning

An indirect enzyme-linked immunosorbent assay (ELISA) was established by optimizing the coating concentrations, blocking times, and antibody dilutions, using a P/N ratio $\geq$2.1 as the positivity threshold for subcloning and detection [13]. SP2/0 cells were thawed, and splenocytes were harvested 3 days post-booster immunization. Cell fusion was performed at a 1:5 ratio (SP2/0–splenocytes) using PEG 4000(Nanjing, China) [11]. Hybridoma subcloning was conducted using the ELISA screening protocol, and stable antibody-secreting cell lines were expanded for ascites production.

Following sensitization, ~$2 \times 10^6$ hybridoma cells were intraperitoneally injected into 6–8-week-old BALB/c mice. Ascites fluid was collected and stored at −20 °C and then purified via IgG affinity chromatography. Antibody titers and subtypes were determined by ELISA; specificity was validated by WB; and reactivity against the virus was confirmed by immunofluorescence assay (IFA).

### 2.7. Identification of Monoclonal Antibody Specificity

MA104 cells were infected with PoRV at a multiplicity of infection (MOI) of 1. Upon the onset of distinct cytopathic effects (CPE), IFA was conducted utilizing the VP6-specific monoclonal antibody (mAb) diluted at 1:100 as the primary antibody and a goat anti-mouse FITC-conjugated antibody as the secondary to evaluate antibody specificity and sensitivity. BHK cells infected with JEV were used as negative controls, and their reactivity was assessed using both the VP6 mAb and a JEV-E-specific antibody.

In parallel, MA104 cells were infected with PoRV G3 and X1 genotypes, while BHK cells were infected with JEV. Uninfected cells served as negative controls. Following cell lysis and protein extraction, Western blotting was performed using the VP6 mAb as the primary antibody and HRP-conjugated secondary antibodies. To further substantiate the assay's specificity, heterologous antibodies and the JEV-NS5 antibody were employed as negative controls.

### 2.8. Construction of VP6 Protein Truncation Mutants

Epitope identification defines critical antigenic regions recognized by mAbs, underpinning affinity optimization, specificity enhancement, and broad-spectrum antiviral development. Following methodologies for truncation variant construction [14], primers were designed using a dichotomy approach to amplify the 1–97 aa (F1) and 87–174 aa (F2) fragments, with a 10 aa overlap. Using pET-28a as a vector and BamHI/XhoI sites, recombinant plasmids were constructed and sequenced (Sangon Biotech, Shanghai, China). Considering typical epitope lengths (6–15 aa), a 10 aa step size balanced mapping efficiency and resolution [15]. Subsequent truncation primers were designed with sequential 10 aa deletions, yielding nine primer pairs (Table S3).

### 2.9. Epitope Identification and Conservation Analysis

Truncated proteins were expressed, and WB was performed using ascites from mAbs 2E7, 2E4, and 3G6 to identify epitope-containing regions.

Full-length VP6 sequences from eight prevalent PoRV strains (Accession Nos. AAB41105.1, AER25319.1, ANQ45164.1, AOW44086.1, AXS77346.1, AYA93307.1, AYA93308.1, AYA93321.1, AYA93335.1) were retrieved from the NCBI and aligned with the VP6 variant (AXF43000.1) using MegAlign Pro 17.1 to evaluate epitope conservation.

The procedural workflow for monoclonal antibody production is illustrated schematically. Please refer to Supplementary Figure S1 for a detailed flowchart.

## 3. Results

### 3.1. Construction, Expression, and Purification of Recombinant pET-28a(+)-VP6

The complete genomic sequence of the RVA/Porcine-tc/KOR/174-1/2006/G8P strain (Accession No. MF940552.1) was retrieved from GenBank. Transmembrane domain analysis using TMHMM-2.0 predicted no transmembrane regions within VP6 (Figure S2). Structural modeling based on the predicted tertiary conformation (Figure S3) guided the selection of an evolutionarily conserved and immunodominant region (2–178 aa) for recombinant expression. The results of target gene amplification and double restriction enzyme digestion confirmed the successful integration of the target gene into the pET-28a(+) vector (Figure 1A,B). SDS-PAGE and WB analyses confirmed the robust expression of the recombinant VP6 protein in the form of inclusion bodies following IPTG induction (1 mM, 37 °C) (Figure 1F,G). Subsequent purification via Ni-NTA affinity chromatography yielded a distinct, homogeneous protein band, which was concentrated to 2.09 mg/mL using sucrose gradient ultrafiltration (Figure 1C). WB analysis utilizing an anti-His tag monoclonal antibody and hyperimmune serum derived from Group A rotavirus-infected swine validated the immunoreactivity and antigenic specificity of the purified VP6 protein (Figure 1D,E).

### 3.2. Generation and Characterization of VP6-Specific Monoclonal Antibodies

The optimization of the indirect ELISA established the following optimal parameters: carbonate buffer coating, antigen concentration at 200 ng/well, blocking at 37 °C for 2 h, and a secondary antibody dilution of 1:5000. Among the immunized mice, Mouse 1 exhibited the most potent humoral response and was subsequently selected for hybridoma production (Figure 2A). Following four cycles of subcloning, three stable hybridoma cell lines—2E7, 2E4, and 3G6—were established, all capable of the continuous secretion of VP6-specific monoclonal antibodies. Isotype characterization revealed that 2E7 secreted IgG2a with a kappa light chain, demonstrating a titer >1:320,000, while 2E4 and 3G6 produced IgG1 isotypes with kappa light chains, each exceeding titers of 1:3,000,000.

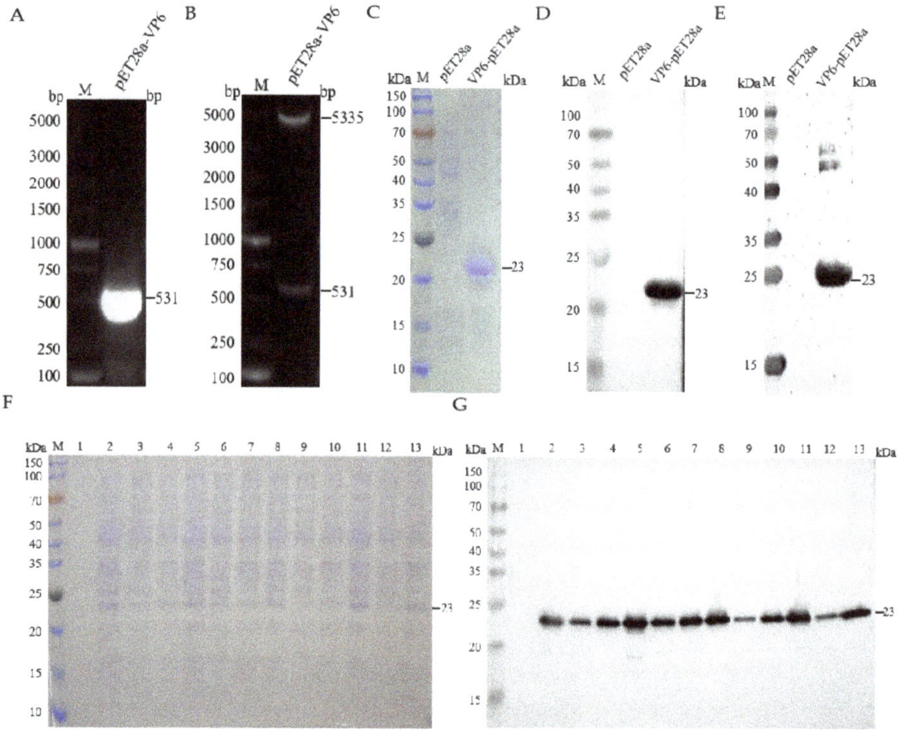


**Figure 1.** Expression and purification of recombinant protein pET-28a(+)-VP6. (**A**) Results of VP6 gene amplification. (**B**) Restriction enzyme digestion identification of recombinant plasmid pET28a(+)-VP6. (**C**) Results of SDS-PAGE for protein-induced expression. (**D**) Results of WB for protein-induced expression. (**E**) Verification of recombinant protein pET-28a(+)-VP6 by SDS-PAGE. (**F**) SDS-PAGE results of induced expression of pET28a(+)-VP6 protein. Lane 1: uninduced bacteria; lanes 2–4: whole bacteria, supernatant, and precipitate induced with 0.5 mM IPTG at 16 °C; lanes 5–7: whole bacteria, supernatant, and precipitate induced with 1 mM IPTG at 16 °C; lanes 8–10: whole bacteria, supernatant, and precipitate induced with 0.5 mM IPTG at 37 °C; lanes 11–13: whole bacteria, supernatant, and precipitate induced with 1 mM IPTG at 37 °C. (**G**) WB results of induced expression of pET28a(+)-VP6 protein. The lanes are the same as in (**E**).

These monoclonal antibodies exhibited exceptional specificity and high-affinity binding, forming a robust foundation for the development of precision diagnostics, the differential detection of co-infections, and advanced immunoassays. WB assays confirmed that all three mAbs specifically recognized the PoRV G3 and X1 genotypes, without cross-reactivity toward heterologous secondary antibodies (Figure 2B–D). Specificity was further substantiated using the JEV-NS5 protein as a negative control (Figure 2E). Consistently, the IFA results corroborated the WB findings, wherein the hybridoma supernatants, diluted at 1:100, retained antigen-binding capabilities in PoRV-infected MA104 cells and exhibited no cross-reactivity with other viruses, such as JEV (Figure 2F).

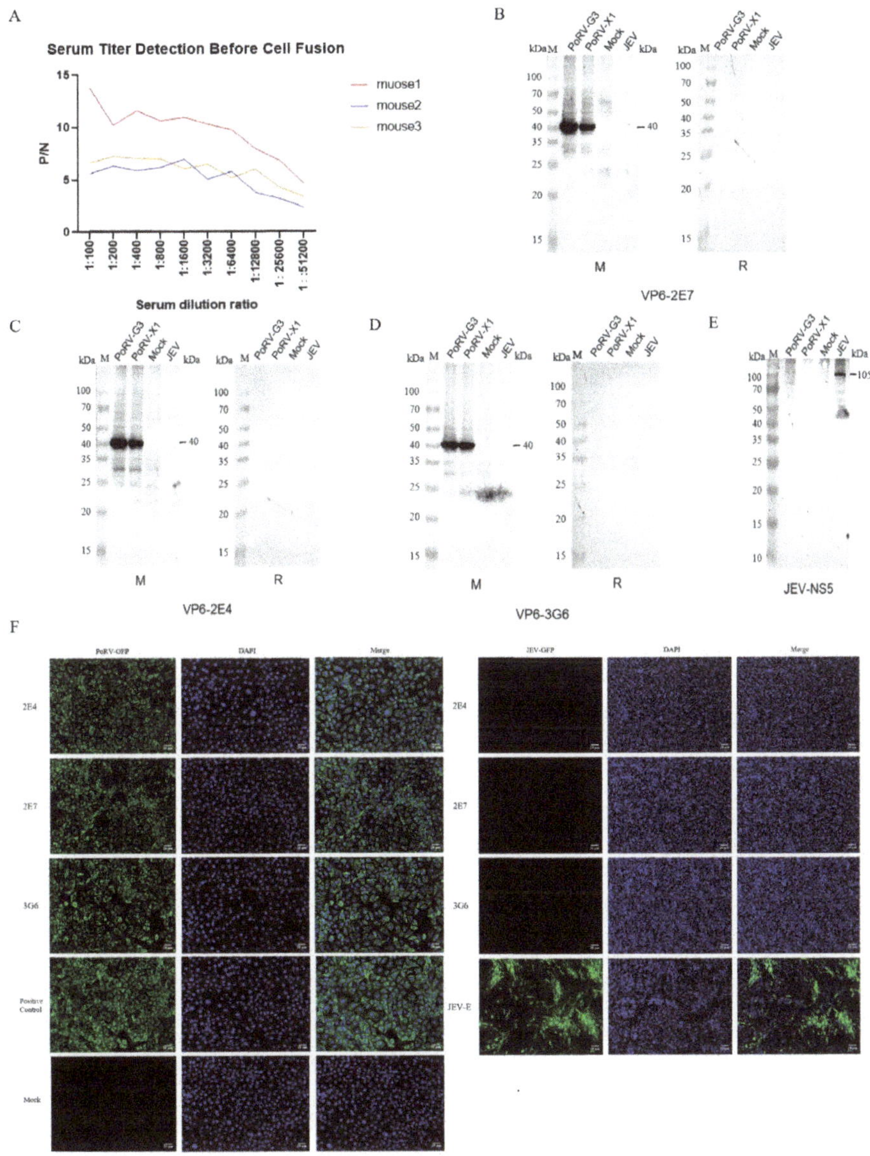


**Figure 2.** Identification of monoclonal antibodies. (**A**) Antibody levels in three mice after immunization. (**B–E**) Specific identification of VP6 mAbs. M: 250 kDa protein marker. PoRV-G3: lysate of cells infected with PoRV-G3. PoRV-X1: lysate of cells infected with PoRV-X1. Mock: lysate of uninfected cells. JEV: lysate of cells infected with Japanese Encephalitis Virus. M refers to mouse-derived secondary antibody, and R refers to rabbit-derived secondary antibody. (**F**) An indirect immunofluorescence assay was performed to determine the reactivity of supernatants from three hybridoma cell lines with both PoRV-infected MA104 cells and JEV-infected BHK cells. Proteins in virus-infected cells were labeled with green fluorescence.

### *3.3. Construction of VP6 Truncation Variants and Epitope Mapping*

A bifurcated truncation strategy was employed to engineer polypeptide fragments spanning amino acids 1–97 and 87–174. Recombinant constructs were successfully assembled and authenticated via Sanger sequencing (Figure 3A). WB analysis demonstrated that all three monoclonal antibodies recognized epitopes localized within the 98–174-amino-acid region (Figure 3C). Subsequently, systematic internal deletions were implemented across this region by sequentially excising 10-amino-acid segments. Each deletion variant was cloned in triplicate, and clones displaying consistent electrophoretic characteristics were selected for commercial sequencing (Sangon Biotech, Shanghai, China) (Figure 3B). The comprehensive epitope mapping workflow is depicted in Figure 3F. Epitope identification revealed that monoclonal antibody 2E7 specifically bound to residues 128–137 (YIKNWNLQNR) (Figure 3E), whereas antibodies 2E4 and 3G6 recognized a contiguous yet distinct epitope encompassing residues 138–147 (RQRTGFVFHK) (Figure 3D).

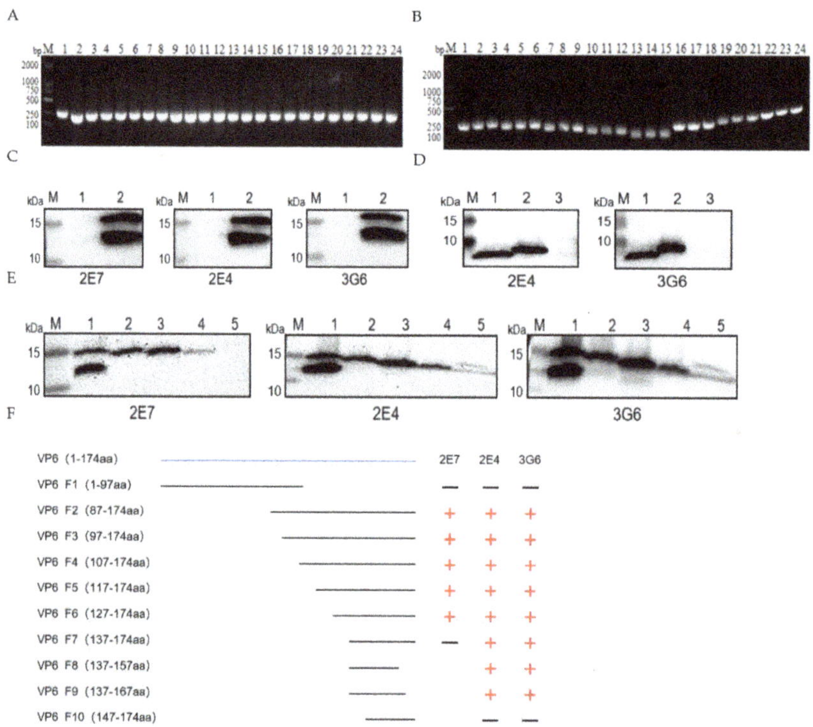


**Figure 3.** Identification of B-cell determinants of mAbs specific to VP6 monoclonal antibodies. (**A**) Construction of VP6 truncation mutants by dichotomy. Lanes 1-12: F1; 13-24: F2. (**B**) PCR amplification of VP6 truncation mutants in the last two rounds. Lanes 1-3: F8; 4-6: F9; 7-9: F10; 10-12: F7; 13-15: F6; 16-18: F5; 19-21: F4; 22-24: F3. (**C**) WB analysis of first-round truncation mutants with antibodies. Lane 1: F1-expressed protein; 2: F2-expressed protein. (**D**) Epitope mapping of 2E4 and 3G6 (antibody incubation). Lane 1: F8-expressed protein; 2: F9; 3: F10. (**E**) Epitope mapping of 2E7 (antibody incubation). Lane 1: F3-expressed protein; 2: F4; 3: F5; 4: F6; 5: F7. (**F**) Flowchart of truncation construct construction.

### 3.4. Epitope Conservation Analysis

Multiple sequence alignment using MegAlign revealed that, within epitope $^{128}$YIKNWNLQNR$^{137}$, residue K130 was substituted with E130 in all analyzed strains except AXF43000.1, while the remaining residues were highly conserved across strains (Figure 4A). In contrast, the epitope $^{138}$RQRTGFVFHK$^{147}$ was completely conserved among the eight representative Porcine Rotavirus isolates evaluated (Figure 4B).

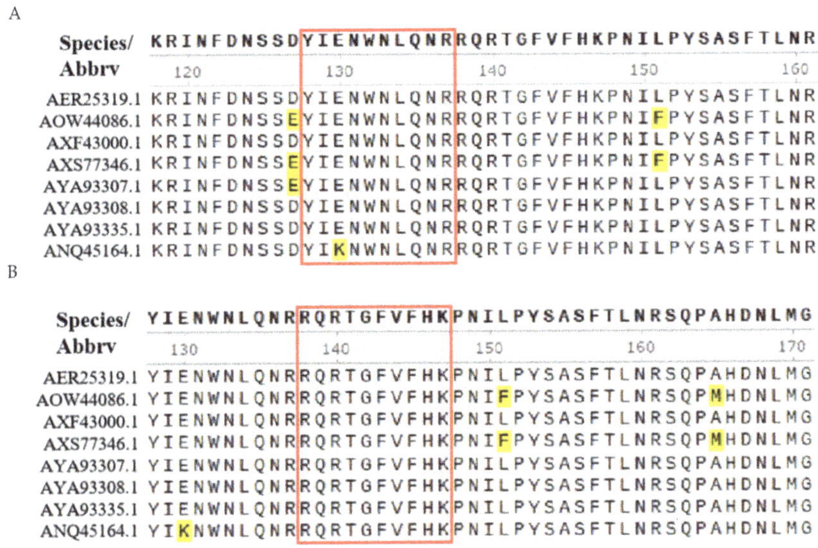


**Figure 4.** Epitope conservation analysis. (**A**) Multiple sequence alignment of the 2E7 epitope $^{128}$YIKNWNLQNR$^{137}$ using MegAlign. (**B**) Multiple sequence alignment of the 2E4/3G6 epitope $^{138}$RQRTGFVFHK$^{147}$ using MegAlign. Note: The region demarcated by the red frame in the figure corresponds to the epitope identified by the antibody, while the highlighted region denotes the differential segment in the sequence alignment.

## 4. Discussion

RV, a member of the Reoviridae family, constitutes a major causative agent of viral gastroenteritis in both young children and livestock worldwide [16,17]. First identified in swine feces in 1975 [18], PoRV is primarily transmitted via the fecal–oral route and causes clinical symptoms such as vomiting, anorexia, dehydration, and severe diarrhea in neonatal piglets, often leading to high morbidity and significant mortality [8]. Endemic diarrhea in piglets continues to impose considerable economic burdens on the swine industry, particularly in China. Contemporary surveillance implicates PoRVA as the predominant viral pathogen responsible for enteric disease in suckling and weaned piglets [8,19]. Furthermore, PoRVA exhibits zoonotic potential, with increasing evidence of cross-species transmission to humans and other economically critical livestock. For instance, epidemiological studies in India have revealed widespread PoRVA dissemination among pig populations, and the genetic characterization of VP7, NSP3, and NSP4 genes underscores its interspecies transmissibility [20]. The close sequence homology between PoRV and human RV strains suggests that the mAbs developed in this study may serve as diagnostic or therapeutic tools in a One Health context—facilitating rapid detection across host species and aiding in integrated control strategies. Notably, the conservation of the identified epitopes may offer a foundation for universal cross-species diagnostics, minimizing missed detections in public health monitoring programs. In China, PoRVA is consistently ranked as a leading etiological agent of piglet diarrhea, second only to Porcine Epidemic Diarrhea Virus (PEDV). Phylogenetic analyses of VP6 genes from Shandong strains revealed five distinct group A RV genotypes [9], underscoring their broad distribution and persistent threat to China's swine sector [19]. Despite the disease burden, there are currently no effective antiviral treatments available for PoRV, rendering immunoprophylaxis the primary strategy to mitigate viral spread and reduce the clinical incidence [21–24]. Hence, the development of efficacious vaccines and precise, field-deployable diagnostic tools is critical to improve disease surveillance and ensure timely outbreak control [25,26].

VP6, the most prevalent genotype circulating in China, is characterized by extensive interstrain conservation, prominent immunogenicity, and unique strain-specific antigenic regions [27]. As the first RV protein to bridge immunoreactivity and genetic variability, VP6 has played a central role in rotavirus taxonomy and is a prime target for diagnostic and vaccine development efforts [28]. Owing to its high expression levels, structural integrity, and immunological prominence, VP6 is ideal for functional analyses and classification-based studies [29–31].

This study hypothesized that "monoclonal antibodies generated against conserved VP6 epitopes can specifically recognize PoRVA and thereby serve as viable tools for diagnostics and vaccine design." Supporting this premise, we successfully produced high-affinity, VP6-specific mAbs with demonstrable specificity and no cross-reactivity to non-rotaviral pathogens. Currently, there is limited research on the VP6 protein of PoRVA and its monoclonal antibodies [32,33]. Despite its significance, research on PoRVA VP6-specific mAbs remains limited. For example, Badillo-Godinez et al. demonstrated that the parenteral administration of VP6 targeting DEC-205 elicited mucosal protection, whereas Atta Muhammad Memon et al. developed VP6-targeted antibodies for antigen capture ELISA without evaluating interstrain cross-reactivity. Notably, the three monoclonal antibodies (mAbs) developed in this study demonstrated high specificity for PoRVA, exhibiting no cross-reactivity with heterologous non-rotaviral pathogens. This result robustly substantiates the hypothesis that VP6 represents a highly specific and diagnostically viable antigenic target for PoRVA detection. For instance, Kohli et al. utilized pepscan technology to identify four antigenic domains on bovine RV VP6, while others characterized five linear B-cell epitopes via truncated expression and synthetic peptide mapping [34]. Analogously, preceding investigations utilized truncated protein expression in conjunction with synthetic peptide methodologies to delineate five linear antigenic epitopes positioned at amino acid residues 6–20, 96–110, 134–144, 302–310, and 361–372 [35]. Nevertheless, the limited conservation of specific epitopes, such as those characterized in the 1E5 reference strain, may undermine diagnostic specificity [35]. Additionally, a B-cell epitope encompassing residues 289–302 was identified through the use of overlapping synthetic peptides [27]. In this study, two novel linear B-cell epitopes ($^{128}$YIKNWNLQNR$^{137}$ and $^{138}$RQRTGFVFHK$^{147}$) within the VP6 protein were elucidated. Conservation analysis revealed that these epitopes are highly conserved among circulating PoRVA strains, underscoring their broad-spectrum diagnostic potential. These findings align with the hypothesis that VP6 conservation enables cross-genotypic detection. While epitope mapping via truncated protein constructs is efficient, technically straightforward, and cost-effective, it is inherently constrained in identifying conformational epitopes, as improper folding may result in epitope loss. Moreover, limitations in terms of expression yield and mapping resolution necessitate integration with complementary approaches to improve the analytical precision. Accordingly, iterative methodological optimization remains critical for the advancement of epitope-based diagnostic platforms and vaccine strategies.

In conclusion, this study successfully generated high-affinity and highly specific monoclonal antibodies against PoRVA VP6 through a prokaryotic expression platform. Two novel, evolutionarily conserved B-cell epitopes were delineated, furnishing a critical theoretical foundation for the development of VP6-oriented diagnostic assays and vaccine candidates. Nonetheless, certain limitations should be acknowledged: specificity assessments were restricted to prototype PoRVA strains and a limited subset of serotypes, with no validation performed against a broader spectrum of emergent variants or clinical isolates, potentially limiting translational applicability. Additionally, the functional activity of the mAbs was evaluated exclusively via Western blotting and indirect ELISA, without in vivo assessment of their neutralizing efficacy; therefore, their neutralizing potential in vivo warrants further investigation. Subsequent investigations will prioritize the elucidation of cross-reactivity profiles against emerging PoRV genotypes (e.g., G9, G12) and geographically diverse clinical strains using molecular virology methodologies to substantiate their utility in epidemiological monitoring. Capitalizing on the conserved nature of the identified epitopes, a double-antibody sandwich ELISA employing these mAbs as capture antibodies will be established and integrated with existing PEDV detection platforms to facilitate rapid, dual-pathogen diagnostic capabilities for piglet enteric diseases. Furthermore, targeting conserved VP6 determinants to enhance diagnostic sensitivity and specificity will bolster early pathogen detection, molecular surveillance, vaccine innovation, and point-of-care diagnostics, thereby contributing substantial practical value to swine health management and public health preparedness.

## 5. Conclusions

Overall, this study successfully harnessed a prokaryotic expression system to produce and purify the VP6 protein of PoRV, laying the groundwork for the generation of three mAbs exhibiting superior specificity and sensitivity. These mAbs effectively recognized VP6 without eliciting non-specific interactions, underscoring their diagnostic potential. Through systematic epitope mapping, two previously uncharacterized linear B-cell epitopes ($^{128}$YIKNWNLQNR$^{137}$ and $^{138}$RQRTGFVFHK$^{147}$) were identified, both highly conserved among diverse circulating PoRVA strains. The elucidation of these conserved epitopes and the development of corresponding high-affinity mAbs provide robust molecular tools and conceptual insights that will advance rapid diagnostic assay development, facilitate epidemiological monitoring, and inform the strategic design of VP6-based vaccines, thereby contributing substantially to PoRV prevention and control efforts.

**Supplementary Materials:** The following supporting information can be downloaded at: https://www.mdpi.com/article/10.3390/vetsci12080710/s1, Figure S1: Flow chart of monoclonal antibody preparation; Figure S2: Prediction of the transmembrane domain of the VP6 gene; Figure S3: Prediction of the 3D modeling structure of the VP6 protein; Table S1: Specific Primers for VP6 Gene Amplification; Table S2: Immunization Protocol (Note: SC: Subcutaneous Injection; IP: Intraperitoneal Injection.); Table S3: Specific Primers for VP6 Gene Amplification.

**Author Contributions:** Conceptualization, B.Z., B.S. and D.M.; methodology, B.S.; software, X.B. and L.Z. (Linke Zou); validation, D.M., B.S. and X.B.; formal analysis, L.Z. (Linhan Zhong) and P.Z.; investigation, J.B., R.L., P.H. and Q.W.; resources, B.Z.; data curation, D.M.; writing—original draft preparation, B.S.; writing—review and editing, B.Z. and J.C.; visualization, J.C.; supervision, B.Z.; project administration, B.Z.; funding acquisition, B.Z. All authors have read and agreed to the published version of the manuscript.

**Funding:** This work was supported by the Jiangsu Province Agricultural Science and Technology Independent Innovation Fund Project (grant number: CX(24)1010).

**Institutional Review Board Statement:** The animal study protocol was approved by the Institutional Review Board of Nanjing Agricultural University (protocol code NJAU.No20220706145 and date 2022-11).

**Informed Consent Statement:** Not applicable.

**Data Availability Statement:** No new data were created or analyzed in this study. Data sharing is not applicable to this article.

**Conflicts of Interest:** The authors declare no conflicts of interest.

## References


1. Mebus, C.A.; Underdahl, N.R.; Rhodes, M.B.; Twiehaus, M.J. Further studies on neonatal calf diarrhea virus. *Proc. Annu. Meet. US Anim. Health Assoc.* **1969**, *73*, 97–99.
2. Woode, G.N.; Bridger, J.; Hall, G.A.; Jones, J.M.; Jackson, G. The isolation of reovirus-like agents (rota-viruses) from acute gastroenteritis of piglets. *J. Med. Microbiol.* **1976**, *9*, 203–209. [CrossRef]
3. Sánchez-Tacuba, L.; Kawagishi, T.; Feng, N.; Jiang, B.; Ding, S.; Greenberg, H.B. The role of the VP4 attachment protein in rotavirus host range restriction in an in vivo suckling mouse model. *J. Virol.* **2022**, *96*, e00550-22. [CrossRef]
4. Mcdonald, S.M.; Nelson, M.I.; Turner, P.E.; Patton, J.T. Reassortment in segmented RNA viruses: Mechanisms and outcomes. *Nat. Rev. Microbiol.* **2016**, *14*, 448–460. [CrossRef] [PubMed]
5. Matthijnssens, J.; Ciarlet, M.; McDonald, S.M.; Attoui, H.; Banyai, K.; Brister, J.R.; Buesa, J.; Esona, M.D.; Estes, M.K.; Gentsch, J.R.; et al. Uniformity of rotavirus strain nomenclature proposed by the rotavirus classification working group (RCWG). *Arch. Virol.* **2011**, *156*, 1397–1413. [CrossRef] [PubMed]
6. Crawford, S.E.; Labbe, M.; Cohen, J.; Burroughs, M.H.; Zhou, Y.J.; Estes, M.K. Characterization of virus-like particles produced by the expression of rotavirus capsid proteins in insect cells. *J. Virol.* **1994**, *68*, 5945–5952. [CrossRef] [PubMed]
7. Yan, W.; Huang, S.; Zhang, L.; Yang, Q.; Liu, S.; Wang, Z.; Chu, Q.; Tian, M.; Zhao, L.; Sun, Y.; et al. Virus-like particles vaccine based on co-expression of g5 porcine rotavirus VP2-VP6-VP7 induces a powerful immune protective response in mice. *Vet. Microbiol.* **2024**, *298*, 110241. [CrossRef]
8. Vlasova, A.N.; Amimo, J.O.; Saif, L.J. Porcine rotaviruses: Epidemiology, immune responses and control strategies. *Viruses* **2017**, *9*, 48. [CrossRef]
9. Xue, R.; Tian, Y.; Zhang, Y.; Zhang, M.; Li, Z.; Chen, S.; Liu, Q. Diversity of group a rotavirus of porcine rotavirus in shandong province China. *Acta Virol.* **2018**, *62*, 229–234. [CrossRef]
10. Jacobson, M. On the infectious causes of neonatal piglet diarrhoea—A review. *Vet. Sci.* **2022**, *9*, 422. [CrossRef]
11. Memon, A.M.; Bhuyan, A.A.; Chen, F.; Guo, X.; Menghwar, H.; Zhu, Y.; Ku, X.; Chen, S.; Li, Z.; He, Q. Development and validation of monoclonal antibody-based antigen capture ELISA for detection of group a porcine rotavirus. *Viral Immunol.* **2017**, *30*, 264–270. [CrossRef]
12. Li, W.; Li, Y.; Li, M.; Zhang, H.; Feng, Z.; Xu, H.; Li, C.; Guo, Z.; Gong, B.; Peng, J.; et al. Development and application of a blocking ELISA based on a n protein monoclonal antibody for the antibody detection against porcine reproductive and respiratory syndrome virus 2. *Int. J. Biol. Macromol.* **2024**, *269*, 131842. [CrossRef]
13. Zhao, Y.; Ren, H.; Lin, Z.; Shi, S.; Zhang, B.; Zhang, Y.; Han, S.; He, W.-R.; Wan, B.; Hu, M.; et al. Identification and characterization of a novel B cell epitope of ASFV virulence protein B125R monoclonal antibody. *Viruses* **2024**, *16*, 1257. [CrossRef] [PubMed]
14. Liang, Y.; Kuang, Q.; Zheng, X.; Xu, Y.; Feng, Y.; Xiang, Q.; Zhang, G.; Zhou, P. Monoclonal antibody development for early detection of ASFV i73r protein: Identification of a linear antigenic epitope. *Virology* **2024**, *597*, 110145. [CrossRef]
15. Gao, C.; Huang, Z.; You, J.; Zhang, W.; Tang, S.; Gong, L.; Zhang, G. Identification of a novel b cell epitope of ASFV pCP312r recognized using a monoclonal antibody. *Vet. Microbiol.* **2024**, *298*, 110247. [CrossRef]
16. Campanha, J.E.T.; Possatti, F.; Lorenzetti, E.; de Almeida Moraes, D.; Alfieri, A.F.; Alfieri, A.A. Longitudinal study of rotavirus c VP6 genotype i6 in diarrheic piglets up to 1 week old. *Braz. J. Microbiol.* **2020**, *51*, 1345–1351. [CrossRef]
17. Wakuda, M.; Ide, T.; Sasaki, J.; Komoto, S.; Ishii, J.; Sanekata, T.; Taniguchi, K. Porcine rotavirus closely related to novel group of human rotaviruses. *Emerg. Infect. Dis.* **2011**, *17*, 1491–1493. [CrossRef] [PubMed]
18. Rodger, S.M.; Craven, J.A.; Williams, I. Letter: Demonstration of reovirus-like particles in intestinal contents of piglets with diarrhoea. *Aust. Vet. J.* **1975**, *51*, 536. [CrossRef] [PubMed]
19. Wenske, O.; Ruckner, A.; Piehler, D.; Schwarz, B.; Vahlenkamp, T.W. Epidemiological analysis of porcine rotavirus a genotypes in germany. *Vet. Microbiol.* **2018**, *214*, 93–98. [CrossRef]
20. Abass, G.; Dubal, Z.B.; Rajak, K.K.; Kale, B.M.; Raorane, A.; Dudhe, N.; Malla, B.A.; Desai, D.; Sinha, D.K.; Kumar, O.R.V.; et al. Molecular characterization of porcine rotavirus a from india revealing zooanthroponotic transmission. *Anim. Biotechnol.* **2022**, *33*, 1073–1085. [CrossRef]
21. Shi, H.; Chen, J.; Li, H.; Sun, D.; Wang, C.; Feng, L. Molecular characterization of a rare G9P [23] porcine rotavirus isolate from China. *Arch. Virol.* **2012**, *157*, 1897–1903. [CrossRef]
22. Azevedo, M.P.; Vlasova, A.N.; Saif, L.J. Human rotavirus virus-like particle vaccines evaluated in a neonatal gnotobiotic pig model of human rotavirus disease. *Expert Rev. Vaccines* **2013**, *12*, 169–181. [CrossRef]
23. Li, F.; Mei, Z.; Ju, N.; Sui, L.; Fan, X.; Wang, Z.; Li, J.; Jiang, Y.; Cui, W.; Shan, Z.; et al. Evaluation of the immunogenicity of auxotrophic lactobacillus with CRISPR-cas9d10a system-mediated chromosomal editing to express porcine rotavirus capsid protein VP4. *Virulence* **2022**, *13*, 1315–1330. [CrossRef] [PubMed]
24. Shao, L.; Fischer, D.D.; Kandasamy, S.; Rauf, A.; Langel, S.N.; Wentworth, D.E.; Stucker, K.M.; Halpin, R.A.; Lam, H.C.; Marthaler, D.; et al. Comparative in vitro and in vivo studies of porcine rotavirus G9P [13] and human rotavirus wa G1P [8]. *J. Virol.* **2016**, *90*, 142–151. [CrossRef] [PubMed]

25. Buttery, J.P.; Danchin, M.H.; Lee, K.J.; Carlin, J.B.; McIntyre, P.B.; Elliott, E.J.; Booy, R.; Bines, J.E.; For the PAEDS/APSU Study Group. Intussusception following rotavirus vaccine administration: Post-marketing surveillance in the national immunization program in australia. *Vaccine* **2011**, *29*, 3061–3066. [CrossRef]
26. Chandler-Bostock, R.; Hancox, L.R.; Payne, H.; Iturriza-Gomara, M.; Daly, J.M.; Mellits, K.H. Diversity of group a rotavirus on a UK pig farm. *Vet. Microbiol.* **2015**, *180*, 205–211. [CrossRef]
27. Banos, D.M.; Lopez, S.; Arias, C.F.; Esquivel, F.R. Identification of a t-helper cell epitope on the rotavirus VP6 protein. *J. Virol.* **1997**, *71*, 419–426. [CrossRef]
28. Nagesha, H.S.; Brown, L.E.; Holmes, I.H. Neutralizing monoclonal antibodies against three serotypes of porcine rotavirus. *J. Virol.* **1989**, *63*, 3545–3549. [CrossRef] [PubMed]
29. Afchangi, A.; Jalilvand, S.; Arashkia, A.; Latifi, T.; Farahmand, M.; Shirazi, M.M.A.; Nasab, S.D.M.; Marashi, S.M.; Roohvand, F.; Shoja, Z. Co-administration of rotavirus nanospheres VP6 and NSP4 proteins enhanced the anti-NSP4 humoral responses in immunized mice. *Microb. Pathog.* **2022**, *163*, 105405. [CrossRef]
30. Shoja, Z.; Jalilvand, S.; Latifi, T.; Roohvand, F. Rotavirus VP6: Involvement in immunogenicity, adjuvant activity, and use as a vector for heterologous peptides, drug delivery, and production of nano-biomaterials. *Arch. Virol.* **2022**, *167*, 1013–1023. [CrossRef]
31. Yuan, J.; Zhang, X.; Shi, H.; Chen, J.; Han, X.; Wei, P.; Feng, L. The interaction of rotavirus a pig/China/NMTL/2008/G9P [23] VP6 with cellular beta-actin is required for optimal RV replication and infectivity. *Vet. Microbiol.* **2016**, *197*, 111–121. [CrossRef] [PubMed]
32. Memon, A.M.; Chen, F.; Khan, S.B.; Guo, X.; Khan, R.; Khan, F.A.; Zhu, Y.; He, Q. Development and evaluation of polyclonal antibodies based antigen capture ELISA for detection of porcine rotavirus. *Anim. Biotechnol.* **2023**, *34*, 1807–1814. [CrossRef] [PubMed]
33. Badillo-Godinez, O.; Gutierrez-Xicotencatl, L.; Plett-Torres, T.; Pedroza-Saavedra, A.; Gonzalez-Jaimes, A.; Chihu-Amparan, L.; Maldonado-Gama, M.; Espino-Solis, G.; Bonifaz, L.; Esquivel-Guadarrama, F. Targeting of rotavirus VP6 to DEC-205 induces protection against the infection in mice. *Vaccine* **2015**, *33*, 4228–4237. [CrossRef]
34. Kohli, E.; Maurice, L.; Bourgeois, C.; Bour, J.B.; Pothier, P. Epitope mapping of the major inner capsid protein of group A rotavirus using peptide synthesis. *Virology* **1993**, *194*, 110–116. [CrossRef] [PubMed]
35. Han, X.; Chen, J.; Shi, H.; Zhang, X.; Shi, D.; Chi, Y.; Li, C.; Feng, L. Preparation of monoclonal antibodies against VP6 protein of porcine group a rotavirus and identification of its antigenic epitopes. *J. Prev. Vet. Med. China* **2014**, *36*, 885–888.

*veterinary sciences*

MDPI

*Article*

# Isolation and Biological Characteristics Study of Porcine Reproductive and Respiratory Syndrome Virus GZ2022 Strain

**Xinmei Yang [1,2,†], Bin Yu [1,2,†], Qing Li [3,†], Hailong Ma [1,2], Zhengjun Yu [4], Pei Ma [4], Shengnan Ruan [1,2], Xuexiang Yu [1,2], Qigai He [1,2,*] and Wentao Li [1,2,*]**

1 National Key Laboratory of Agricultural Microbiology, College of Veterinary Medicine, Huazhong Agricultural University, Wuhan 430070, China; xinmei@webmail.hzau.edu.cn (X.Y.); yu940262935@163.com (B.Y.); littlelate@163.com (H.M.); ruan@webmail.hzau.edu.cn (S.R.); yuxuexiang@mail.hzau.edu.cn (X.Y.)

2 Key Laboratory of Preventive Veterinary Medicine in Hubei Province, The Cooperative Innovation Center for Sustainable Pig Production, Wuhan 430070, China

3 Wuhan Golden Dragon Group, Wuhan 430015, China; liqing@whjinlong.com

4 Hunan Provincial Engineering Research Center for Biosecurity and Disease Control in Swine Production, Changsha 410125, China; yuzhengjun_scbio@163.com (Z.Y.); mapei_scbio@163.com (P.M.)

* Correspondence: he628@mail.hzau.edu.cn (Q.H.); wentao@mail.hzau.edu.cn (W.L.)

† These authors contributed equally to this work.

**Simple Summary:** Porcine reproductive and respiratory syndrome (PRRS) is a significant infectious disease threatening swine production. In China, emerging NADC30-like PRRSV strains exhibit concerning features including frequent recombination events and partial immune evasion capabilities. In this study, we isolated and characterized a novel recombinant NADC30-like PRRSV strain, designated GZ2022, from a swine farm in Guizhou Province. The biological and pathogenic properties of this strain were systematically evaluated. Phylogenetic and recombination analyses revealed that GZ2022 is a recombinant virus, with NADC30 serving as the major parental strain and HuN4 as the minor contributor. In vivo experiments demonstrated that GZ2022 exhibits moderate virulence, inducing typical PRRS clinical signs in piglets, but causes less severe symptoms than highly pathogenic strains. These findings provide important insights into the molecular epidemiology and pathogenicity of PRRSV in China.

**Abstract:** PRRSV continues to evolve, complicating its epidemiological landscape in China. In this study, we isolated a novel PRRSV strain, GZ2022, from a swine farm in Guizhou Province. Subsequent analyses performed on this isolate included complete genome sequencing, phylogenetic analysis, recombination assessment, and characterization of its biological properties. Phylogenetic analysis revealed that GZ2022 clusters within Lineage 1 (NADC30-like) and features a 131-amino-acid deletion in NSP2, consistent with NADC30-derived strains. Recombination analysis identified NADC30 as the major parental strain (75% genomic contribution), with a minor recombinant region (25%) derived from the highly pathogenic HuN4 strain. In vitro growth kinetics revealed peak viral titers in Marc-145 cells at 72 h post infection (hpi). Pathogenicity was evaluated in 21-day-old piglets infected with GZ2022, the highly pathogenic PRRSV strain WUH3, or negative controls. Both infected groups exhibited typical PRRS clinical signs (fever, respiratory distress) and histopathological lesions (interstitial pneumonia, pulmonary consolidation). However, GZ2022-infected piglets exhibited attenuated virulence compared to WUH3, with reduced pulmonary hemorrhage and 0% mortality compared to 80% in the WUH3 group. Seroconversion (N-protein antibodies) was observed at 14 dpi (days post inoculation) in GZ2022-infected animals, persisting throughout the 28-day trial. Viral shedding dynamics aligned with moderate pathogenicity. These findings classify GZ2022 as a moderately virulent NADC30-like recombinant strain with partial HuN4-derived genomic regions. The emergence of such strains underscores the need for sustained surveillance of PRRSV genetic diversity and systematic evaluation of the biological properties of novel variants to refine control measures and inform vaccine development.

**Keywords:** PRRSV; GZ2022; complete genome sequencing; genetic variation; phylogenetic analysis; pathogenicity

## 1. Introduction

PRRS is a major infectious disease affecting the swine industry in China [1–3]. It is characterized by reproductive disorders in sows and respiratory problems in pigs of all ages. Following infection with PRRSV, pigs often experience immune suppression, leading to secondary infections that severely impact herd health and productivity [4,5]. Furthermore, PRRSV evolves continuously through mutation and recombination, resulting in new strains that diminish the effectiveness of existing vaccines, complicating disease control efforts [6,7]. Despite ongoing management strategies, significant challenges in controlling PRRS persist in China.

The PRRSV genome is a single-stranded, positive-sense RNA molecule approximately 15 kb in length [8,9], characterized by a high mutation rate that enables rapid adaptation to the host and evasion of immune detection. Genetic variability and low homology between different strains can result in up to 60% genetic divergence, with divergence reaching about 10% even within the same strain [2]. PRRSV strains are diverse and are primarily categorized into European (PRRSV-1) and American (PRRSV-2) types [10,11], which differ significantly in biological characteristics and pathogenicity, complicating control strategies. Moreover, PRRSV has developed several mechanisms to escape host immune surveillance, leading to immune suppression after infection and facilitating secondary or mixed infections, particularly with other viruses or bacteria, which significantly increase mortality rates [12,13].

Recent years have witnessed frequent reports in China of a novel PRRSV strain belonging to Lineage 1. This strain shares high sequence similarity with the NADC30 strain first isolated in the United States in 2008, particularly in its amino acid (aa) deletion pattern within the nonstructural protein 2 (NSP2) region, leading to its designation as an NADC30-like strain [14–16]. NADC30-like strains exhibit three characteristic deletions in the NSP2 coding region: a 111-aa deletion spanning aa positions 322–432; a single-aa deletion at position 483; and a 19-aa deletion spanning positions 504–522, for a total of 131 noncontiguous aa deletions [17]. These deletions serve as key molecular markers distinguishing NADC30-like strains from other PRRSV lineages. However, both NADC30-like and highly pathogenic PRRSV (HP-PRRSV) strains are prone to recombination events, resulting in variants with significantly altered biological properties and virulence [18]. Since their emergence, NADC30-like strains have rapidly recombined with domestic PRRSV isolates and accumulated mutations, gradually replacing HP-PRRSV as the predominant lineage in Chinese swine herds [17,19]. Given the limited cross-protection afforded by commercial vaccines against these recombinant variants [20], isolating and characterizing currently circulating PRRSV strains remains essential for refining control measures and guiding vaccine development in China.

To better understand the biological characteristics and pathogenicity of currently circulating PRRSV strains, this study focused on the genetic evolution of the NADC30-like strain. A recombinant strain, designated PRRSV GZ2022, was isolated and identified, and its pathogenicity was systematically evaluated in comparison with the highly pathogenic PRRSV strain WUH3. PRRSV GZ2022 and WUH3 strains were separately inoculated into weaned piglets to investigate their pathogenicity, infection patterns, and the cross-protective ability of different sera against these strains. This research aims to provide a theoretical foundation for exploring the pathogenic mechanisms, immune protection, and vaccine development for recombinant strains of NADC30 and highly pathogenic PRRSV strains.

## 2. Materials and Methods

### 2.1. Cells, Viruses, Animals, and Ethics Statement

Porcine alveolar macrophages (PAMs) and Marc-145 cell lines, cryopreserved in liquid nitrogen, were obtained from our laboratory. Viral isolation was performed using PRRSV-positive lung tissue samples collected from pigs clinically suspected of being affected by PRRS at a swine farm in Jianhe County, Guizhou Province. The PRRSV GZ2022 strain (GenBank accession no. OR369723.1) was isolated, identified, and preserved in this study. The classical PRRSV strain R98 (accession number: DQ355796.1) and the highly pathogenic PRRSV strain WUH3 (accession number: HM853673.2) were maintained in our repository. Reference strains for phylogenetic analysis were retrieved from the NCBI GenBank database and peer-reviewed publications (full metadata are provided in Supplementary Table S1). Fifteen, 21-day-old, PRRSV antigen-negative healthy male piglets were procured from Hubei Jinlin Original Breeding Livestock Co., Ltd., Wuhan, China. All animal experiments were conducted under Animal Experimental License SYXK (E) 2020-0084 and Ethics Approval No. HZAUSW-2024-0038, in compliance with institutional guidelines for animal welfare and biosafety protocols.

### 2.2. Isolation and Identification of PRRSV Field Strains

Lung tissues from PRRS-suspected pigs were homogenized in DMEM, clarified by centrifugation (12,000× $g$, 3 min), and filtered (0.22 μm) for RNA extraction and cDNA synthesis. *ORF5* and *ORF7* genes were amplified via PCR (primers listed in Supplementary Table S2) and sequenced. PAMs were inoculated with a PCR-positive supernatant along with controls (WUH3 strain, DMEM). After a 2 h adsorption period, cells were maintained at 37 °C in 5% $CO_2$ and monitored for cytopathic effects (CPEs) at 24 hpi. Virus isolation was validated through three blind passages in PAMs, followed by serial propagation in Marc-145 cells, with cryopreservation at −80 °C after CPE stabilization.

### 2.3. Viral Plaque Purification and Growth Curve Detection

Plaque assays were performed as described previously [21]. Cells were fixed with 0.1% crystal violet in 20% ethanol for 4 h at room temperature (RT), and plaque morphology was imaged under standardized conditions. All experiments were performed in triplicate. PFU/mL and MOI were calculated based on PFU and cell counts.

For viral growth curve analysis, Marc-145 cells cultured in 6-well plates were infected with PRRSV isolates at a multiplicity of infection (MOI) of 0.1. Following infection, cell supernatants were collected at 12, 24, 36, 48, 60, 72, 84, 96, 108, and 120 hpi. All experiments were performed in triplicate. Viral titers were quantified using 50% tissue culture infectious dose ($TCID_{50}$) assays, with calculations performed according to the Reed–Muench method [22].

### 2.4. Indirect Immunofluorescence Assay (IFA)

Indirect immunofluorescence assays were performed as described by Ma et al. [23]. The cell culture supernatant exhibiting CPE was inoculated onto Marc-145 cells. At 48 hpi, cells were incubated with a murine anti-PRRSV N monoclonal antibody (this antibody was generated by the research group of Professor Zili Li at Huazhong Agricultural University), followed by an Alexa Fluor 488-conjugated goat anti-mouse IgG secondary antibody. Nuclei were counterstained with DAPI, and fluorescence signals were captured using a Nikon Eclipse Ti2 microscope under standardized exposure parameters.

### 2.5. *Phylogenetic and Evolutionary Analysis*

Overlapping primers (Supplementary Table S2) for full-genome amplification of strain GZ2022 were designed based on the NADC30-like PRRSV reference genome (GenBank: JN654459). Total RNA was reverse-transcribed and amplified into full-length PRRSV cDNA. Primer synthesis and Sanger sequencing were performed by Beijing Tsingke Biotechnology Co., Ltd. (Beijing, China). Genome assembly and annotation were performed using SeqMan Pro v17.0 and EditSeq v12.1.

The GZ2022 genome was phylogenetically analyzed against reference PRRSV strains (Supplementary Table S1) using (1) maximum-likelihood tree construction, (2) *NSP2* deletion profiling, and (3) comparative genomic/GP5 amino acid analyses. Evolutionary distances were derived from pairwise genome comparisons (1000 bootstrap replicates). To characterize recombination events in strain GZ2022, full-genome sequences were analyzed using SimPlot v3.5.1 and RDP5 Beta 4.101. SimPlot sliding window analysis (2000 bp window, 200 bp step) identified recombination signals between GZ2022 and potential parental strains. RDP5 was used to predict recombination sites within the full genomic sequence of the GZ2022 strain.

### 2.6. *Animal Experiments*

#### 2.6.1. Virus Inoculation in Piglets and Clinical Monitoring

A total of 15 piglets were randomly assigned to three groups ($n$ = 5 per group). Piglets in Group 1 were simultaneously inoculated with 2 mL of the PRRSV GZ2022 strain ($10^6$ $TCID_{50}$/mL) via the intranasal route and 2 mL via intramuscular injection. Piglets in Group 2 received an identical inoculation regimen with the PRRSV WUH3 strain ($10^6$ $TCID_{50}$/mL). Group 3 served as the negative control and received 2 mL of Marc-145 cell supernatant via both the intranasal and intramuscular routes. Rectal temperatures were recorded daily for 28 dpi; body weights and clinical signs (scored daily using a standardized system: see Supplementary Table S3 for detailed criteria) were assessed at 1, 3, 5, 7, 9, 11, 13, 15, 17, and 19 dpi [24].

#### 2.6.2. Viral Load Quantification and Humoral Immune Response Assessment

Serum samples collected at 0, 1, 3, 7, 14, 21, and 28 dpi were analyzed for viral RNA via qRT-PCR (primers listed in Supplementary Table S2). Piglets that succumbed during the trial underwent immediate necropsy, while survivors were euthanized at 28 dpi. Tissues (lungs, spleen, kidneys, tonsils, and inguinal lymph nodes) were harvested for viral load quantification by qRT-PCR [25]. Serum samples (0–28 dpi) were analyzed using a PRRSV-specific ELISA kit (BioChek, Reeuwijk, The Netherlands) following the method of Lu et al. [26], with antibody levels quantified via standard curves (seropositivity cutoff: 0.4). Virus neutralization assays [27] tested GZ2022-infected piglet sera (14–28 dpi) against R98, WUH3 (HP-PRRSV), and NADC30-like strains, with neutralizing titers calculated by the Reed–Muench method.

#### 2.6.3. Necropsy and Gross Pathological Examination

Piglets that died during the experimental period underwent immediate necropsy, with major gross pathological lesions systematically documented. At the study endpoint (28 dpi), all surviving piglets were euthanized and subjected to necropsy to evaluate the pathological features of major organs, such as the lungs, in the GZ2022 infection group, WUH3 infection group, and control group.

### 2.7. Statistical Analyses

Statistical analyses were performed using Student's $t$-test for independent samples in GraphPad Prism 10 (GraphPad Software Inc., San Diego, CA, USA). Data are expressed as mean $\pm$ standard deviation (mean $\pm$ SD). Statistical significance was defined as follows: * $p < 0.05$ (significant), ** $p < 0.01$, *** $p < 0.001$, and **** $p < 0.0001$ (highly significant); differences with $p > 0.05$ were considered nonsignificant.

## 3. Results

### 3.1. Virus Isolation and Identification

Specific amplification of the *ORF5* (957 bp) and *ORF7* (671 bp) genes via PCR and agarose gel electrophoresis confirmed PRRSV infection, with banding patterns consistent with positive controls (Figure 1A). Distinct CPEs were observed in PRRSV-inoculated cells. PAM cells exhibited pronounced shrinkage and membrane rupture, while Marc-145 cells developed characteristic grape-like clusters, a hallmark of PRRSV infection. In contrast, negative control cells retained intact morphology without pathological alterations (Figure 1B). These results demonstrate that the GZ2022 strain retains cytopathogenic capability in both PAM and Marc-145 cells, confirming its infectivity and replication efficiency in vitro.

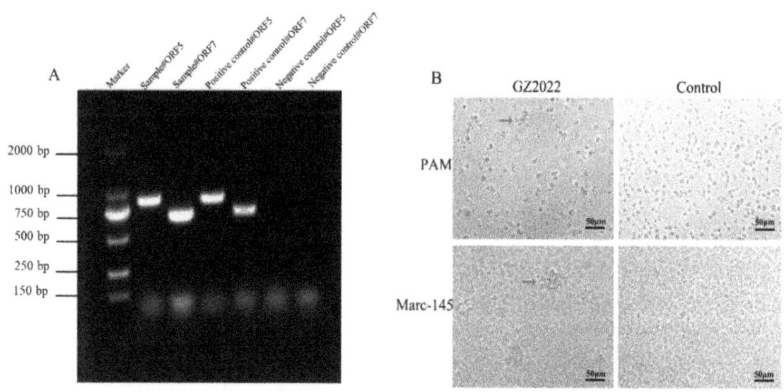


**Figure 1.** Isolation and identification of PRRSV GZ2022. (**A**) PCR identification of PRRSV GZ2022. PCR detection of PRRSV-ORF7 and PRRSV-ORF5 genes was performed on the cDNA of the samples. (**B**) CPE identification of the GZ2022 strain in PAM and Marc-145 cells. The GZ2022 strain was inoculated into PAM and Marc-145 cells, with a blank control group included. The red arrow indicates the site of cell lesion. Cytopathic effects were observed under an optical microscope at 3 to 4 dpi (CPE; 40 $\times$ magnification).

### 3.2. Plaque Purification and Growth Curve Assay of GZ2022

Serial 10-fold dilutions ($10^{-2}$–$10^{-6}$) of GZ2022 were subjected to plaque assays on Marc-145 monolayers. After three rounds of plaque purification, uniform plaque morphology was observed across generations. At the $10^{-6}$ dilution, a single plaque was isolated, while no plaques formed in negative controls. Crystal violet staining revealed distinct plaque boundaries (Figure 2A).

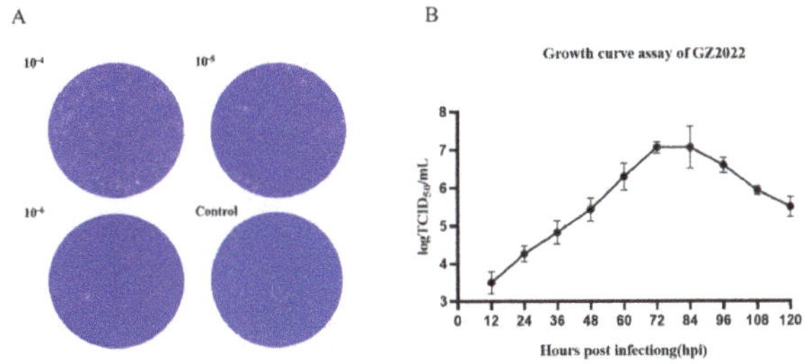


**Figure 2.** (**A**) Plaque assays of GZ2022 strain. Plaque assays were conducted using three experimental groups at dilutions of $10^{-4}$, $10^{-5}$, and $10^{-6}$, alongside a control group (Scale bar = 200 μm). (**B**) One step growth curve of PRRSV GZ2022 strain. The experiment was repeated three times at each time point.

The 50% tissue culture infectious dose ($TCID_{50}$) of GZ2022 was determined to be $10^{-6}$/mL via the Reed–Muench method. A one-step growth curve revealed peak viral titers ($10^{7.1}$ $TCID_{50}$/mL) at 72 hpi in Marc-145 cells, followed by a gradual decline (Figure 2B).

### 3.3. IFA Results of GZ2022

Immunofluorescence analysis using a monoclonal antibody targeting the PRRSV N protein revealed intense green fluorescence in Marc-145 cells infected with the GZ2022 strain (Figure 3), confirming active viral replication. DAPI counterstaining confirmed the integrity of the nuclei, with fluorescence predominantly localized to the perinuclear regions.

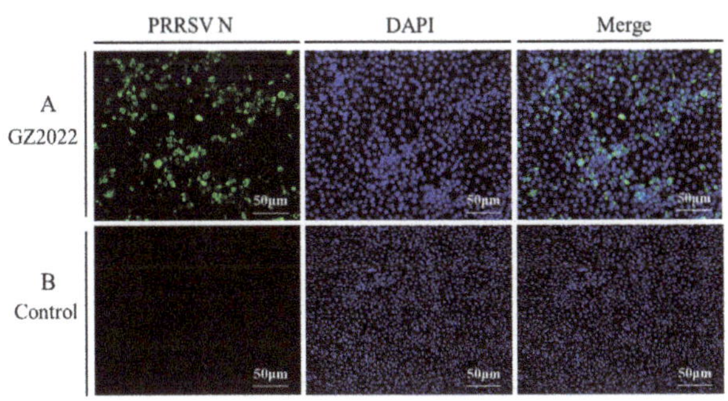


**Figure 3.** Immunofluorescence analysis of PRRSV infection in Marc-145 cells at 48 hpi. (**A**) Marc-145 cells infected with the PRRSV GZ2022 strain exhibited positive immunofluorescence staining (green) accompanied by CPEs. (**B**) No specific immunofluorescence signal was detected in mock-inoculated cells (CPE; 40 × magnification).

### 3.4. Genomic Characterization and Recombination Analysis of the GZ2022 Strain

Whole-genome sequencing of the GZ2022 strain generated a 14,638 bp sequence, which was deposited in the NCBI GenBank under accession number OR369723.1. Phylogenetic analysis of the *ORF5* (Figure 4A), *NSP2* (Figure 4B), and complete genome (Figure 4C) confirmed its classification within Lineage 1 (NADC30-like strains). We performed sequence alignment of NSP2 amino acid sequences from the GZ2022 isolate and reference strains, revealing three discontinuous deletions in GZ2022 totaling 131 amino acids relative to classical VR-2332: a 111-aa deletion (positions 322–432), a 1-aa deletion (position 483), and a 19-aa deletion (positions 504–522), with this deletion topology being identical to the NADC30 reference strain as shown in Figure 4D. GP5 amino acid mutation analysis identified critical epitope mutations through comparison of GP5 protein sequences between GZ2022 and NADC30 strains (Figure 4E). Functional mutations occurred in predicted B-cell epitope regions (9–13 aa, 32–34 aa, 158 aa, 168–172 aa), with six consecutive mutations in the 9–26 aa region. Three consecutive mutations were present in both 32–34 aa and 168–172 aa epitopes. Key T-cell epitope mutations clustered in the 121–158 aa region, while the linear antigenic epitope at 58–61 aa contained three consecutive mutations. Nucleotide and amino acid homology analysis of all ORFs revealed that GZ2022 shared the highest sequence identity with NADC30 (Supplementary Table S4). Recombination analysis using RDP5 (Figure 4F) and SimPlot (Figure 4G) conclusively identified GZ2022 as an NADC30-like recombinant strain, with NADC30 (Lineage 1) as the major parental strain and the highly pathogenic HuN4 (Lineage 8) as the minor parental strain.

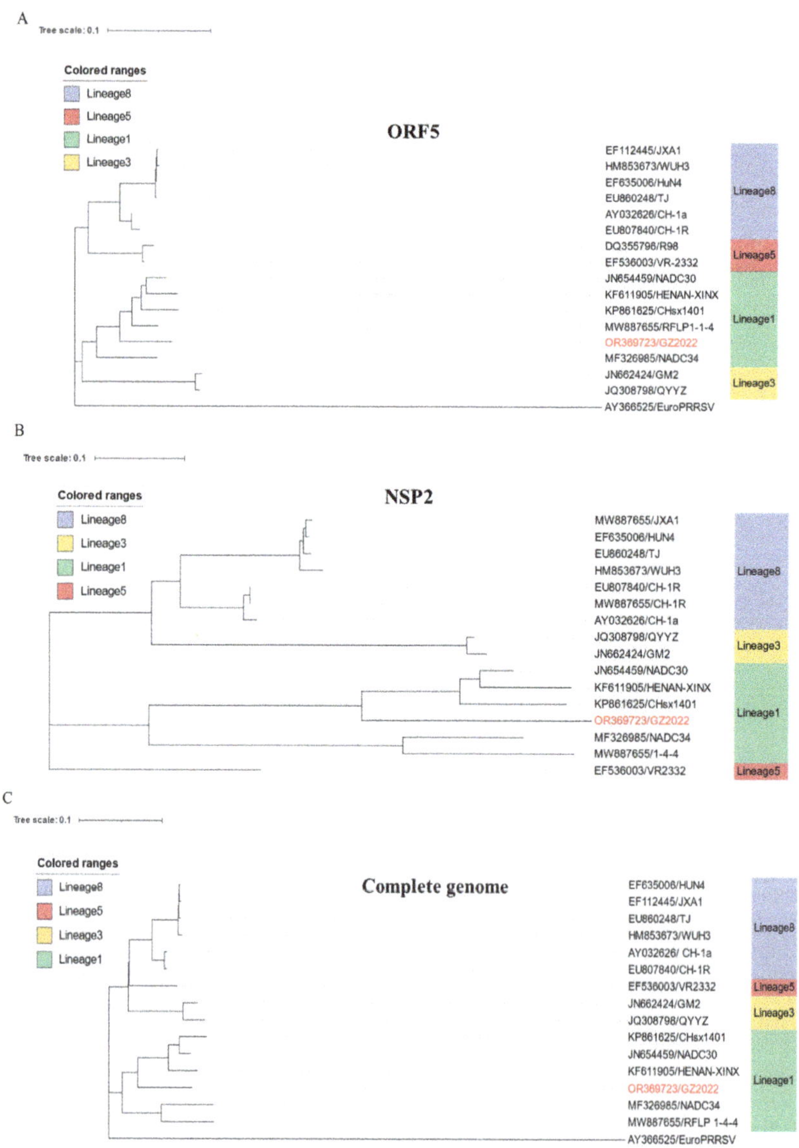


**Figure 4.** *Cont.*

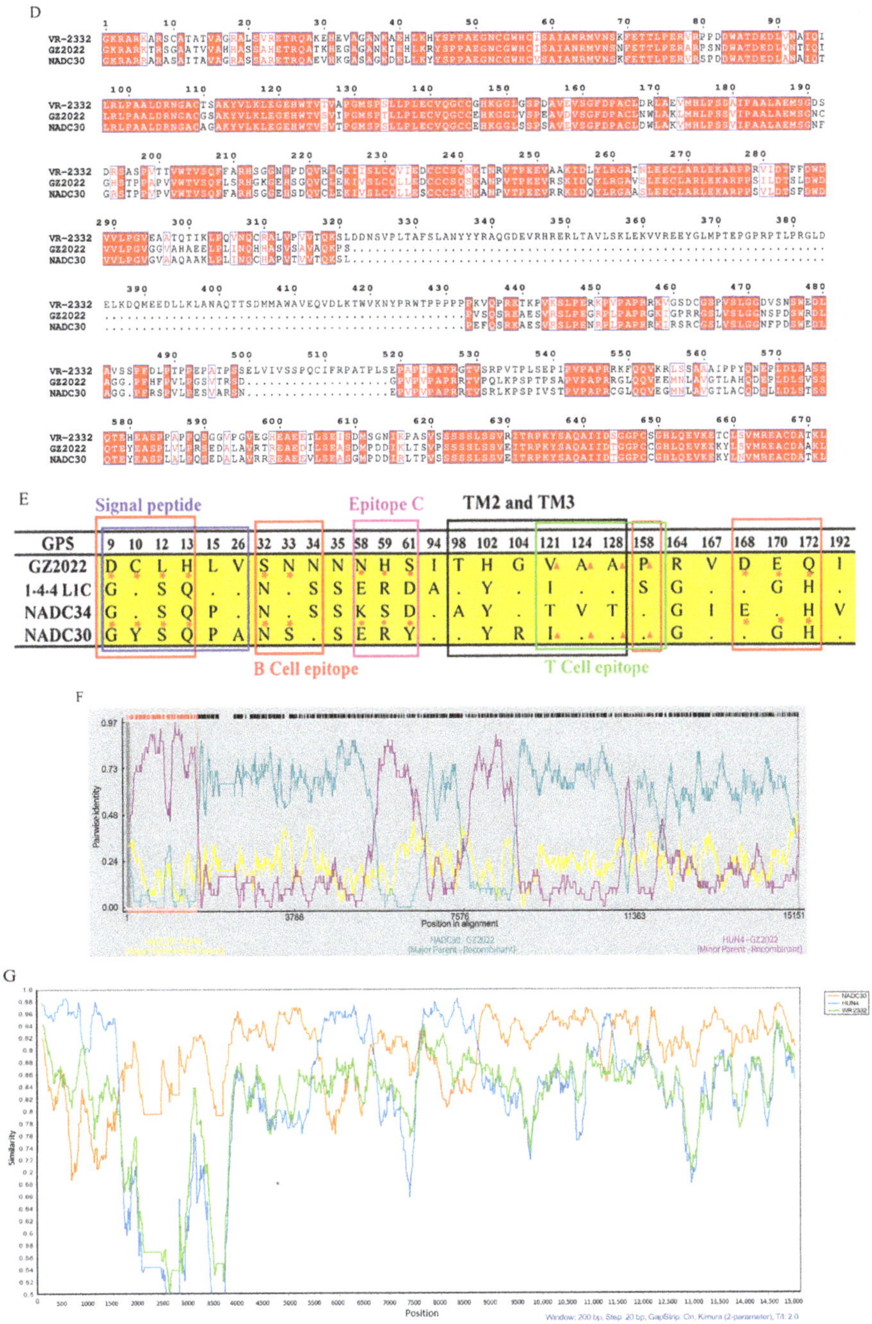


**Figure 4.** (**A**) ORF5 phylogenetic tree. (**B**) NSP2 phylogenetic tree. (**C**) Full-length genome phylogenetic tree. (**D**) NSP2 region deletion pattern analysis. Multiple sequence alignment of NSP2 sequences was conducted to identify strain-specific deletions. (**E**) GP5 amino acid analysis. The asterisk (*) denotes B-cell epitopes, the circle (○) denotes T-cell epitopes, and the triangle (Δ) denotes linear antigenic epitopes. (**F**) Recombination breakpoints identified using the RDP5 algorithm. (**G**) SimPlot analysis showing genome-wide recombination signals.

### 3.5. Divergent Virulence, Clinical Dynamics, and Growth Impacts of GZ2022

The progression of clinical signs in PRRSV-infected piglets is summarized in Figure 5A. By 3 dpi, both infected groups exhibited early symptoms, including fever (>40.0 °C) and huddling behavior. The GZ2022-infected group reached peak clinical severity at 7 dpi, characterized by sustained hyperthermia (>40.5 °C), lethargy, anorexia, and respiratory manifestations (cough, nasal discharge, conjunctivitis), followed by complete resolution of clinical signs occurring by 17 dpi, manifested as absence of fever and restoration of normal appetite, indicative of a self-limiting disease course. In contrast, the WUH3-infected group displayed more severe pathogenic phenotypes, with exacerbated respiratory distress (abdominal breathing), hypersecretion, and neurological signs (tremors) emerging by 9 dpi. Cumulative mortality in the WUH3 group reached four cases (1, 2, and 1 death at 10, 13, and 17 dpi, respectively). The single surviving piglet in this group recovered from the observed PRRS clinical signs (including respiratory distress and fever) by 17 dpi. Control piglets remained clinically normal throughout the observation period. These findings demonstrate significant heterogeneity in virulence and disease progression between PRRSV strains.

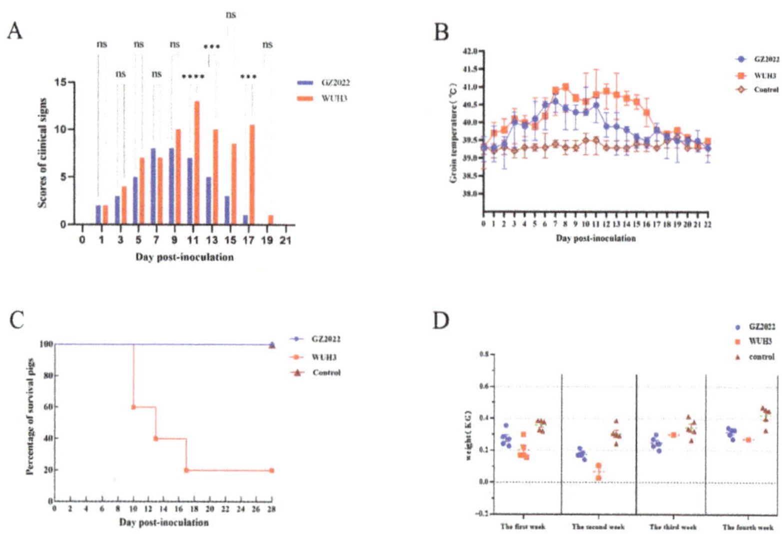


**Figure 5.** (**A**) Clinical signs score. (**B**) Temperature changes of piglets. Note: At 17–22 dpi, only one piglet survived in the WUH3 group, and thus no variation could be calculated ($n$ = 1). (**C**) Survival rate of PRRSV-infected piglets. (**D**) Weekly mean weight gain of PRRSV-infected piglets. Values in are shown as mean ± SD. *** $p < 0.001$; **** $p < 0.0001$; ns: no significance.

Rectal temperatures were recorded daily from 0 to 28 dpi (Figure 5B). In the infected groups, body temperatures rose significantly starting at 1 dpi. The GZ2022-infected group reached a peak temperature of 40.6 °C at 7 dpi, followed by a gradual decline to baseline by 17 dpi. The WUH3-infected group exhibited a higher peak of 41.0 °C at 8 dpi, returning to normal levels by 20 dpi. No mortality occurred in the GZ2022 group, while the WUH3 group experienced cumulative mortality (2, 1, and 1 piglet at 10, 13, and 16 dpi, respectively; the survival curve is shown in Figure 5C). Control group temperatures remained within physiological ranges throughout the trial.

Body weight was monitored weekly until 28 dpi (Figure 5D). During the first week, daily weight gain (DWG) in the GZ2022- and WUH3-infected groups was significantly lower than in the control group, with values of 0.27 kg and 0.20 kg, respectively, compared to 0.36 kg in the controls. By week 2, DWG further decreased in the infected groups, with GZ2022 at 0.17 kg and WUH3 at 0.20 kg, compared to 0.30 kg in the control group. From week 3 onward, DWG gradually recovered in the infected groups, with GZ2022 at 0.30 kg and WUH3 at 0.24 kg, while the control group maintained a DWG of 0.34 kg. By week 4, DWG in the infected groups approached near-normal levels, with GZ2022 at 0.31 kg and WUH3 at 0.27 kg, compared to 0.42 kg in the control group.

### *3.6. Divergent Viral Shedding, Tissue Tropism, and Antibody Profiles of PRRSV Strains*

Viral RNA shedding in nasal/oral and rectal swabs from infected piglets was quantitatively monitored via qPCR at multiple time points (0, 1, 3, 5, 7, 14, 21, and 28 dpi). Low levels of viral RNA were detected in the infected groups as early as 1 dpi. Nasal/oral shedding peaked at 7 dpi (Ct value = 22.1 ± 0.8, Figure 6A), while rectal shedding reached its maximum at 8 dpi (Ct value = 24.5 ± 1.2, Figure 6B). Both routes exhibited a gradual decline in viral load thereafter. Control groups remained negative throughout the experiment (Ct values > 38).

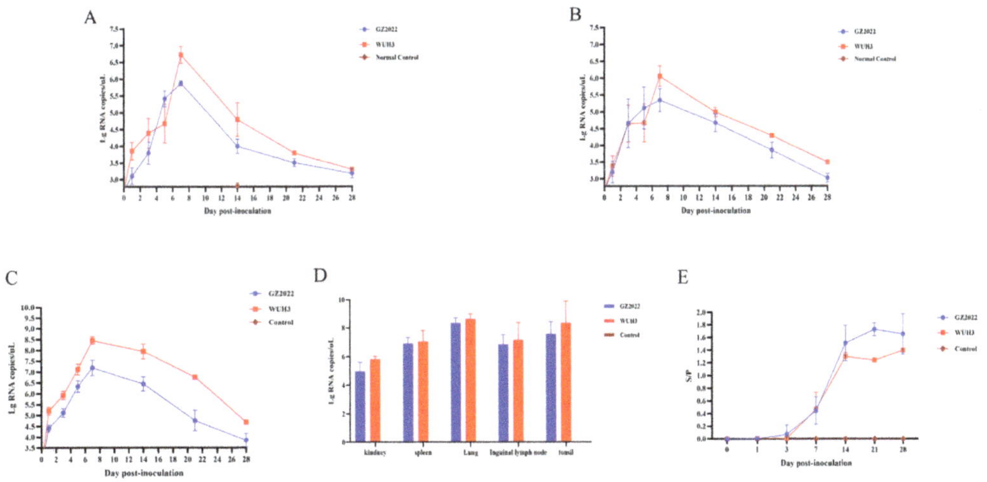


**Figure 6.** (**A**) Detection of virus shedding via nasal swabs. (**B**) Detection of virus shedding via rectal swabs. (**C**) Viral load in serum. (**D**) Viral load in tissues. Note: Viral loads in the WUH3-infected group represent individual measurements from both deceased and surviving piglets. (**E**) PRRSV antibody levels in serum. Values in are shown as mean ± SD.

Serum viral RNA loads (Figure 6C) peaked at 7 dpi in both infected groups, with GZ2022-infected piglets showing consistently lower titers than the WUH3 group ($p < 0.01$). Viral clearance began after 7 dpi in both cohorts. Tissue-specific analysis (Figure 6D) revealed the highest viral burden in lung tissue (mean $10^{7.5}$ copies/g), followed by tonsils ($10^{6.9}$ copies/g), spleen ($10^{6.3}$ copies/g), inguinal lymph nodes ($10^{5.8}$ copies/g), and kidney ($10^{5.2}$ copies/g), indicating distinct organotropism of the virus.

Serum antibodies against the PRRSV N protein were quantified using a commercial ELISA kit at 0, 1, 3, 7, 14, 21, and 28 dpi (Figure 6E). Both GZ2022- and WUH3-infected groups seroconverted by 7 dpi, with mean sample-to-positive (S/P) ratios of 0.45 and 0.47, respectively (cutoff S/P ≥ 0.4). Antibody levels in the GZ2022 group peaked at 21 dpi (mean S/P = 1.7), while the WUH3 group reached maximal titers at 14 dpi (mean S/P = 1.3), followed by stabilization. Throughout the observation period, GZ2022-infected piglets exhibited higher antibody levels than the WUH3 group. All control samples remained seronegative (S/P < 0.2).

### 3.7. Gross Lesions in Piglet Lungs

Necropsy of surviving piglets at 28 dpi revealed marked pulmonary lesions in both infected groups (Figure 7). The cardiac and apical lobes of the lungs exhibited parenchymal consolidation with hemorrhagic foci; however, hemorrhagic manifestations were milder in the GZ2022 group compared with the WUH3-infected animals. The lungs of the control group maintained a pinkish hue and spongy texture, with no macroscopic abnormalities detected in other organs.

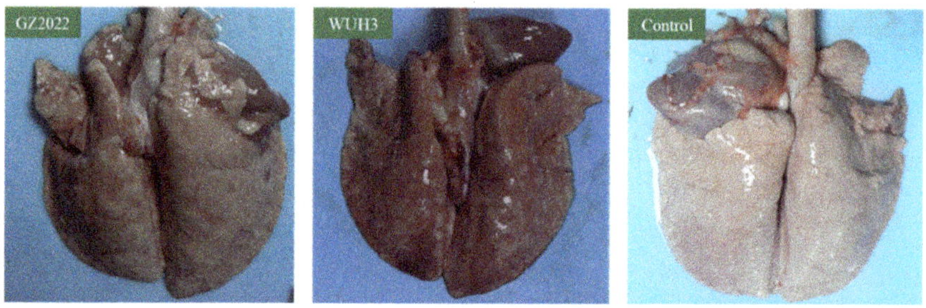


**Figure 7.** Gross lesions in piglet lungs. At 28 dpi, all surviving piglets were euthanized and necropsied to observe gross pathological lesions in lungs among the GZ2022-infected, WUH3-infected, and control groups.

### 3.8. Pathological Analysis of Lungs and Hilar Lymph Nodes in Piglets

Serum neutralization assays were performed using sera collected from GZ2022-infected piglets at 14, 21, and 28 dpi to evaluate neutralizing antibody titers against the PRRSV classical strain (R98), the highly pathogenic strain (WUH3), and the NADC30-like strain (GZ2022) (Table 1). The results indicated that, for the NADC30-like strain (GZ2022), serum collected at 21 dpi from one piglet demonstrated a neutralizing titer of 1: 4, while at 28 dpi, sera from three piglets reached a titer of 1: 4, and one piglet's serum reached a titer of 1: 8. For the highly pathogenic strain (WUH3) and the classical strain (R98), one piglet in each group showed a neutralizing titer of 1: 4 at 28 dpi. For the classical strain (R98), neutralizing titers remained below 1: 4 across all time points. These results indicate varying neutralizing responses to different PRRSV strains in the infected piglets.

**Table 1.** Positive rate of the neutralization test against different strains of PRRSV.

| Day | NADC30-Like Strain (GZ2022) | | | Highly Pathogenic Strain (WUH3) | | | PRRSV Classical Strain | | |
|---|---|---|---|---|---|---|---|---|---|
| | 1: 2 | 1: 4 | 1: 8 | 1: 2 | 1: 4 | 1: 8 | 1: 2 | 1: 4 | 1: 8 |
| 14 | 0 | 0 | 0 | 0 | 0 | 0 | 0 | 0 | 0 |
| 21 | 40% (2/5) | 20% (1/5) | 0 | 20% (1/5) | 0 | 0 | 20% (1/5) | 0 | 0 |
| 28 | 60% (3/5) | 40% (2/5) | 20% (1/5) | 20% (1/5) | 20% (1/5) | 0 | 20% (1/5) | 0 | 0 |

## 4. Discussion

The emergence of novel recombinant PRRSV strains poses significant challenges to global swine health management. Since the introduction of NADC30-like strains into China in 2014, their rapid dissemination and recombination dynamics have significantly reshaped the epidemiological landscape of PRRSV [28–30]. This study reports the isolation and characterization of a NADC30-like recombinant strain, GZ2022, which exhibits distinctive biological properties and investigates its pathogenicity. The findings provide critical insights into the evolutionary dynamics and pathogenic mechanisms of PRRSV.

In this study, a PRRS outbreak on a pig farm in Guizhou Province was confirmed to be caused by PRRSV infection through an *ORF7* gene RT-PCR assay. Further evolutionary analysis of the ORF5 gene revealed that the strain belongs to Lineage 1 (NADC30-like clade) [31], showing the closest genetic similarity to NADC30-like strains (93.7% nucleotide identity). Notably, this strain exhibited unique adaptability to Marc-145 cells, overcoming the in vitro cultivation limitations observed in most conventional NADC30-like strains. This characteristic supports the "functional modular recombination" hypothesis proposed by Kong et al. [32], which suggests that PRRSV acquires advantageous genomic segments through cross-lineage recombination to optimize host adaptation. After three rounds of plaque purification to obtain a monoclonal virus population, whole-genome recombination analysis confirmed that GZ2022 originated from recombination between NADC30 (major parent) and HuN4 (minor parent) within the ORF1a/b region (14.3% of the genome), suggesting that structural protein exchange during recombination may enhance cellular tropism [33,34].

The *NSP2* region of GZ2022 harbors a 131-amino-acid discontinuous deletion identical to that of NADC30 [35], a feature that may attenuate its ability to activate host inflammatory responses while enhancing immune evasion [13,36]. Mutation analysis of the *ORF5* gene revealed critical substitutions in the GP5 protein at the signal peptide (G9 $\rightarrow$ D9), B-cell epitope (H172 $\rightarrow$ Q172), and neutralizing epitope (S33 $\rightarrow$ N33). These mutations likely alter antigenic epitope conformation [37] or charge distribution [17], thereby impairing the efficiency of neutralizing antibodies and limiting cross-protection [38,39]. Similar observations have underscored *ORF5* genetic variation as a key mechanism of PRRSV immune evasion [40,41].

Animal challenge experiments demonstrated that GZ2022 infection induced persistent viremia and characteristic interstitial pneumonia, yet the clinical severity was significantly milder than that induced by the highly pathogenic strain WUH3. Evaluation of pulmonary lesions remains critical for understanding PRRSV pathogenesis, as lung pathology directly reflects viral cytopathic effects and immunopathological sequelae [42]. GZ2022-infected piglets exhibited attenuated clinical signs, whereas WUH3 infection induced severe diffuse pulmonary consolidation accompanied by hemorrhage. The enhanced pathogenicity by WUH3-infected groups was attributed to accelerated replication kinetics and amplified proinflammatory cytokine induction [43], which exacerbate tissue damage despite comparable levels of viral persistence. Additionally, 7 dpi was identified as a critical window for intervention, coinciding with peak viral replication and hyperthermia, consistent with the viral kinetics model proposed by Labarque et al. [44]. ELISA detected protective N-protein antibody levels by 14 dpi; however, neutralizing antibody responses were delayed, likely due to antigenic drift resulting from GP5 epitope variation [13]. Serological analysis further revealed limited cross-neutralizing activity (maximum titer: 1: 8) against heterologous strains, reinforcing the role of antigenic drift in immune evasion [45]. Although no statistically significant differences were detected in neutralization titers between strains, likely due to the limited sample size ($n$ = 5) and early-stage immune response dynamics, the observed numerical trends suggest strain-dependent variation in immunogenicity.

The Marc-145 cell adaptability of GZ2022 offers a novel resource for vaccine seed virus development; however, its antigenic diversity underscores the limitations of conventional monovalent vaccines. Designing multivalent or chimeric vaccines targeting recombination hotspots (e.g., NSP2 and ORF3) may represent a pivotal future strategy [46]. Furthermore, the establishment of a molecular surveillance network to monitor recombinant strains and track viral evolution in real time [47], coupled with reverse genetics approaches to dissect the functional contributions of specific recombinant fragments [48,49], is strongly recommended.

In conclusion, the discovery of GZ2022 not only enriches the genetic database of PRRSV recombinant strains but also provides novel insights into viral evolution and host–

pathogen interactions. Future studies should prioritize elucidating the mechanisms of immune evasion and cross-lineage transmission risks associated with recombinant strains, in order to develop precision-based interventions and mitigate the global threat of PRRS to swine production.

## 5. Conclusions

This study suggests that recombination events contribute to the genetic diversification of NADC30-like PRRSV strains in China. The recombinant GZ2022 strain, classified within Lineage 1, retained key NADC30-derived genomic features (e.g., a 131-amino-acid deletion in *NSP2*) while acquiring HuN4-derived segments that enhance its adaptability to Marc-145 cells. In vivo experiments demonstrated moderate virulence, inducing typical PRRS clinical manifestations in infected piglets, albeit with attenuated severity compared to highly pathogenic strains. Although GZ2022 infection elicited neutralizing antibodies in piglets by 21 dpi, sera exhibited limited cross-neutralization against heterologous PRRSV isolates, underscoring the strain-specific nature of immune responses. These findings highlight the recombinant genomic architecture and intermediate pathogenicity of GZ2022, providing valuable insights for refining vaccine development strategies against evolving PRRSV variants.

**Supplementary Materials:** The following supporting information can be downloaded at: https://www.mdpi.com/article/10.3390/vetsci12070651/s1, Table S1: Information of PRRSV reference strains; Table S2: Primers and probes used in this study; Table S3: Scoring criteria of clinical symptoms; Table S4: The whole genome nucleotide sequence homology analysis of GZ2022 strain.

**Author Contributions:** X.Y. (Xinmei Yang), B.Y. and Q.L.: conceptualization, writing—original draft preparation, methodology, formal analysis, H.M., Z.Y., P.M., S.R., X.Y. (Xuexiang Yu), Q.H. and W.L.: resources, validation and software, visualization. Provided reagents and material: Q.H. and W.L. All authors have read and agreed to the published version of the manuscript.

**Funding:** This research was supported by the National Key Research and Development Program of China (2023YFF1000901), the China Agriculture Research System of MOF and MARA (No. CARS-35), the Fundamental Research Funds for the Central Universities, National Key Laboratory of Agricultural Microbiology (AML2023A02), 2025 Huazhong Agricultural University Self-directed Science and Technology Innovation Fund Project—Research on Structure and Function of Important Animal Pathogens and Novel Veterinary Drugs (2662025DKPY007), and the "Yingzi Tech & Huazhong Agricultural University Intelligent Research Institute of Food Health" (No. IRIFH202209).

**Institutional Review Board Statement:** All the piglets were cared for in accordance with the guidelines of the Regulations for the Administration of Affairs Concerning Experimental Animals established by the Ministry of Science and Technology of China. The experimental procedures were conducted following protocols approved by the Scientific Ethics Committee of Huazhong Agricultural University (Approval Number: HZAUSW-2024-0086).

**Informed Consent Statement:** Informed consent was obtained from all subjects involved in the study.

**Data Availability Statement:** The data presented in this study are available on request from the corresponding author.

**Acknowledgments:** Thank National Key Laboratory of Agricultural Microbiology Core Facility and the Cooperative Innovation Center for Sustainable Pig Production for assistance during the experiments.

**Conflicts of Interest:** The authors declare that there are no conflicts of interest.

## References

1. Guo, Z.; Chen, X.X.; Li, R.; Qiao, S.; Zhang, G. The prevalent status and genetic diversity of porcine reproductive and respiratory syndrome virus in China: A molecular epidemiological perspective. *Virol. J.* **2018**, *15*, 2. [CrossRef] [PubMed]
2. Sun, Q.; Xu, H.; An, T.; Cai, X.; Tian, Z.; Zhang, H. Recent Progress in Studies of Porcine Reproductive and Respiratory Syndrome Virus 1 in China. *Viruses* **2023**, *15*, 1528. [CrossRef] [PubMed]
3. Zhou, L.; Han, J.; Yang, H. The evolution and diversity of porcine reproductive and respiratory syndrome virus in China. *Vet. Microbiol.* **2024**, *298*, 110252. [CrossRef]
4. Amadori, M.; Razzuoli, E. Immune Control of PRRS: Lessons to be Learned and Possible Ways Forward. *Front. Vet. Sci.* **2014**, *1*, 2. [CrossRef]
5. Li, J.; Wang, S.; Li, C.; Wang, C.; Liu, Y.; Wang, G.; He, X.; Hu, L.; Liu, Y.; Cui, M.; et al. Secondary Haemophilus parasuis infection enhances highly pathogenic porcine reproductive and respiratory syndrome virus (HP-PRRSV) infection-mediated inflammatory responses. *Vet. Microbiol.* **2017**, *4*, 35–42. [CrossRef] [PubMed]
6. Chae, C. Commercial PRRS Modified-Live Virus Vaccines. *Vaccines* **2021**, *9*, 185. [CrossRef]
7. Huang, B.; Deng, L.; Xu, T.; Jian, Z.; Lai, S.; Ai, Y.; Xu, Z.; Zhu, L.; Jones, C.J. Isolation and pathogenicity comparison of two novel natural recombinant porcine reproductive and respiratory syndrome viruses with different recombination patterns in Southwest China. *Microbiol. Spectr.* **2024**, *12*, e0407123. [CrossRef]
8. Music, N.; Gagnon, C.A. The role of porcine reproductive and respiratory syndrome (PRRS) virus structural and non-structural proteins in virus pathogenesis. *Anim. Health Res. Rev.* **2010**, *11*, 135–163. [CrossRef]
9. Snijder, E.J.; Meulenberg, J.J. The molecular biology of arteriviruses. *J. Gen. Virol.* **1998**, *79*, 961–979. [CrossRef]
10. Murtaugh, M.P.; Elam, M.R.; Kakach, L.T. Comparison of the structural protein coding sequences of the VR-2332 and Lelystad virus strains of the PRRS virus. *Arch. Virol.* **1995**, *140*, 1451–1460. [CrossRef]
11. Nelsen, C.J.; Murtaugh, M.P.; Faaberg, K.S. Porcine reproductive and respiratory syndrome virus comparison: Divergent evolution on two continents. *J. Virol.* **1999**, *73*, 270–280. [CrossRef]
12. Yoo, D.; Song, C.; Sun, Y.; Du, Y.; Kim, O.; Liu, H.C. Modulation of host cell responses and evasion strategies for porcine reproductive and respiratory syndrome virus. *Virus Res.* **2010**, *154*, 48–60. [CrossRef] [PubMed]
13. Lunney, J.K.; Fang, Y.; Ladinig, A.; Chen, N.; Li, Y.; Rowland, B.; Renukaradhya, G.J. Porcine Reproductive and Respiratory Syndrome Virus (PRRSV): Pathogenesis and Interaction with the Immune System. *Annu. Rev. Anim. Biosci.* **2016**, *4*, 129–154. [CrossRef]
14. Bian, T.; Sun, Y.; Hao, M.; Zhou, L.; Ge, X.; Guo, X.; Han, J.; Yang, H. A recombinant type 2 porcine reproductive and respiratory syndrome virus between NADC30-like and a MLV-like: Genetic characterization and pathogenicity for piglets. *Infect. Genet. Evol.* **2017**, *54*, 279–286. [CrossRef] [PubMed]
15. Liu, J.; Liu, C.; Xu, Y.; Yang, Y.; Li, J.; Dai, A.; Huang, C.; Luo, M.; Wei, C.; Jones, C.J. Molecular Characteristics and Pathogenicity of a Novel Recombinant Porcine Reproductive and Respiratory Syndrome Virus Strain from NADC30-, NADC34-, and JXA1-Like Strains That Emerged in China. *Microbiol. Spectr.* **2022**, *10*, e0266722. [CrossRef] [PubMed]
16. Zhou, L.; Wang, Z.; Ding, Y.; Ge, X.; Guo, X.; Yang, H. NADC30-like Strain of Porcine Reproductive and Respiratory Syndrome Virus, China. *Emerg. Infect. Dis.* **2015**, *21*, 2256–2257. [CrossRef]
17. Tian, K. NADC30-Like Porcine Reproductive and Respiratory Syndrome in China. *Open Virol. J.* **2017**, *11*, 59–65. [CrossRef]
18. Li, X.; Wu, J.; Tan, F.; Li, Y.; Ji, G.; Zhuang, J.; Zhai, X.; Tian, K. Genome characterization of two NADC30-like porcine reproductive and respiratory syndrome viruses in China. *Springerplus* **2016**, *5*, 1677. [CrossRef]
19. Ruan, S.; Yu, X.; Wu, H.; Lei, M.; Ku, X.; Ghonaim, A.H.; Li, W.; Jiang, Y.; He, Q. Assessing the antiviral activity of antimicrobial peptides Caerin1.1 against PRRSV in Vitro and in Vivo. *Vet. Microbiol.* **2024**, *297*, 110210. [CrossRef]
20. Zhou, L.; Yang, B.; Xu, L.; Jin, H.; Ge, X.; Guo, X.; Han, J.; Yang, H. Efficacy evaluation of three modified-live virus vaccines against a strain of porcine reproductive and respiratory syndrome virus NADC30-like. *Vet. Microbiol.* **2017**, *207*, 108–116. [CrossRef]
21. Kumar, N.; Barua, S.; Riyesh, T.; Chaubey, K.K.; Rawat, K.D.; Khandelwal, N.; Mishra, A.K.; Sharma, N.; Chandel, S.S.; Sharma, S.; et al. Complexities in Isolation and Purification of Multiple Viruses from Mixed Viral Infections: Viral Interference, Persistence and Exclusion. *PLoS ONE* **2016**, *11*, e0156110. [CrossRef] [PubMed]
22. Pizzi M: Sampling variation of the fifty percent end-point, determined by the Reed-Muench (Behrens) method. *Hum. Biol.* **1950**, *22*, 151–190.
23. Ma, N.; Zhang, M.; Zhou, J.; Jiang, C.; Ghonaim, A.H.; Sun, Y.; Zhou, P.; Guo, G.; Evers, A.; Zhu, H.; et al. Genome-wide CRISPR/Cas9 library screen identifies C16orf62 as a host dependency factor for porcine deltacoronavirus infection. *Emerg. Microbes Infect.* **2024**, *13*, 2400559. [CrossRef]
24. Li, Y.; Zhou, L.; Zhang, J.; Ge, X.; Zhou, R.; Zheng, H.; Geng, G.; Guo, X.; Yang, H.; Murtaugh, M.P. Nsp9 and Nsp10 contribute to the fatal virulence of highly pathogenic porcine reproductive and respiratory syndrome virus emerging in China. *PLoS Pathog.* **2014**, *10*, e1004216. [CrossRef]

25. Sun, Y.; Li, C.; Liu, Z.; Zeng, W.; Ahmad, M.J.; Zhang, M.; Liu, L.; Zhang, S.; Li, W.; He, Q. Chinese herbal extracts with antiviral activity: Evaluation, mechanisms, and potential for preventing PRV, PEDV and PRRSV infections. *Anim. Dis.* **2023**, *3*, 35. [CrossRef]
26. Lu, Y.; Gillam, F.; Cao, Q.M.; Rizzo, A.; Meng, X.J.; Zhang, C. Hepatitis B core antigen-based vaccine demonstrates cross-neutralization against heterologous North American Porcine Reproductive and Respiratory Syndrome Virus (PRRSV-2) strains. *J. Virol. Methods* **2020**, *285*, 113945. [CrossRef] [PubMed]
27. Liu, B.; Zheng, X.; Sun, X.; Wan, B.; Dong, J.; Zhou, Z.; Nan, Y.; Wu, C. Characterization of in vitro viral neutralization targets of highly pathogenic porcine reproductive and respiratory syndrome virus (HP-PRRSV) in alveolar macrophage and evaluation of protection potential against HP-PRRSV challenged based on combination of HP-PRRSV-structure proteins in vitro. *Vet. Microbiol.* **2024**, *292*, 110035.
28. Huang, Y.; Zhai, W.; Wang, Z.; He, Y.; Tao, C.; Chu, Y.; Pang, Z.; Zhu, H.; Jia, H. Analysis of the Immunogenicity of African Swine Fever F317L Protein and Screening of T Cell Epitopes. *Animals* **2024**, *14*, 1331. [CrossRef]
29. Ouyang, Y.; Du, Y.; Zhang, H.; Guo, J.; Sun, Z.; Luo, X.; Mei, X.; Xiao, S.; Fang, L.; Zhou, Y. Genetic Characterization and Pathogenicity of a Recombinant Porcine Reproductive and Respiratory Syndrome Virus Strain in China. *Viruses* **2024**, *16*, 993. [CrossRef]
30. Zhao, J.; Zhu, L.; Huang, J.; Yang, Z.; Xu, L.; Gu, S.; Huang, Y.; Zhang, R.; Sun, X.; Zhou, Y.; et al. Genetic characterization of a novel recombined porcine reproductive and respiratory syndrome virus 2 among Nadc30-like, Jxa1-like and TJ-like strains. *Vet. Med. Sci.* **2021**, *7*, 697–704. [CrossRef]
31. Yim-Im, W.; Anderson, T.K.; Paploski, I.A.D.; VanderWaal, K.; Gauger, P.; Krueger, K.; Shi, M.; Main, R.; Zhang, J.; Chao, D.-Y. Refining PRRSV-2 genetic classification based on global ORF5 sequences and investigation of their geographic distributions and temporal changes. *Microbiol. Spectr.* **2023**, *11*, e0291623. [CrossRef] [PubMed]
32. Kong, C.; Li, D.; Hu, Y.; Gao, P.; Zhang, Y.; Zhou, L.; Ge, X.; Guo, X.; Han, J.; Yang, H.; et al. The Genetic Variation of Porcine Reproductive and Respiratory Syndrome Virus Replicase Protein nsp2 Modulates Viral Virulence and Persistence. *J. Virol.* **2023**, *97*, e0168922. [CrossRef]
33. An, T.Q.; Li, J.N.; Su, C.M.; Yoo, D. Molecular and Cellular Mechanisms for PRRSV Pathogenesis and Host Response to Infection. *Virus Res.* **2020**, *286*, 197980. [CrossRef]
34. Qiu, M.; Li, S.; Li, S.; Sun, Z.; Lin, H.; Yang, S.; Cui, M.; Qiu, Y.; Qi, W.; Yu, X.; et al. The GP2a 91/97/98 amino acid substitutions play critical roles in determining PRRSV tropism and infectivity but do not affect immune responses. *J. Virol.* **2025**, *99*, e0004825. [CrossRef]
35. Zhou, L.; Kang, R.; Yu, J.; Xie, B.; Chen, C.; Li, X.; Xie, J.; Ye, Y.; Xiao, L.; Zhang, J.; et al. Genetic Characterization and Pathogenicity of a Novel Recombined Porcine Reproductive and Respiratory Syndrome Virus 2 among Nadc30-Like, Jxa1-Like, and Mlv-Like Strains. *Viruses* **2018**, *10*, 551. [CrossRef] [PubMed]
36. Huang, C.; Du, Y.; Yu, Z.; Zhang, Q.; Liu, Y.; Tang, J.; Shi, J.; Feng, W.-H. Highly Pathogenic Porcine Reproductive and Respiratory Syndrome Virus Nsp4 Cleaves VISA to Impair Antiviral Responses Mediated by RIG-I-like Receptors. *Sci. Rep.* **2016**, *6*, 28497. [CrossRef]
37. Chang, C.C.; Yoon, K.J.; Zimmerman, J.J.; Harmon, K.M.; Dixon, P.M.; Dvorak, C.M.; Murtaugh, M.P. Evolution of porcine reproductive and respiratory syndrome virus during sequential passages in pigs. *J. Virol.* **2002**, *76*, 4750–4763. [CrossRef] [PubMed]
38. Barfoed, A.M.; Blixenkrone-Møller, M.; Jensen, M.H.; Bøtner, A.; Kamstrup, S. DNA vaccination of pigs with open reading frame 1–7 of PRRS virus. *Vaccine* **2004**, *22*, 3628–3641. [CrossRef]
39. Hu, W.; Tang, D.; Zeng, Z.; Wang, B.; Zhou, M.; Mao, Y.; Zhou, P.; He, S. Research progress on the molecular mechanism of immune escape of porcine reproductive and respiratory syndrome virus. *Virology* **2025**, *602*, 110298. [CrossRef]
40. Chen, N.; Trible, B.R.; Kerrigan, M.A.; Tian, K.; Rowland, R.R.R. ORF5 of porcine reproductive and respiratory syndrome virus (PRRSV) is a target of diversifying selection as infection progresses from acute infection to virus rebound. *Infect. Genet. Evol.* **2016**, *40*, 167–175. [CrossRef]
41. Faaberg, K.S.; Hocker, J.D.; Erdman, M.M.; Harris, D.H.; Nelson, E.A.; Torremorell, M.; Plagemann, P.G. Neutralizing antibody responses of pigs infected with natural GP5 N-glycan mutants of porcine reproductive and respiratory syndrome virus. *Viral Immunol.* **2006**, *19*, 294–304. [CrossRef] [PubMed]
42. Sun, W.; Wu, W.; Jiang, N.; Ge, X.; Zhang, Y.; Han, J.; Guo, X.; Zhou, L.; Yang, H. Highly Pathogenic PRRSV-Infected Alveolar Macrophages Impair the Function of Pulmonary Microvascular Endothelial Cells. *Viruses* **2022**, *14*, 452. [CrossRef] [PubMed]
43. Zhao, K.; Gao, J.C.; Xiong, J.Y.; Guo, J.C.; Yang, Y.B.; Jiang, C.G.; Tang, Y.-D.; Tian, Z.-J.; Cai, X.-H.; Tong, G.-Z.; et al. Two Residues in NSP9 Contribute to the Enhanced Replication and Pathogenicity of Highly Pathogenic Porcine Reproductive and Respiratory Syndrome Virus. *J. Virol.* **2018**, *92*, e02209-17. [CrossRef]

44. Labarque, G.G.; Nauwynck, H.J.; Van Reeth, K.; Pensaert, M.B. Effect of cellular changes and onset of humoral immunity on the replication of porcine reproductive and respiratory syndrome virus in the lungs of pigs. *J. Gen. Virol.* **2000**, *81*, 1327–1334. [CrossRef]
45. Shi, M.; Lam, T.T.; Hon, C.C.; Murtaugh, M.P.; Davies, P.R.; Hui, R.K.; Li, J.; Wong, L.T.; Yip, C.W.; Jiang, J.W.; et al. Phylogeny-based evolutionary, demographical, and geographical dissection of North American type 2 porcine reproductive and respiratory syndrome viruses. *J. Virol.* **2010**, *84*, 8700–8711. [CrossRef]
46. Liu, D.; Zhou, R.; Zhang, J.; Zhou, L.; Jiang, Q.; Guo, X.; Ge, X.; Yang, H. Recombination analyses between two strains of porcine reproductive and respiratory syndrome virus in vivo. *Virus Res.* **2011**, *155*, 473–486. [CrossRef] [PubMed]
47. Sánchez-Carvajal, J.M.; Rodríguez-Gómez, I.M.; Ruedas-Torres, I.; Zaldívar-López, S.; Larenas-Muñoz, F.; Bautista-Moreno, R.; Garrido, J.J.; Pallarés, F.J.; Carrasco, L.; Gómez-Laguna, J.; et al. Time Series Transcriptomic Analysis of Bronchoalveolar Lavage Cells from Piglets Infected with Virulent or Low-Virulent Porcine Reproductive and Respiratory Syndrome Virus 1. *J. Virol.* **2022**, *96*, e0114021. [CrossRef] [PubMed]
48. Chaudhari, J.; Vu, H.L.X. Porcine Reproductive and Respiratory Syndrome Virus Reverse Genetics and the Major Applications. *Viruses* **2020**, *12*, 1245. [CrossRef]
49. Guo, J.; Li, C.; Lu, H.; Wang, B.; Zhang, L.; Ding, J.; Jiao, X.; Li, Q.; Zhu, S.; Wang, A.; et al. Reverse genetics construction and pathogenicity of a novel recombinant NADC30-like PRRSV isolated in China. *Front. Vet Sci.* **2024**, *11*, 1434539. [CrossRef]

*veterinary sciences*

MDPI

*Article*

# In Silico Designed Multi-Epitope Vaccine Based on the Conserved Fragments in Viral Proteins for Broad-Spectrum Protection Against Porcine Reproductive and Respiratory Syndrome Virus

**Shaukat Ullah [1,2], Hikmat Ullah [2,3], Kainat Fatima [1,2] and Tan Lei [4,5,6,*]**

1 Key Laboratory of Quantitative Synthetic Biology, Shenzhen Institute of Synthetic Biology, Shenzhen Institutes of Advanced Technology, Chinese Academy of Sciences, Shenzhen 518055, China; ullah@siat.ac.cn (S.U.); kainat@siat.ac.cn (K.F.)
2 University of Chinese Academy of Sciences, Beijing 100049, China; hikmat@siat.ac.cn
3 Center for Energy Metabolism and Reproduction, Institute of Biomedicine and Biotechnology, Shenzhen Institutes of Advanced Technology, Chinese Academy of Sciences, Shenzhen 518000, China
4 State Key Laboratory of Reproductive Medicine and Offspring Health, Nanjing Medical University, Nanjing 211166, China
5 Innovation Center of Suzhou Nanjing Medical University, Suzhou 215000, China
6 National Center of Technology Innovation for Biopharmaceuticals, Suzhou 215000, China
* Correspondence: lei.tan19800707@njmu.edu.cn; Tel.: +86-178-1271-7184

**Simple Summary:** Porcine reproductive and respiratory syndrome virus (PRRSV) represents a significant threat to the global pig farming industry, primarily due to its high mutation rate and genetic variability, which have hindered the development of effective vaccines. Consequently, we present a groundbreaking approach to vaccine design, using conserved protein fragments from multiple PRRSV strains, afterward predicted optimized epitopes combined that could stimulate B cells, helper T lymphocytes (HTLs), and cytotoxic T lymphocytes (CTLs). Subsequently, our designed vaccine candidate, PRRSV-V-2, demonstrated higher physicochemical properties and effectively bound to immune receptors (TLR3 and TLR8), enhancing immune system activation. Moreover, the recombinant PRRSV-V-2 plasmid, designed for expression in *E. coli*, offers a promising, scalable solution for controlling PRRSV across diverse strains. This novel vaccine could revolutionize swine health management by providing broad-spectrum protection against PRRSV.

**Abstract:** Porcine reproductive and respiratory syndrome virus (PRRSV) is a major viral threat to swine, causing significant economic loss in the global pig farming industry. This virus includes two major genotypes, PRRSV1 and PRRSV2, both characterized by high mutation rates and genetic variability, complicating the development of a universally effective vaccine and disease control. To address this challenge, this study utilizes immunoinformatics tools to identify conserved epitopes and design a multi-epitope vaccine candidate against PRRSV based on reverse vaccinology. The complete sequences of PRRSV-encoded proteins were retrieved worldwide, and the conserved fragments were identified through the alignment of polypeptide sequences. Subsequent screening was conducted to screen epitopes for their potential to be safe and to activate B cells, HTLs (helper T cells), and CTLs (cytotoxic T cells). By conjugating the selected epitopes with distinct adjuvant proteins, three vaccine candidates were designed and termed PRRSV-vaccine (PRRSV-V-1, PRRSV-V-2, and PRRSV-V-3, respectively). Furthermore, systematic evaluations of their physicochemical properties, structural stability, binding with pattern recognition receptors, and induction of the host immune system were performed. PRRSV-V-2 had the most promising physicochemical and structural characteristics, strong binding with toll-like receptors (TLR3 and TLR8), and the most vigorous reactions to host immune responses. As the most promising candidate, the recombinant PRRSV plasmid was in silico designed for expression in *Escherichia coli*. Our study proposed a novel approach to PRRSV vaccine development against PRRSV, offering a promising strategy for controlling the infection across diverse PRRSV strains in swine. Despite providing significant insights into vaccine design through computational methods, the results of this study remain predictive. So, it is open for the experimental validations of the scientific community to ensure its actual immunological properties, especially the safety and efficacy.

**Keywords:** porcine reproductive and respiratory syndrome virus; PRRSV vaccine; in silico; multi-epitopes; immune response

## 1. Introduction

Porcine reproductive and respiratory syndrome (PRRS), caused by the PRRS virus (PRRSV), was first identified in the United States in 1987. Since then, PRRSV has become a global threat, inflicting severe economic losses in the swine industry [1]. Recent estimates show that PRRSV losses the swine industry approximately USD 1.2 billion annually [2,3]. In Germany, financial losses due to PRRSV resulted in a 19.1% average reduction in farm profit, reaching up to 41% under severe conditions in 2021 [4]. One of the key challenges in controlling PRRSV lies in its genetic and antigenic variability [5–7], which complicates disease tracking and vaccine development. The virus's high mutation rate has hindered the making of a broadly protective vaccine, prolonging the fight against this devastating pathogen [8,9].

PRRSV is a member of the Arterivirus genus within the Arteriviridae family, with a genome of approximately 15.4 kb and encodes 10 open reading frames (ORFs) [10]. This virus produces two large polyproteins from ORFs 1a and 1b, which are cleaved into 14 nonstructural proteins [11]. Additionally, 7 structural proteins are encoded by ORFs 2–7, including glycoproteins (GP) (GP2a, GP3, GP4, GP5, GP2b or E protein, GP6 or M protein, and GP7 or N protein) [12]. These structural proteins play key roles in virus–host interactions and are highly immunogenic, making them key targets for vaccine development. Currently, 365 distinct PRRSV strains have been isolated worldwide, with significant genetic diversity due to ongoing recombination and mutations in the viral genome. PRRSV is classified into two major genotypes based on these variations: PRRSV1 (European) and PRRSV2 (North American). Notably, the GP2, GP3, GP4, and GP5 proteins are hotspots for mutations, while the GP6 (M protein) and GP7 (N protein) sequences are relatively conserved [13,14]. The emergence of new strains continues to challenge the development of vaccines capable of providing broad protection across diverse genotypes.

In response to this constant evolution, a variety of vaccine types have been developed, including 17 modified live virus (MLV) vaccines, 3 killed virus (KV)vaccines, 5 adenovirus-vectored vaccines, 2 synthetic peptide vaccines, and 14 subunit vaccines, as of 2024 [15,16]. Although MLV and KV vaccines have shown practical effectiveness, each has significant limitations. MLV vaccines are effective against homologous strains but pose safety concerns, such as the risk of virulent reversion and recombination with field strains [17–19]. Currently, a total of five MLV vaccines are used in the U.S., seven in Europe, and four in Asia (Supplementary Table S1). However, significant safety concerns are always accompanied due to the possibility of virulence reversion and recombination with virulent field strains [19,20]. On the contrary, KV vaccines are valued for safety, and no adverse effects have been reported [21], which benefits from the killed nature and consequently eliminates risks of replication, mutation, or spreading among vaccinated animals. However, it's regrettable that KV vaccines are often considered less effective than MLV vaccines [17,19], which limits their market capacity (Supplementary Table S2). As reported, the enhancement of humoral immune response was constantly disabled with the variety of virus concentrations and inactivation methods [22]. Since then, KV vaccines have not yet been approved in the United States, despite being approved in Europe and other regions.

To provide effective and wide-spectrum immunoprotection, identifying and targeting protective epitopes is crucial in vaccine development. Several neutralizing epitopes have been identified in PRRSV structural proteins, particularly GP3, GP4, GP5, and M [23,24]. GP5, as a major envelope glycoprotein, plays a key role in viral entry and assembly [25], with neutralizing epitopes found in its N-terminus ectodomain [25,26]. Similarly, epitopes in the M proteins have been identified as important for susceptibility to neutralization antibodies [24,27,28]. In M protein, the pattern of GP5, a fragment of 70 amino acid residues, was determined as the crucial modulator of the susceptibility to neutralization antibody in porcine serum [29]. However, partial protection has been observed with immunization using the GP5-M ectodomain, suggesting that additional epitopes are needed for full viral neutralization. Alternatively, neutralizing epitopes have also been shown in GP2, GP3, and GP4 [30–32], with another complex formed as an essential factor in receptor binding and viral infection [33,34]. With the progression of epitope discovery, new candidates were continuously identified. For instance, 37–45 amino acids in GP5 of both genotypes [27] and 57–68 amino acids in GP4 of type I [32,35] have been reported as neutralizing epitopes. T cell epitopes were also highlighted from GP4 and GP5 [36,37]. In the N protein, four to five antigenic regions are included [12] and are essential to induce protective B- and T-cell responses [38]. Despite these findings, cross-genotype protection remains a major hurdle due to the high mutation rate of PRRSV antigens, which allows for immune evasion and complicates vaccine development. Optimized epitopes must be identified to update the vaccine formulation and address this challenge.

To discover effective epitopes with high efficiency in current vaccine strategies, an in silico approach is proposed for identifying novel epitopes and designing a multi-epitope vaccine. With the current advancement of immunoinformatics, the identification of epitopes and formulation of vaccines has become feasible and efficacious [39]. Under the in silico strategy, vaccines have been designed successfully for pathogens, such as the human coronavirus [40], dengue virus [41], influenza virus [42], etc. In the current study, we conducted an in silico design to target either type of PRRSV. The effective and non-risky epitopes were screened from the viral protein fragments conserved across the globally isolated strains. Subsequently, the vaccine candidate was constructed by assembling selected epitopes with different adjuvant proteins. With the evaluation of candidates in physiochemistry, structural biology, and immunology, one promising candidate was ultimately highlighted. This study aims to provide a promising strategy for overcoming the challenges of PRRSV vaccine development, offering cross-genotype protection despite the genetic diversity of PRRSV.

## 2. Materials and Methods

As illustrated in Figure 1, the PRRSV vaccine design was carried out stepwise in three stages. In the first stage, viral protein sequences were retrieved worldwide and then used to extract the conserved fragments inside as the basis of epitope screening. The second stage aimed to obtain the candidate vaccine, including the steps of epitope primary screening, evaluation, selection, and assembly. For the third and last stage, candidate vaccines were subjected to a panel of evaluations to visualize their properties in physiochemistry, structure, host interaction, and immune induction capacity. The web links of databases and online platforms used in this study were listed in Supplementary Materials Table S3. All the Supplementary Tables are included in Supplementary Material File Table S1, and all the figures are included in Supplementary Material File Figure S2.

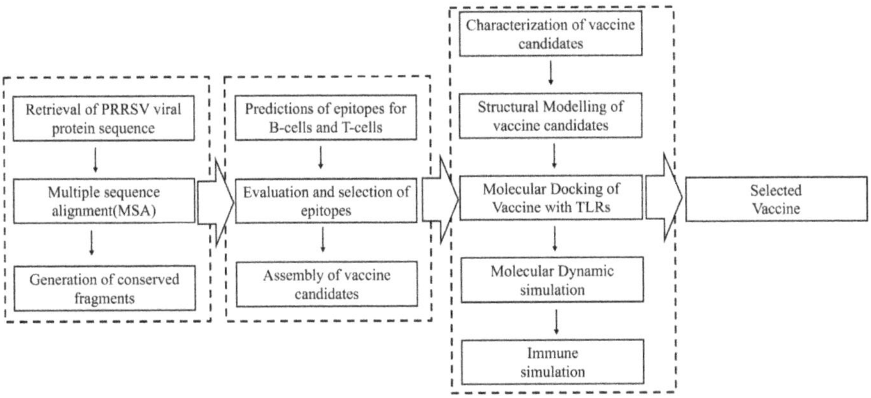


**Figure 1.** An overview of the reverse vaccinology process used to design the Porcine reproductive and respiratory syndrome virus vaccine. Data collection, epitope prediction, vaccine candidate construction, and in silico validation are highlighted.

### 2.1. *Retrieval of PRRSV Protein Sequence*

To initiate the design, full-length sequences of all nonstructural and structural proteins from both PRRSV genotypes (PRRSV1 and PRRSV2) were retrieved from NCBI (Supplementary Tables S4 and S5), focusing on sequences deposited between 2018 and 2023. Conserved fragments within each protein were extracted using multiple sequence alignment (MSA). Initial alignment was conducted using MUSCLE [43], followed by confirmation with CLUSTALW version 2.0 [44]. Conserved sequences shared across retrieved strains were extracted using BioEdit version 7.2 [45] and are listed in Supplementary Table S6.

### 2.2. *Step-by-Step Selection of Epitopes*

Identification of epitopes for B cells, HTLs (helper T cells), and CTLs (cytotoxic T cells) was carried out using IEDB (Immune Epitope Database) tools based on the conserved protein fragments in each viral protein.

The Bepipred Linear Epitope Prediction 2.0 method of IEDB was applied for linear B-cell epitopes. The minimum length of the epitope was set as 6 amino acids. The threshold of 0.5 was used to determine the likelihood of an immunogenic epitope [46]. The NetMHCIIPan 4.1 EL recommended method of IEDB was employed to predict HTL epitopes. Since there were no SLA options for MHCII prediction, the HLA reference was set as below: HLA-DPA1*02:01/DPB1*05:01, HLA-DRB4*01:04, HLA-DRB1*12:04, HLA-DPA1*02:01/DPB1*04:01, HLA-DRB5*01:03, HLA-DPA1*01:03/DPB1*01:01, HLA-DPA1*02:01/DPB1*02:01, HLA-DRB1*04:04) [47]. A fixed length of 15 amino acids was used in the prediction of HTL, and the top 10% of affinity was considered for further analysis, among which %Rank < 2% were regarded as strong binders (SB) and %Rank < 10% for weak binders (WBs). For the prediction of CTL epitopes, the NetMHCpan EL4.1 recommended method of IEDB was used with an MHCI source specified as "Pigs". Epitope lengths were defined as 9 amino acids, targeting the most common SLA (swine leukocyte antigen) class I alleles [48]: SLA-1*0702, SLA-1*1101, SLA-2*0101, SLA-2*1101, SLA-2*1201, SLA-3*0101, SLA-3*0401, SLA-3*0501, and SLA-3*0701. For CTL peptides ranked within the top 1% (%Rank < 0.5% were considered as strong binders (SB), and %Rank < 2% thresholds are considered for weak binders (WBs)), affinity was selected for further analysis [47].

Selected epitopes for all three types of B cells, HTLs, and CTLs were further assessed for conservancy, antigenicity, allergenicity, and toxicity with the IEDB analysis resource [49], VaxiJen v2.0 bioinformatics tool for protein antigenicity prediction with a threshold of 0.4 for viruses that excludes non-antigen epitopes from vaccine design [50], AllerTOP v.2.0 bioinformatics tool for protein allergy prediction that excludes allergenic epitopes from vaccine design [51] and the ToxinPred bioinformatics tool for protein toxicity prediction that excludes toxic epitopes from vaccine design [52], respectively. The inclusion criteria required sequence identity ≥80%, antigenicity score >0.5, and non-allergic, non-toxic properties. Additional scoring of immunogenicity was conducted for CTL epitopes with the Class I Immunogenicity server of IEDB [49], and the ones who scored ≤0.2 were excluded. For HTL epitopes, further predictions were made to identify potential inducers of IFN-$\gamma$ and IL-4 using the IFNepitope [53] and IL4pred [54] servers.

### 2.3. Design and Assembly of Candidate Vaccines

The candidate epitopes that passed the stepwise selection process were then used for vaccine construction. Initially, the epitopes were categorized based on their reactivity with B cells, HTLs, and CTLs, and subsequently assembled into a core antigen. Distinct adjuvant domains were attached to enhance the immune response, as shown below:

S50 L7/12 ribosomal protein, isolated from *Mycobacterium tuberculosis*, is known for its ability to provoke an immune response [55].

$\beta$-defensin, recognized for its antimicrobial and immunomodulatory properties [56].

Heparin-binding hemagglutinin (HBHA), sourced from *Mycobacterium* species, is known for its immunogenic properties [57].

For vaccine assembly and further reinforcement of immunogenicity, the following linkers were used to organize the immune elements [55]:

KK: linker to connect B-cell epitopes [58].
AAY: linker used for the organization of CTL epitopes [59].
GPGPG: linker to group HTL epitope [41].
EAAAK: a rigid linker to fuse the adjuvant domains to the N-terminus of the core antigen.

### 2.4. Evaluation of Candidate Vaccines in Physicochemical and Immune Properties

The Expasy ProtParam Tool [60] was employed to predict a range of physicochemical properties for the candidate vaccines, including molecular weight, isoelectric point, half-life, stability, and other essential characteristics. A protein whose instability index is smaller than 40 is predicted as stable. The solubility of these molecules was predicted using SOLpro of SCRATCH [61]. A threshold of >0.5 confirms that the chosen candidates remained soluble and functional under physiological conditions. Additionally, the immune properties of the candidate's vaccine were assessed using immunoinformatic tools, including VaxiJen v2.0 for antigenicity, AllerTop for allergenicity assessment, and ToxinPred to evaluate toxicity.

### 2.5. Prediction and Validation of Molecular Structure for Candidate Vaccines

The secondary structure of the vaccine molecules, including $\alpha$-helices, $\beta$-strands, and coils, was predicted using the PSIPRED server [62] based on their polypeptide sequences.

Three-dimensional (3D) structures of the designed vaccine construct were predicted using the PHYRE2 server [63], a widely used platform for protein modeling and analysis. With PHYRE2, you can construct 3D models, predict ligand-binding sites, and evaluate the effect of amino acid variants on protein function and structure. With a confidence level of 90%, the model demonstrated high-confidence structural predictions suitable for downstream structural and immunological analyses. The predicted model by PHYRE2 was further refined with the Galaxy Refine server [64]. Five revised models were suggested for each candidate based on GDT-HA, RMSD, MolProbity, Clash Score, Poor Rotamers, and Rama Favoritism. Subsequently, to validate the refined models, ProSA-web [65] and PROCHECK [66] were employed, generating a Z-score and Ramachandran plot, which were used to assess the overall quality of vaccine proteins.

### 2.6. Molecular Docking and Dynamic Analysis of Vaccine-TLR (Toll-like Receptors) Complexes

To evaluate the potential recognition of vaccine molecules by the host immune system, molecular docking was performed between the candidate vaccines and two pattern recognition receptors (TLRs), TLR3 (PDB ID: D0V2A2) and TLR8 (PDB ID: Q865R7) of swine were retrieved from the UniProt protein data bank [67,68]. HDOCK [69] was used for docking, which uses a hybrid algorithm to combine template-based modeling and free docking. In the docking procedure, HDOCK used its default rigid-body protocol and ranked and refined docked conformations through iterative global as well as local search optimization. The multi-receptor docking approach provided a comprehensive assessment of the vaccine constructs' potential to engage key pattern recognition receptors and trigger innate immune responses. For each vaccine–TLR complex, model 1 with the lowest binding energy was selected for analysis of non-covalent interactions, including hydrogen bonds, salt bridges, and interface areas, using the PDBsum [70] server. The 3D structures with related interactions were then visualized by PyMOL [71].

For the top-ranked docking models, PRRSV vaccine with -TLR complexes, molecular dynamics (MD) simulations were performed using the AMBER [72]. A 14-angstrom padding around the protein and water box was applied. The AMBERS FF14SB force field [73,74] was used for both vaccines and TLRs. The vaccine molecules were placed in a TIP3P water box with counterions added to neutralize the system. To minimize structural collisions, steepest descent or conjugate gradient minimization methods were employed, with transition steps every 1000 cycles up to 5000 cycles. Thermal equilibrium was reached at 300 K after 250 picoseconds (ps) of system heating. A production run of 100 nanoseconds (ns) was conducted to explore the dynamic behavior of the vaccine–receptor complexes. In the canonical ensemble, periodic boundary conditions were applied to maintain the rigidity of the covalently bound hydrogen atoms using the SHAKE algorithm [75]. In these simulations, the temperature was regulated at 300 K using a Langevin thermostat with a 10 Å cutoff for non-bounded interactions and Ewald simulations for long-range interactions [76]. Statistical parameters such as the root mean square deviation (RMSD), root mean square fluctuations (RMSF), and radius of gyration (Rg) of vaccine–receptor complexes were analyzed using the CPPTRAJ module [77]. The results were visualized in PyMOL [78] to assess the structural stability and dynamic properties of the vaccine-receptor.

### 2.7. Immune Simulation of Candidate Vaccines

Immune simulations were conducted using the C-ImmSim server [79] to model the immune responses elicited after vaccine administration according to a specified immunization schedule. The simulation protocol involved three doses of immunization with a two-week interval, corresponding to simulation steps 1, 84, and 168, where each step represented eight hours of real time. Default settings were used for all additional parameters to ensure consistency and standardization. The dynamics of innate and immune response properties were visualized as curves.

### 2.8. In Silico Cloning of Selected Vaccine Candidate

To generate the DNA coding sequence of the vaccine protein, JCat servers were used for reverse translation and codon optimization based on *E. coli* codon usage bias. The codon adaptation index (CAI) and GC content percentage of the resulting sequence were calculated to assess the protein expression level [80]. A good outcome was defined as a CAI > 0.8, while GC content ranged from 30% to 70%. Furthermore, in silico cloning was performed in SnapGene 8.0.2 using the pET-28a (+) vector with restriction sites XhoI and BamHI.

## 3. Results

### 3.1. Extraction of Conserved Fragments from PRRSV Protein Sequences Collected Worldwide

Due to the nature of rapid evolution, PRRSV protein sequences deposited from 2018~2023 were taken as the starting material for the current design. Approximately 7315 full-length sequences were downloaded in total from NCBI in FASTA format, including 90 proteins for PRRSV1 and 7225 for PRRSV2. Protein accession numbers are listed in Table S4. Based on the alignments, conserved fragments were identified for each strain, and those with at least 10 amino acids were used for epitope screening. Finally, 30 conserved fragments were obtained for PRRSV1 and 77 for PRRSV2 (Table S6).

### 3.2. Prediction and Evaluation of Epitopes for B Cells, CTLs, and HTLs

B-cell epitope prediction was performed based on structural proteins of PRRSV, including GP2a, GP3, GP4, GP5, M, and N. Primarily, 68 epitopes were generated for PRRSV1 (European genotype) and 97 for PRRSV2 (American genotype). Epitopes were then evaluated for their antigenicity, allergenicity, and toxicity. Subsequently, the ones that were antigenic, non-allergic, and non-toxic were reserved for further usage, including 06 for PRRSV1 and 07 for PRRSV2. The conservancy of epitopes was also analyzed using the epitope conservancy analysis algorithm in the IEDB platform, and the ones scored above 50% were highlighted among the reserved epitopes. The final selection of B-cell epitopes that were present on the surface of each protein is included in Table 1. The procedure of epitope selection is presented in Figure S1 and Tables S7 and S8.

**Table 1.** Prediction of B-cell epitopes that are antigenic, non-allergic, and non-toxic.

| Genotype | Protein | Start –End | Epitopes | Length | Antigenicity Score | Allergenicity /Toxicity | Intra/Inter Conservancy |
|---|---|---|---|---|---|---|---|
| PRRSV1 | ORF2A | 57–66 | APRYSVRALP | 10 | 0.8784 | Non/Non | 75.00% /20.00% |
| PRRSV2 | GP3 | 38–45 | LVRGNFSF | 8 | 1.7037 | Non/Non | 100.00% /33.33% |
| PRRSV2 | GP4 | 44–57 | QHTQQHHLVIDHIR | 14 | 0.5457 | Non/Non | 100.00% /33.33% |
| PRRSV1 | GP5 | 49–61 | LNGTDWLDKRFDW | 13 | 1.7333 | Non/Non | 75.00% /20.00% |
| PRRSV2 | M | 131–144 | RKPGLTSVNGTLV | 13 | 0.8659 | Non/Non | 100.00% /33.33% |
| PRRSV1 | N | 6–13 | HFPLAAED | 8 | 0.9417 | Non/Non | 66.00% /22.22% |

Epitopes for T lymphocytes were predicted from all PRRSV proteins, and 4560 epitopes were generated for HTLs and 4208 for CTLs (Figures S2 and S3). The first selection was then carried out based on binding affinity between candidate epitopes and MHC II or I alleles, with the top 10% and the top 1% as the threshold, respectively [81]. Further selection of epitopes was conducted for both HTLs and CTLs to highlight the most promising inducer of T cell reaction. For CTLs, epitopes were chosen once antigenicity scored >0.5 and immunogenicity > 0.1, non-allergenic and non-toxic, while excluded if allergenic, toxic, or conservancy was lower than 50% (Tables S9 and S10). Similar filters were also set up for HTLs, but the capacity of IFN-$\gamma$ and IL-4 induction was evaluated instead of immunogenicity scoring (Tables S11 and S12). Resultantly, 10 HTL epitopes and 10 CTL epitopes were selected for the final vaccine design (Tables 2 and 3).

**Table 2.** Prediction of HTL epitopes: Predicted antigenicity, allergenicity, interferon induction, toxicity, and conservancy analysis of shortlisted HTL (MHC-II) epitopes.

| Genotype | Protein | HLA- Allele | Epitope | Antigenicity | Allergenicity /Toxicity | IL-4/IFN- $\gamma$ Induction | Intra/Inter Conservancy |
|---|---|---|---|---|---|---|---|
| PRRSV1 | ORF1a | DPA1*02:01 DPB1*05:01 | GVAPAVRIAERYRGR | 0.9596 | Non/Non | Inducer /Inducer | 100.00% /13.33% |
| PRRSV1 | ORF1a | DRB4*01:04 | KPIAYANLDEKKISA | 0.9123 | Non/Non | Inducer /Inducer | 100.00% /13.33% |
| PRRSV2 | ORF1b | DRB1*12:04 | CLGDFKQLHPVGFDS | 1.2326 | Non/Non | Inducer /Inducer | 95.24% /20.00% |
| PRRSV2 | GP2a | DPA1*02:01 DPB1*04:01 | LSFASDWFAPRYSVR | 0.5461 | Non/Non | Inducer /Inducer | 75.00% /20.00% |
| PRRSV2 | GP2b | DRB1*04:04 | VFCIRLVCSAIHRSR | 1.3347 | Non/Non | Inducer /Inducer | 100.00% /20.00% |
| PRRSV2 | GP3 | DPA1*02:01 DPB1*04:01 | NWFHLEWLRPFFSSW | 0.4583 | Non/Non | Inducer /Inducer | 75.00% /20.00% |
| PRRSV1 | GP4 | DRB5*01:03 | ACVNFTDYVAHVTQH | 1.0257 | Non/Non | Inducer /Inducer | 75.00% /13.33% |
| PRRSV2 | GP5 | DPA1*01:03 DPB1*01:01 | WRYSCTRYTNFLLDT | 0.5114 | Non/Non | Inducer /Inducer | 100.00% /13.33% |
| PRRSV1 | M | DPA1*02:01 DPB1*02:01 | LAFSITYTPIIYALK | 1.4017 | Non/Non | Inducer /Inducer | 100.00% /13.33% |
| PRRSV2 | N | DQA1*01:01 DQB1*02:01 | PHFPLATEDDVRHHF | 0.4807 | Non/Non | Inducer /Inducer | 66.67% /26.67% |

**Table 3.** Prediction of CTL epitopes. Shortlisted CTL (MHC-I) epitopes along with their strain, predicted immunogenicity, antigenicity, toxicity, and conservancy analysis.

| Genotype | Protein | SLA- Allele | Epitope | Antigenicity /Immunogenicity | Allergenicity /Toxicity | Intra/Inter Conservancy |
|---|---|---|---|---|---|---|
| PRRSV1 | ORF1a | 1*1101 | AALTGRTL | 1.2071 /0.14994 | Non/Non | 100.00% /25.00% |
| PRRSV1 | ORF1a | 3*0501 | HQKPIAYANL | 1.1782 /0.1123 | Non/Non | 100.00% /20.00% |
| PRRSV2 | ORF2b | 3*0701 | TRARHAIF | 1.0977 /0.31383 | Non/Non | 95.24% /25.00% |
| PRRSV2 | GP2a | 3*0401 | SVRALPFTL | 1.5046 /0.15641 | Non/Non | 75.00% /22.22% |
| PRRSV1 | GP2b | 2*0101 | IFLAILFGF | 0.7123 /0.27353 | Non/Non | 75.00% /22.22% |
| PRRSV2 | GP3 | 2*0101 | YAWLAFLSF | 1.1215 /0.10302 | Non/Non | 100.00% /22.22% |
| PRRSV2 | GP4 | 1*0702 | SACVNFTDY | 1.6035 /0.17366 | Non/Non | 66.67% /22.22% |
| PRRSV2 | GP5 | 3*0101 | TRYTNFLL | 1.2753 /0.1324 | Non/Non | 66.67% /25.00% |
| PRRSV1 | M | 2*1201 | FSITYTPII | 1.4228 /0.1836 | Non/Non | 66.67% /22.22% |
| PRRSV1 | N | 2*1101 | FPLATEDDVRHHF | 0.4645 /0.34939 | Non/Non | 50.00% /38.46% |

### 3.3. Design of Three Candidate Vaccines

To construct the candidate vaccines, the core antigen was first organized via the assembly of selected epitopes for B cells, HTLs, and CTLs, with the help of linker sequences. KK, GPGPG, and AAY linkers were employed to group three classes of epitopes (Figure 2A). After that, three distinct adjuvant proteins, L7/12 S50 ribosomal protein, β-defensin, and HBHA, were added to enhance the boosting of the host immune system. The adjuvant proteins were attached with the core antigen at the N-terminus with EAAAK, a rigid linker. Correspondingly, three candidate vaccines were generated and termed PRRSV-V-1, PRRSV-V-2, and PRRSV-V-3, as shown in Figure 2B–D, while protein sequences of vaccine candidates were contained in Figure S4:

PRRSV-V-1: Core antigen attached with S50 L7/12 ribosomal protein on N-terminal.
PRRSV-V-2: Core antigen attached with β-defensin on the N-terminal.
PRRSV-V-3: Core antigen attached with HBHA adjuvant on N-terminal.

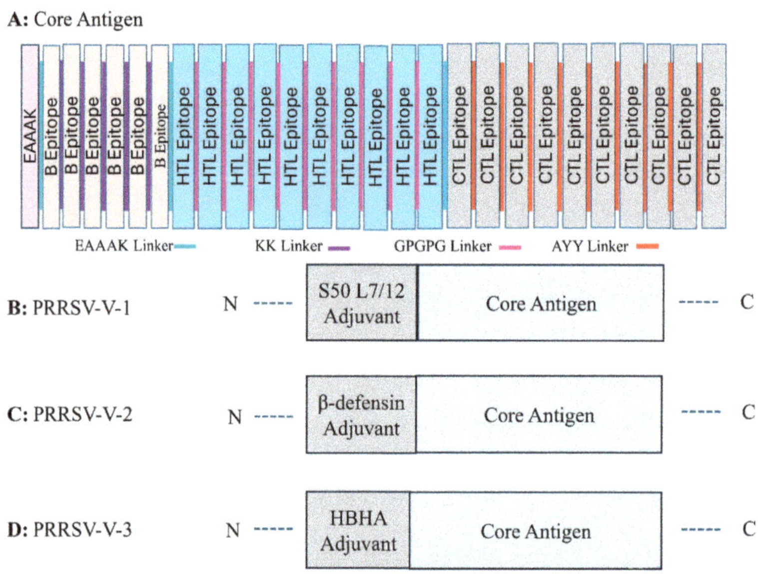


**Figure 2.** Organization of multi-epitope vaccine peptides. (**A**) Graphic representation of the vaccine design approach that includes the core antigen and a summary of the approach taken to create the multi-epitope vaccine peptide. This design includes specific B cells, CTLs, and HTLs epitopes that are bordered by adjuvant and connected by peptide linkers. (**B–D**) Vaccine candidates of PRRSV-V-1, PRRSV-V-2, PRRSV-V-3 (**B**), PRRSV-V-1, Core antigen conjugated with S50 L7/12 ribosomal protein at the N-terminal, (**C**) PRRSV-V-2, at the N-terminal, β-defensin is attached to the core antigen, (**D**) HBHA adjuvant is conjugated with core antigen at the N-terminal.

### 3.4. Properties of Candidate Vaccines in Physicochemical and Immunology

To ensure the qualification of candidate vaccines primarily, their antigenicity, allergenicity, and toxicity were analyzed before any other characteristics (Table 4). The physicochemical properties were analyzed thereafter to assess their molecular size, stability, solubility, and hydropathicity, which were correlated with the feasibility of the future preparation and application of candidate vaccines. For the molecular size, all three candidate vaccines, PRRSV-V-1~3, contained approximately 531 amino acids, 448 amino acids, and 551 amino acids (line 1 in Table 4) and weighed around 57 kDa, 49 kDa, and 60 kDa (line 2 in Table 4). To ensure good stability in vitro, instability indexes were calculated much lower than 40 (line 3 in Table 4). The vaccines' aliphatic index ranged between 69 and

82, indicative of a stable protein (line 4, Table 4). For intracellular stability, the protein half-life was estimated to be more than 30 h in mammalian cells, beyond 20 h in yeasts, and in *E. coli*, no less than 10 h (line 5 in Table 4). The water solubility of candidates was scored with SOLpro. Resultantly, the solubility scores were around 0.9 (line 6 in Table 4), GRAVY slightly below 0 (line 7 in Table 4), and pI generally above 9 (line 8 in Table 4). In one word, these three candidate vaccines were suggested as hydrophilic and soluble in water at physiological pH. All three molecules showed non-allergenicity and non-toxicity and scored >0.6 in antigenicity (lines 9–11 in Table 4), which is deemed to be safe and effective in immunization.

**Table 4.** Properties of PRRSV-V-1~3 vaccines in physiochemistry and immunology.

| Parameters | PRRSV-V-1 Vaccine | PRRSV-V-2 Vaccine | PRRSV-V-3 Vaccine |
|---|---|---|---|
| **No. of amino acids** | 531 | 448 | 551 |
| **Molecular weight** | 57,526.93 kDa | 49,446.89 kDa | 60,733.93 kDa |
| **Instability index** | 22.04 | 25.72 | 28.66 |
| **Aliphatic index** | 81.02 | 80.91 | 77.04 |
| **Half-life** | 30 h (mammalian reticulocytes, in vitro). >20 h (yeast, in vivo). >10 h (Escherichia coli, in vivo) | 30 h (mammalian reticulocytes, in vitro). >20 h (yeast, in vivo). >10 h (Escherichia coli, in vivo) | 30 h (mammalian reticulocytes, in vitro). >20 h (yeast, in vivo). >10 h (Escherichia coli, in vivo) |
| **Solubility** | 0.979905 | 0.975094 | 0.986037 |
| **GRAVY** | −0.054 | −0.191 | −0.249 |
| **Theoretical pI** | 9.12 | 9.82 | 9.17 |
| **Antigenicity** | 0.61 | 0.68 | 0.63 |
| **Allergenicity** | Non-allergenic | Non-allergenic | Non-allergenic |
| **Toxicity** | Non-toxic | Non-toxic | Non-toxic |

### *3.5. Prediction and Refinement of Secondary and Tertiary Structure*

To determine the composition of secondary structural elements in candidate vaccines, the PSIPRED server was used (Figure 3A–C). PRRSV-V-1 contains 44.06% of alpha helix, 11.11% of beta-strand, and 44.82% of random coil, as plotted along its polypeptide sequence. The alpha helix, beta-strand, and random coil in PRRSV-V-2 were 43.30%, 12.05%, and 44.64%, while 57.16%, 6.71%, and 36.11% in PRRSV-V-3, respectively (Figure 3D).

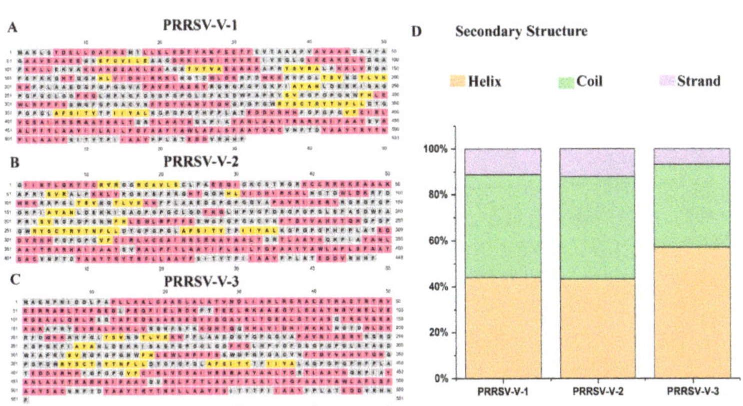


**Figure 3.** Secondary structure of vaccine candidate. (**A–C**) Illustration of the secondary structure for the PRRSV-V-1~PRRSV-V-3 vaccine candidate. Strand, helix, and coil are indicated with distinct colors: yellow, pink, and gray, respectively. (**D**) Histogram plot of secondary structure for PRRSV-V-1~PRRSV-V-3, showing the percentage distribution of coils, helices, and β-strands in each protein.

For the modeling of tertiary structure, templates c8yeqA, c1kj6A, and c2ch7B were employed on the Phyre2 platform corresponding to PRRSV-V-1, 2, and 3 candidates (Figure 4A–C). The levels of confidence and coverage are shown in Figure 4D,E.

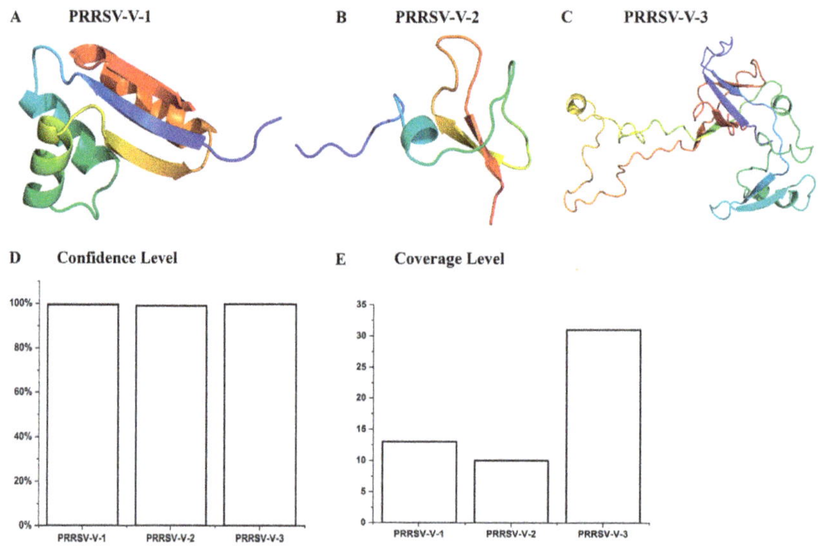


**Figure 4.** Tertiary structure of vaccine candidate. (**A–C**) The tertiary structure of the template model was used to model the PPRRSV-V-1~PRRSV-V-3 tertiary structure. (**D**) Confidence level for the tertiary structure models of PRRSV-V-1~PRRSV-V-3. (**E**) Coverage level of tertiary structure models PRRSV-V-1~PRRSV-V-3.

The structural models were then verified with Procheck, and the Ramachandran plots were generated subsequently. As visualized in the Ramachandran plot, 96.7% of residues in PRRSV-V-1 were located in the favored region, 3.3% in the allowed region, and 0% in the disallowed region. Similarly, the residues of PRRSV-V-2 and 3 were mainly in the favored region, which is 84.2% and 99.4%, and the allowed region was 15.8% and 0.6%, respectively (Figure 5A–D). On the other side, ProSA-web showed Z-scores of −5.98, −4.80, and −1.24 for vaccine candidates PRRSV-V-1, 2, and 3, respectively, with which high quality was indicated for all three models (Figure 5E–H).

*3.6. Molecular Docking and Dynamics Simulation for Complexes of Vaccine-Immune Receptor*

Molecular docking was performed using the HDock server to evaluate the interaction between the vaccine protein and pattern recognition receptors. Two receptor molecules of swine, TLR3 and TLR8, were used in docking with three candidate vaccines, and then 6 complexes were generated. For each complex, 10 models were constructed and ranked by docking score from low to high, among which model 1 exhibited the strongest binding.

Tight binding between vaccines and receptors was suggested, with the lowest binding energy score ranging from −100~−350 kJ/mol (Table S13). The interactions were illustrated in detail, while the number of hydrogen bonds, salt bridges, non-ionic interactions, and the size of the interface were listed. To PRRSV-V-1, complexes with TLR3 and 8 were scored as −196.74 and −190.43 kJ/mol, respectively. The interface between PRRSV-V-1 and TLR3 was composed of 15 and 20 residues on either the vaccine or receptor side, respectively. The area of this interface reached 1104 Å2, within which 3 hydrogen bonds, 2 salt bridges, and 121 non-covalent interactions were formed. An interface of a much smaller scale was formed between PRRSV-V-1 and TLR8, which contained a much smaller number of connections, including 2 hydrogen bonds, 0 salt bridges, and 87 non-covalent interactions. PRRSV-V-2 exhibited strong and consistent binding across TLR3 and TLR8. It achieved the best docking score with TLR3 (–308.16 kJ/mol) and TLR8 (−263.17 kJ/mol), forming a stable interface with 6, 9 hydrogen bonds, 3, 1 salt bridges, and 155,176 non-covalent interactions, as illustrated in Figure 6. The other PRRSV-V-3 docking analysis showed strong interactions (−260.99 kJ/mol with TLR3 and −335.61 kJ/mol with TLR8), along with a stable interface with 3,3 hydrogen bonds, 4, 1 salt bridges, and 316,282 non-covalent interactions. All these docking analyses of PRRSV-V-1~3 are presented in Table 5. PRRSV-V-1 and PRRSV-V-3 docking interactions, number of hydrogen bonds, salt bridges, and non-covalent interactions were supplemented in Figure S5. Molecular docking analysis revealed that PRRSV-V-2 and PRRSV-V-3 bind most favorably to innate immune receptors TLR3 and TLR8, resulting in significantly lower docking scores and enhanced interfacial interactions. Based on these results, their potential as highly immunogenic vaccine candidates is highlighted by the likelihood of receptor engagement and downstream immune activation. In contrast, PRRSV-V-1 exhibited significantly lower binding affinity and fewer interaction interfaces, suggesting a limited ability to trigger an immune response.

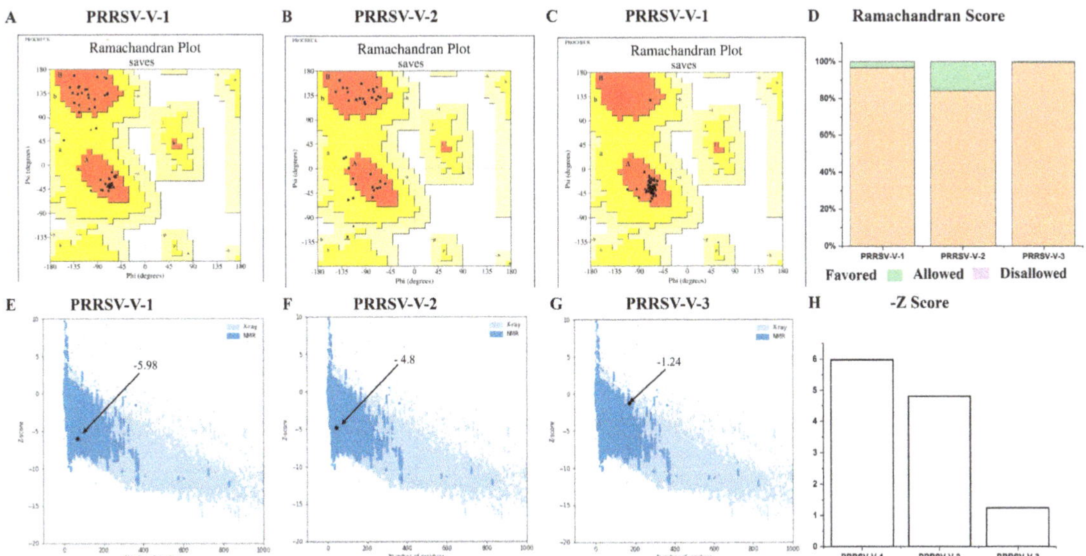


**Figure 5.** Validation of vaccine candidates' molecular structure. (**A–C**) The Ramachandran plot of PRRSV-V-1~PRRSV-V-3 emphasizes the residues in favorable A, B, and L regions displayed with red colors. (**D**) Histogram plot of Ramachandran, the distribution of residues in favored, allowed, and disallowed regions of PRRSV-V-1~PRRSV-V-3. (**E–G**) The Z-score plot of PRRSV-V-1 was scored as −5.98, PRRSV-V-1 score −4.8, and PRRSV-V-3 score as −1.24. (**H**) Histogram plot of Z-score of PRRSV-V-1~3, according to their Z-score plots.

The conformational stability and dynamic behavior of vaccine–receptor complexes were further evaluated using molecular dynamics simulations (MD). A complex's overall structural stability was determined by RMSD, while its compactness was determined by Rg. In the PRRSV-V-2–TLR3 complex, the RMSD value consistently remained within an acceptable range below 4, while in the PRRSV-V-2–TLR8 complex, equilibrium was reached approximately 40 ns after reaching equilibrium, indicating that conformational behavior had remained stable throughout the simulation (Figure 7A). The Rg values for PRRSV-V-2–TLR3 and PRRSV-V-2–TLR8 complexes range from 32.0 to 33.5, indicating compact and stable conformations throughout simulation (Figure 7B). Furthermore, the number of intermolecular hydrogen bonds fluctuated between 8 and 16 in the complexes of PRRSV-V-2 with TLR3 or TLR8; therefore, persistent and favorable vaccine–receptor interactions were suggested (Figure 7C). However, with overranged RMSD and Rg indicated during simulation, PRRSV-V-1 and PRRSV-V-3 molecules were suggested to be unable to form stable complexes with TLR3 and TLR8. RMSD values of the PRRSV-V-1–TLR3 and PRRSV-V-1–TLR8 complexes exceeded 5 Å and 30 Å, respectively. There were also pronounced fluctuations in the radius of gyration (Rg) of PRRSV-V-1 and PRRSV-V-3 with TLR3/TLR8 complexes, ranging from 54 Å to 34 Å, suggesting partial unfolding and loss of structural compactness (Figures S6 and S7). Based on the above analysis and comparison with PRRSV-V-1 and PRRSV-V-3, such variability of key structural parameters implies weakened and less stable interactions. Based on the analysis above, the variability of key structural parameters among the three candidate vaccine molecules suggested PRRSV-V-2, the one with robust interactions with TLR3 and TLR8.

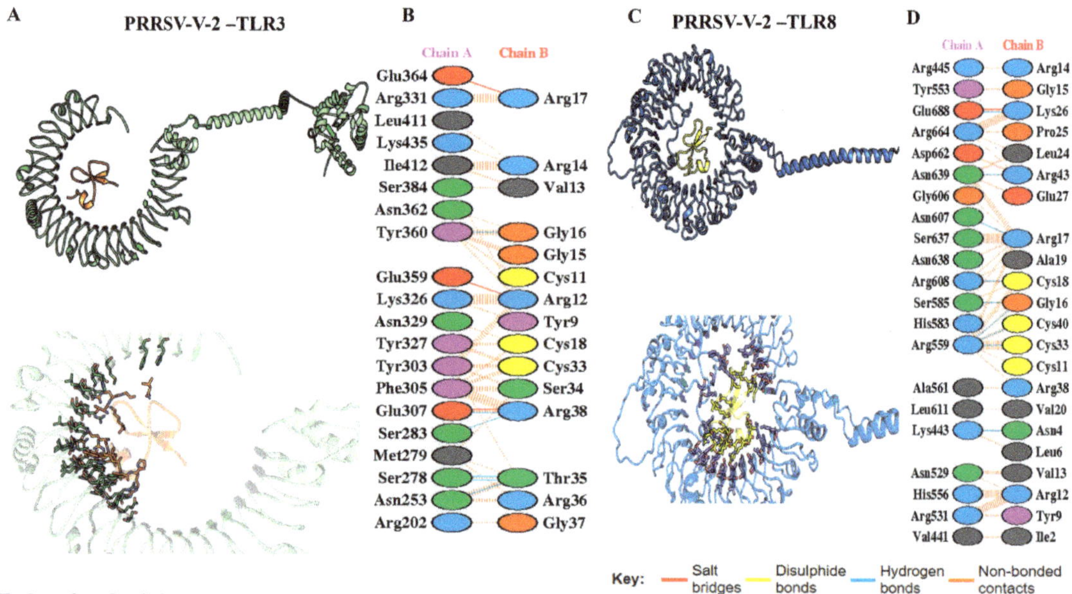


**E Interface Statistics**

| Chain | No of interface residues | Interface residues (Å²) | No. of Salt bridges | No of hydrogen bonds | No of Non-bonded contacts |
|---|---|---|---|---|---|
| PRRSV-V-2<br>TLR3 | 20<br>15 | 939<br>1063 | 3 | 6 | 155 |
| PRRSV-V-2<br>TLR8 | 21<br>22 | 1010<br>975 | 1 | 9 | 176 |

**Figure 6.** A molecular docking analysis of PRRSV-V-2-TLR3/8 complexes. PRRSV-V-2 docking with (**A**) TLR3 and (**C**) TLR8. The salt bridges (red), hydrogen bonds (blue), disulfide bonds (yellow), and non-bonded contacts (brown dots) formed in the interface between PRRSV-V-2 with (**B**) LR3 and (**D**) TLR8. (**E**) An analysis of the interface statistics properties of PRRSV-V-2 with TLR3 and TLR8.

**Table 5.** Molecular docking and interaction analysis of PRRSV-V-1~PRRSV-V-3 with swine TLR2, TLR3, TLR4, and TLR8 complexes. Each vaccine–receptor complex is summarized by docking scores, the number of interface residues, interface area, salt bridges, hydrogen bonds, and non-covalent interactions.

| Vaccine–Receptor Complex | Docking Score | Number of Residues in Interface | Area of Interface (Å2) | Number of Salt Bridges | Number of Hydrogen Bonds | Number of Non-Covalent Interactions |
|---|---|---|---|---|---|---|
| PRRSV-V-1<br>TLR-3 | −196.74 | 15<br>20 | 1104<br>1114 | 2 | 3 | 121 |
| PRRSV-V-1<br>TLR-8 | −190.43 | 10<br>14 | 791<br>870 | 0 | 2 | 87 |
| PRRSV-V-2<br>TLR-3 | −308.16 | 20<br>15 | 939<br>1063 | 3 | 6 | 155 |
| PRRSV-V-2<br>TLR-8 | −263.17 | 21<br>22 | 1010<br>975 | 1 | 9 | 176 |
| PRRSV-V-3<br>TLR-3 | −260.99 | 30<br>37 | 2129<br>1943 | 4 | 3 | 316 |
| PRRSV-V-3<br>TLR-8 | −335.61 | 34<br>37 | 2340<br>2211 | 1 | 3 | 282 |

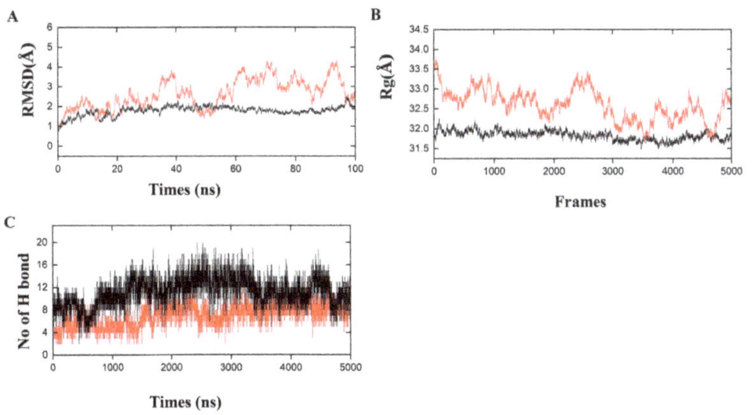


**Figure 7.** Molecular dynamic simulations of PRRSV-V-2 with TLR3/TLR8. (**A**) Root mean square deviation (RMSD): shows the RMSD for the complexes of PRRSV-V-2 with TLR3 and TLR8. (**B**) Radius of Gyration (Rg): Rg is the compactness of PRRSV-V-2 with TLR3 and TLR8. (**C**) Hydrogen Bonds: the count of intra-molecular hydrogen bonds formed by PRRSV-V-2 with TLR3 and TLR8. Graphical representations of the dynamic simulation of PRRSV-V-2 with TLR3 are indicated in red, whereas the dynamics simulation of PRRSV-V-2 with TLR8 is shown in black.

### *3.7. Simulation of Immune Responses Post-Vaccination*

To demonstrate the efficacy of vaccine candidate PRRSV-V-2 to induce host immunity, a simulation was then conducted on the platform C-ImmSim with three doses of vaccination at 4-week intervals till 350 days post-immunization. Dendritic cells, macrophage cells, and natural killer cells were monitored in the simulation to see the status of innate immunity. A significant activation of macrophage cells was demonstrated following each dose of immunization. In addition, cytokines like IFN-γ and IL-2 were also rapidly elevated up to 400,000 and 700,000 ng/mL, respectively. Thus, it can be seen that vaccine candidate PRRSV-V-2 can initiate the innate part of the host immune system, as shown in Figure 8.

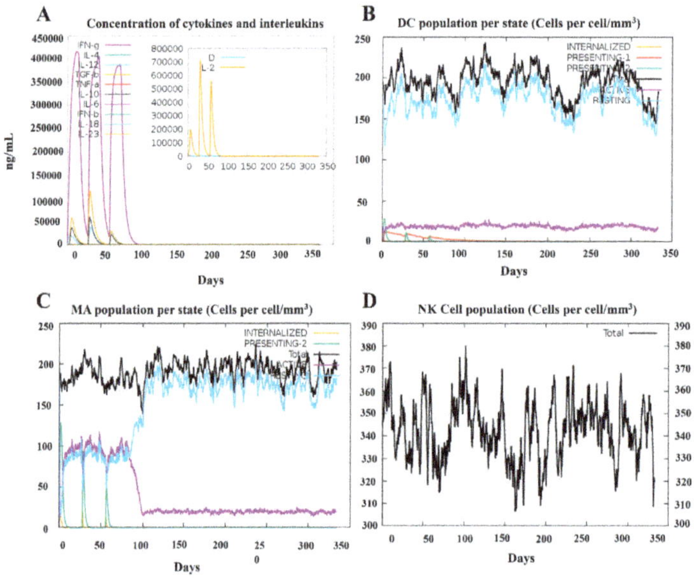


**Figure 8.** Simulation of innate immune response spots PRRSV-V-2 vaccination. (**A**) Kinetics of important cytokine expression levels after three PRRSV-V-2 dosages. (**B**) Dendritic cell (DC) responses following three PRRSV-V-2 doses. (**C**) Macrophages (MA) reactions to three doses of PRRSV-V-2 vaccination. (**D**) Natural killer (NK) cell dynamics after vaccination.

The adaptive immune response was also analyzed in this simulation to assess the activation of B cells, T cells, and immunoglobulins, with a particular focus on the B- and T-cell-related immune memory.

Following three vaccine doses, a robust and long-lasting activation of B cells was observed, with notable increases in memory B cells and IgM isotypic B cells, which persisted till one year after the initial immunization. Production of antibodies was observed in a pattern following the dynamics pattern corresponding to B cell activation. The total levels of IgM and IgG peaked at nearly 200,000, with individual levels reaching around 100,000 for each, suggesting a potent humoral response comparable to other published vaccines designed in silico [41]. T-cell activation was also evident, with strong responses in both cytotoxic and helper T-cell populations. Notably, memory helper T cells displayed high dynamic activity, maintaining a level above 600 at 350 days post-immunization, corresponding to 30% of the peak response. The peak immune response across different parameters was achieved approximately 50 days after the first dose, coinciding with the period immediately following the third vaccination (Figure 9).

These findings highlight the vaccine's robust immunogenic potential, suggesting that it can induce strong and systematic adaptive immune protection against PRRSV infection. Especially, prolonged memory B- and T-cell responses indicate the potential for long-term immunity following immunization.

### *3.8. Codon Optimization of the Final Vaccine Candidate*

Its coding sequence was first generated using the JCat server to produce the designed vaccine. Codon optimization was then performed with JCat to align the codon usage bias of *E. coli* K12. The optimized DNA sequence, consisting of 1386 bp, had a GC content of 56.02% and a 0.90 CAI, indicating efficient expression potential. This sequence was subsequently cloned into the pET-28a (+) vector between *XhoI* and *BamHI* restriction sites.

As a result, a recombinant plasmid for vaccine expression was successfully constructed, ready for application in the *E. coli* expression system, as shown in Figure S8.

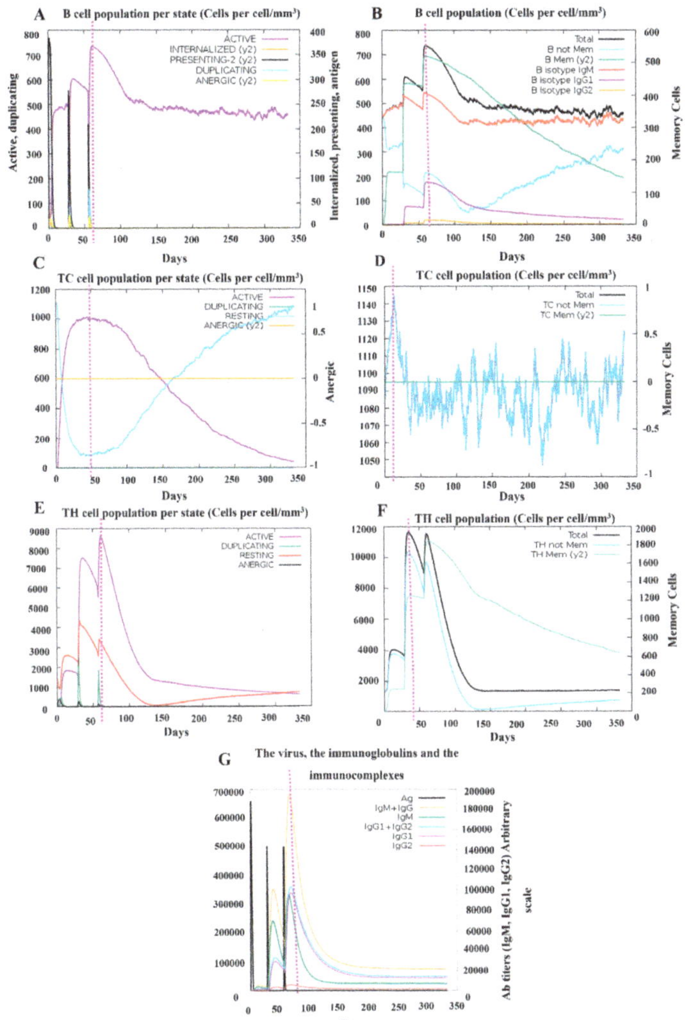


**Figure 9.** Simulation of the adaptive immune response following PRRSV-V-2 vaccination. (**A**) The status of the B-cell population post-vaccination is represented. (**B**) Distribution of B-cell subpopulations based on the antibody production status, (**C**,**D**) Dynamics of CD8+ cytotoxic T cells post-vaccination is shown, with distinct activation statuses (**C**), Such as active, duplicating, resting, (**D**) anergic (y2), and memory conditions. (**E**,**F**) Post-immunization trend CD4+ helper T cells are displayed, categorized by activation status (**E**), such as active, duplicating, resting, and (**F**) energic (y2), and memory conditions. (**G**) Immunoglobulin production following three PRRSV-V-2 doses shows antigen (Ag) levels.

## 4. Discussion

Porcine reproductive and respiratory syndrome (PRRS) has been a significant veterinary disease for over three decades, causing substantial economic losses and affecting pig farms worldwide. Despite extensive research, specific therapeutic drugs remain ineffective, and current vaccines offer only limited protection. Moreover, current commercially available vaccines offer just modest protection. Even worse, among the technical platforms ever tested, only modified live virus [19] or killed virus [18]-based vaccines have been proven to be effective. These challenges have made the control of PRRS exceptionally difficult.

Several key issues complicate the development of an effective PRRSV vaccine [82]. First, the extensive genetic variations in PRRSV genomes complicated the host immune response, making it difficult to achieve full protection across strains, which is only achievable with strains homologous to the vaccine [83]. Second, PRRSV is prone to genetic recombination and replacement, particularly among RNA viruses. This poses a substantial risk of virulence reversal or genetic recombination with field strains when live attenuated vaccines are used [84]. Although inactivated vaccines can mitigate this risk, they do not effectively promote cellular immunity. Furthermore, certain viral proteins interact with the host immune system, evading both innate and adaptive immunity, which either prolongs infection or reduces the immunogenicity of vaccines [85]. Several neutralizing epitopes have been identified in PRRSV structural proteins, particularly GP3, GP4, GP5, and M [23,24]. GP5, as a major envelope glycoprotein, plays a key role in viral entry and assembly [25], with neutralizing epitopes found in its N-terminus ectodomain [25,26]. Similarly, epitopes in the M proteins have been identified as important for susceptibility to neutralization antibodies [24,27,28]. In M protein, the pattern of GP5, a fragment of 70 amino acid residues, was determined as the crucial modulator of the susceptibility to neutralization antibody in porcine serum [29]. However, partial protection has been observed with immunization using the GP5-M ectodomain, suggesting that additional epitopes are needed for full viral neutralization. Alternatively, neutralizing epitopes have also been shown in GP2, 3, and 4 [30–32], with which another complex is formed as an essential player in receptor binding and viral infection [33,34]. Thus, in-depth investigation and continuous development of new vaccines with great potency will provide targeted opportunities for preventing and controlling PRRSV. However, a long way of struggle is still required to bring forth the vaccine design.

With the advancement of immunoinformatic platforms, a new path for vaccine development has emerged, distinct from the traditional approaches. Instead of relying on a small number of representative strains in the wet lab, large panels of strains from the same species can be utilized to identify epitopes. This approach can enhance the quality and coverage of vaccines. By predicting and scoring epitopes using various bioinformatics tools, promising epitopes can be selected, while suboptimal or risky ones are excluded. Subsequently, this method enables the quick construction of candidate vaccines by combining selected epitopes with adjuvant proteins, followed by multidisciplinary evaluations. Based on the predicted structural and immunogenic properties, the range of optimized proposals is narrowed down from a panel of candidates before conducting actual wet lab assays. This approach has facilitated the de novo design of vaccines for a wide range of pathogens, particularly those with significant complexity and challenges in vaccine development. With the benefit brought out with such an effective strategy, the previous design of the PRRSV vaccines in 2024 and 2025 dates [81,86]. The studies acknowledged certain limitations that need further investigation. In our study, a multi-epitope vaccine was developed to protect the swine from either PRRSV1 or 2, eliciting both humoral and cellular immune responses. In this vaccine, epitopes were screened from 30 and 77 conserved fragments of viral proteins in PRRSV1 and PRRSV2, respectively, and 6, 10, and 10 epitopes were selected to target B cells, CTLs, and HTLs. Stringent criteria based on antigenicity, immunogenicity, allergenicity, toxicity, conservancy, and the capacity of cytokine induction were applied. Three candidate vaccines were generated by organizing the selected epitopes and conjugating them with adjuvant proteins such as S50 L7/12 ribosomal protein, β-defensin, and HBHA. In silico evaluations demonstrated good structural properties, tight interactions with host pattern recognition receptors, and strong immune stimulation.

PRRSV-V-2 emerged as the most suitable candidate due to its favorable interaction with TLR3 and TLR8. The MD-derived structural stability metrics also have direct implications for vaccine effectiveness. For sustained receptor engagement and effective innate immune activation, vaccine–TLR interactions must be stable. Most residues of the complex exhibit a few atomic fluctuations, particularly at key contact regions, as indicated by the low Rg values across most residues. As a result of this structural rigidity, receptor engagement may be consistent, leading to reliable downstream signaling and cytokine activation, resulting in an enhanced ability to elicit a strong and sustained adaptive immune response. To effectively defend against viruses, vaccine constructs that interact with TLRs may enhance Th1-skewed immune responses [87].

When compared to earlier in silico PRRSV vaccine design [81,86], our approach differed significantly. While the previous design was based on pathogenic isolates specific to his research group, our study included PRRSV protein sequences from global strains, providing a broader representation. Additionally, we used all viral proteins, focusing on conserved fragments across strains, whereas Dr. Liu's design was restricted to GP3 and GP5 molecules. This difference potentially offers a wider range of epitope sources for our design and potentially improves the vaccine's coverage in a broader array of PRRSV strains. At the next stage, screening of epitopes was carried out in previous work; however, the evaluations of allergenicity and toxicity were absent in the earlier design [81,86]. In speculation, these were due to the limited supply of source antigens; however, potential risks were then introduced. Unlike that, with plenty of sources of viral antigens, enough resultant epitopes were still selected in our work, even after a stringent evaluation of risks to each single epitope predicted originally. With the abundance of outcomes, more CTL and HTL epitopes were selected to construct the vaccine candidates with a reasonable length sequence of 448 amino acids, as compared to the previous PRRSV in silico design vaccines' lengths of 688 and 455 amino acids [81,86]. Possibly, this will be a positive factor to reinforce the T-cell immunity, which is equally important in the control of disease and restraint of severity post-infection. In the stages, after epitopes were obtained, more adjuvant proteins and pattern recognition receptors were tried in our work. Especially, rather than human TLR3 and TLR4 used in previous studies [81,86], docking of our candidate vaccines was performed with TLR3 and TLR8 of swine, the prospective host. This approach offers more accuracy in predicting the real-world interaction between vaccines and receptors in the host. Additionally, our immune simulation was extended to 350 days, rather than 35 days in Dr. Liu's study, which provided a more comprehensive understanding of immune persistence. With longer observation, the persistence of immune protection was demonstrated in candidates of our selection. The scale of total and active B cell populations was maintained at a high level for up to one year, while the production of immunoglobulins was also retained to a certain degree. Notably, about 30% of peak levels were reserved for the memory B and T cells until the end of the simulation, consolidating the confidence in our strategy and choice in the current design.

## 5. Conclusions

By using comprehensive in silico approaches, we designed a novel multi-epitope vaccine candidate to address PRRSV's high mutation rate and genetic variability. We identified conserved antigenic regions and systematically screened them for B cell, helper T lymphocyte (HTL), and cytotoxic T lymphocyte (CTL) epitopes by integrating genomic and proteomic data from multiple PRRSV strains. Epitopes were selected based on antigenicity, immunogenicity, non-allergenicity, and non-toxicity criteria. Among the three vaccine candidates evaluated, PRRSV-V-2 displayed acceptable physicochemical properties, stable interactions with TLRs, and consequently, robust induction of host immunity.

In addition, an expression plasmid for PRRSV-V-2 was constructed in silico, facilitating downstream studies.

Using computational tools for high-throughput epitope identification and validation, our approach streamlines vaccine development at the early stage. Through molecular docking and molecular dynamics simulations, we gained further insight into the conformational stability and binding affinity of vaccine–receptor complexes, increasing our confidence in the vaccine candidate's ability to elicit robust immune responses. With this strategy, time, cost, and resource burden can be significantly saved when compared with traditional experimental approaches. Even so, this study's findings are inherently predictive and limited by in silico methodologies. Although computational tools are powerful, they cannot capture the immunological complexity of a living host or replicate the dynamic interaction between immune cells and signaling molecules. It is therefore necessary to experimentally validate the predicted immunogenicity, safety, and efficacy of the proposed vaccine construct.

To evaluate cellular immune responses, future directions will involve expressing and purifying the PRRSV-V-2 construct in vitro in a bacterial expression system, followed by immunological assays, including ELISpot, T-cell activation profiling, and cytokine release tests. In vivo studies will be essential to assess protective efficacy, immunological memory induction, and overall safety. Experimental validation will be crucial for translating our computational findings into a preclinical vaccine candidate with real-world application potential.

**Supplementary Materials:** The following supporting information can be downloaded at: https://www.mdpi.com/article/10.3390/vetsci12060577/s1.

**Author Contributions:** Conceptualization, T.L. and S.U.; methodology, S.U.; software, S.U.; validation, T.L. and S.U.; formal analysis, S.U.; investigation, T.L.; resources, H.U. and K.F.; data curation, S.U.; writing—original draft preparation, S.U., H.U. and T.L.; writing—review and editing, T.L.; visualization, T.L.; supervision, T.L. All authors have read and agreed to the published version of the manuscript.

**Funding:** This research project was supported by the Startup Foundation for New Faculty at the State Key Laboratory of Reproductive Medicine and Offspring Health, Nanjing Medical University (Grant No NMUR20240018) (Title: Diagnosis and therapy of viral infectious diseases related to reproduction).

**Institutional Review Board Statement:** Not applicable.

**Informed Consent Statement:** Not applicable.

**Data Availability Statement:** All data are available in the manuscript and Supplementary Materials.

**Acknowledgments:** This project was supported by the One Hundred Talented Youth Program of the Chinese Academy of Sciences, as well as the startup foundation for new faculty in the State Key Laboratory of Reproductive Medicine and Offspring Health of Nanjing Medical University. We provide our thanks to the Alliance of International Science Organizations (ANSO) for their support of scholarships to international students involved in this research.

**Conflicts of Interest:** The authors declare no conflicts of interest.

## Abbreviations

| | |
|---|---|
| PRRSV | Porcine respiratory and reproductive syndrome virus |
| PRRSV-V | Porcine respiratory and reproductive syndrome virus-vaccine |
| SLA | Swine Leukocyte Antigen |
| HLA | Human Leukocyte Antigen |

## References


1. Albina, E. Epidemiology of porcine reproductive and respiratory syndrome (PRRS): An overview. *Vet. Microbiol.* **1997**, *55*, 309–316. [CrossRef] [PubMed]
2. Lalonde, C.; Provost, C.; Gagnon, C.A. Whole-genome sequencing of porcine reproductive and respiratory syndrome virus from field clinical samples improves the genomic surveillance of the virus. *J. Clin. Microbiol.* **2020**, *58*, 10–1128. [CrossRef]
3. Neumann, E.J.; Kliebenstein, J.B.; Johnson, C.D.; Mabry, J.W.; Bush, E.J.; Seitzinger, A.H.; Green, A.L.; Zimmerman, J.J. Assessment of the economic impact of porcine reproductive and respiratory syndrome on swine production in the United States. *J. Am. Vet. Med. Assoc.* **2005**, *227*, 385–392. [CrossRef]
4. Renken, C.; Nathues, C.; Swam, H.; Fiebig, K.; Weiss, C.; Eddicks, M.; Ritzmann, M.; Nathues, H. Application of an economic calculator to determine the cost of porcine reproductive and respiratory syndrome at farm-level in 21 pig herds in Germany. *Porc. Health Manag.* **2021**, *7*, 3. [CrossRef]
5. Sun, Q.; Xu, H.; Li, C.; Gong, B.; Li, Z.; Tian, Z.-J.; Zhang, H. Emergence of a novel PRRSV-1 strain in mainland China: A recombinant strain derived from the two commercial modified live viruses Amervac and DV. *Front. Vet. Sci.* **2022**, *9*, 974743. [CrossRef]
6. Yu, F.; Liu, L.; Tian, X.; Chen, L.; Huang, X.; Sun, Y.; Yan, Y.; Tian, Z.; Cai, X.; Liu, D. Genomic analysis of porcine reproductive and respiratory syndrome virus 1 revealed extensive recombination and potential introduction events in China. *Vet. Sci.* **2022**, *9*, 450. [CrossRef]
7. Yu, F.; Yan, Y.; Shi, M.; Liu, H.-Z.; Zhang, H.-L.; Yang, Y.-B.; Huang, X.-Y.; Gauger, P.C.; Zhang, J.; Zhang, Y.-H. Phylogenetics, genomic recombination, and NSP2 polymorphic patterns of porcine reproductive and respiratory syndrome virus in China and the United States in 2014–2018. *J. Virol.* **2020**, *94*, 10–1128. [CrossRef]
8. Wang, L.-j.; Wan, B.; Guo, Z.; Qiao, S.; Li, R.; Xie, S.; Chen, X.-x.; Zhang, G. Genomic analysis of a recombinant NADC30-like porcine reproductive and respiratory syndrome virus in China. *Virus Genes* **2018**, *54*, 86–97. [CrossRef]
9. Zhao, K.; Ye, C.; Chang, X.-B.; Jiang, C.-G.; Wang, S.-J.; Cai, X.-H.; Tong, G.-Z.; Tian, Z.-J.; Shi, M.; An, T.-Q. Importation and recombination are responsible for the latest emergence of highly pathogenic porcine reproductive and respiratory syndrome virus in China. *J. Virol.* **2015**, *89*, 10712–10716. [CrossRef]
10. Ruedas-Torres, I.; Rodríguez-Gómez, I.; Sánchez-Carvajal, J.M.; Larenas-Muñoz, F.; Pallarés, F.; Carrasco, L.; Gómez-Laguna, J. The jigsaw of PRRSV virulence. *Vet. Microbiol.* **2021**, *260*, 109168. [CrossRef]
11. Fang, Y.; Snijder, E.J. The PRRSV replicase: Exploring the multifunctionality of an intriguing set of nonstructural proteins. *Virus Res.* **2010**, *154*, 61–76. [CrossRef] [PubMed]
12. Dea, S.; Gagnon, C.; Mardassi, H.; Pirzadeh, B.; Rogan, D. Current knowledge on the structural proteins of porcine reproductive and respiratory syndrome (PRRS) virus: Comparison of the North American and European isolates. *Arch. Virol.* **2000**, *145*, 659–688. [CrossRef]
13. Nan, Y.; Wu, C.; Gu, G.; Sun, W.; Zhang, Y.-J.; Zhou, E.-M. Improved vaccine against PRRSV: Current progress and future perspective. *Front. Microbiol.* **2017**, *8*, 1635. [CrossRef]
14. Guan, Z.; Pang, L.; Ouyang, Y.; Jiang, Y.; Zhang, J.; Qiu, Y.; Li, Z.; Li, B.; Liu, K.; Shao, D. Secondary Highly Pathogenic Porcine Reproductive and Respiratory Syndrome Virus (HP-PRRSV2) Infection Augments Inflammatory Responses, Clinical Outcomes, and Pathogen Load in Glaesserella-parasuis-Infected Piglets. *Vet. Sci.* **2023**, *10*, 365. [CrossRef]
15. Zhang, H.; Luo, Q.; He, Y.; Zheng, Y.; Sha, H.; Li, G.; Kong, W.; Liao, J.; Zhao, M. Research Progress on the Development of Porcine Reproductive and Respiratory Syndrome Vaccines. *Vet. Sci.* **2023**, *10*, 491. [CrossRef]
16. Wang, H.; Feng, W. Current Status of Porcine Reproductive and Respiratory Syndrome Vaccines. *Vaccines* **2024**, *12*, 1387. [CrossRef]
17. Charerntantanakul, W. Porcine reproductive and respiratory syndrome virus vaccines: Immunogenicity, efficacy and safety aspects. *World J. Virol.* **2012**, *1*, 23. [CrossRef]
18. Renukaradhya, G.J.; Meng, X.-J.; Calvert, J.G.; Roof, M.; Lager, K.M. Inactivated and subunit vaccines against porcine reproductive and respiratory syndrome: Current status and future direction. *Vaccine* **2015**, *33*, 3065–3072. [CrossRef]
19. Chae, C. Commercial PRRS modified-live virus vaccines. *Vaccines* **2021**, *9*, 185. [CrossRef]
20. Lee, M.-A.; Jayaramaiah, U.; You, S.-H.; Shin, E.-G.; Song, S.-M.; Ju, L.; Kang, S.-J.; Hyun, B.-H.; Lee, H.-S. Molecular characterization of porcine reproductive and respiratory syndrome virus in Korea from 2018 to 2022. *Pathogens* **2023**, *12*, 757. [CrossRef]
21. Papatsiros, V. Impact of a killed PRRSV vaccine on sow longevity in a PRRSV infected swine herd. *J. Appl. Anim. Res.* **2012**, *40*, 297–304. [CrossRef]
22. Kim, H.; Kim, H.K.; Jung, J.H.; Choi, Y.J.; Kim, J.; Um, C.G.; Hyun, S.B.; Shin, S.; Lee, B.; Jang, G. The assessment of efficacy of porcine reproductive respiratory syndrome virus inactivated vaccine based on the viral quantity and inactivation methods. *Virol. J.* **2011**, *8*, 323. [CrossRef]

23. Van Nieuwstadt, A.; Meulenberg, J.; van Essen-Zanbergen, A.; Petersen-den Besten, A.; Bende, R.; Moormann, R.; Wensvoort, G. Proteins encoded by open reading frames 3 and 4 of the genome of Lelystad virus (Arteriviridae) are structural proteins of the virion. *J. Virol.* **1996**, *70*, 4767–4772. [CrossRef]
24. Cancel-Tirado, S.M.; Evans, R.B.; Yoon, K.-J. Monoclonal antibody analysis of porcine reproductive and respiratory syndrome virus epitopes associated with antibody-dependent enhancement and neutralization of virus infection. *Vet. Immunol. Immunopathol.* **2004**, *102*, 249–262. [CrossRef]
25. Delputte, P.; Meerts, P.; Costers, S.; Nauwynck, H. Effect of virus-specific antibodies on attachment, internalization and infection of porcine reproductive and respiratory syndrome virus in primary macrophages. *Vet. Immunol. Immunopathol.* **2004**, *102*, 179–188. [CrossRef]
26. Delputte, P.L.; Van Breedam, W.; Delrue, I.; Oetke, C.; Crocker, P.R.; Nauwynck, H.J. Porcine arterivirus attachment to the macrophage-specific receptor sialoadhesin is dependent on the sialic acid-binding activity of the N-terminal immunoglobulin domain of sialoadhesin. *J. Virol.* **2007**, *81*, 9546–9550. [CrossRef]
27. Ostrowski, M.; Galeota, J.; Jar, A.; Platt, K.; Osorio, F.A.; Lopez, O. Identification of neutralizing and nonneutralizing epitopes in the porcine reproductive and respiratory syndrome virus GP5 ectodomain. *J. Virol.* **2002**, *76*, 4241–4250. [CrossRef]
28. Plagemann, P.; Rowland, R.; Faaberg, K. The primary neutralization epitope of porcine respiratory and reproductive syndrome virus strain VR-2332 is located in the middle of the GP5 ectodomain. *Arch. Virol.* **2002**, *147*, 2327–2347. [CrossRef]
29. Fan, B.; Liu, X.; Bai, J.; Zhang, T.; Zhang, Q.; Jiang, P. Influence of the amino acid residues at 70 in M protein of porcine reproductive and respiratory syndrome virus on viral neutralization susceptibility to the serum antibody. *Virol. J.* **2016**, *13*, 1–11. [CrossRef]
30. Vanhee, M.; Costers, S.; Van Breedam, W.; Geldhof, M.F.; Van Doorsselaere, J.; Nauwynck, H.J. A variable region in GP4 of European-type porcine reproductive and respiratory syndrome virus induces neutralizing antibodies against homologous but not heterologous virus strains. *Viral Immunol.* **2010**, *23*, 403–413. [CrossRef]
31. Costers, S.; Lefebvre, D.J.; Van Doorsselaere, J.; Vanhee, M.; Delputte, P.L.; Nauwynck, H.J. GP4 of porcine reproductive and respiratory syndrome virus contains a neutralizing epitope that is susceptible to immunoselection in vitro. *Arch. Virol.* **2010**, *155*, 371–378. [CrossRef] [PubMed]
32. Costers, S.; Vanhee, M.; Van Breedam, W.; Van Doorsselaere, J.; Geldhof, M.; Nauwynck, H.J. GP4-specific neutralizing antibodies might be a driving force in PRRSV evolution. *Virus Res.* **2010**, *154*, 104–113. [CrossRef]
33. Wissink, E.; Kroese, M.; Van Wijk, H.; Rijsewijk, F.; Meulenberg, J.; Rottier, P. Envelope protein requirements for the assembly of infectious virions of porcine reproductive and respiratory syndrome virus. *J. Virol.* **2005**, *79*, 12495–12506. [CrossRef]
34. Das, P.B.; Dinh, P.X.; Ansari, I.H.; De Lima, M.; Osorio, F.A.; Pattnaik, A.K. The minor envelope glycoproteins GP2a and GP4 of porcine reproductive and respiratory syndrome virus interact with the receptor CD163. *J. Virol.* **2010**, *84*, 1731–1740. [CrossRef]
35. Meulenberg, J.; Van Nieuwstadt, A.; van Essen-Zandbergen, A.; Langeveld, J. Posttranslational processing and identification of a neutralization domain of the GP4 protein encoded by ORF4 of Lelystad virus. *J. Virol.* **1997**, *71*, 6061–6067. [CrossRef]
36. Vashisht, K.; Goldberg, T.L.; Husmann, R.J.; Schnitzlein, W.; Zuckermann, F.A. Identification of immunodominant T-cell epitopes present in glycoprotein 5 of the North American genotype of porcine reproductive and respiratory syndrome virus. *Vaccine* **2008**, *26*, 4747–4753. [CrossRef]
37. Díaz, I.; Pujols, J.; Ganges, L.; Gimeno, M.; Darwich, L.; Domingo, M.; Mateu, E. In silico prediction and ex vivo evaluation of potential T-cell epitopes in glycoproteins 4 and 5 and nucleocapsid protein of genotype-I (European) of porcine reproductive and respiratory syndrome virus. *Vaccine* **2009**, *27*, 5603–5611. [CrossRef]
38. Luo, Q.; Zheng, Y.; Zhang, H.; Yang, Z.; Sha, H.; Kong, W.; Zhao, M.; Wang, N. Research progress on glycoprotein 5 of porcine reproductive and respiratory syndrome virus. *Animals* **2023**, *13*, 813. [CrossRef]
39. Gong, H.-R.; Hu, Y.-f.; Li, X.; Yau, T.; Zhang, B.-Z.; Huang, J.-D. Non-neutralizing epitopes shade neutralizing epitopes against omicron in a multiple epitope-based vaccine. *ACS Infect. Dis.* **2022**, *8*, 2586–2593. [CrossRef]
40. Devi, A.; Chaitanya, N.S. In silico designing of multi-epitope vaccine construct against human coronavirus infections. *J. Biomol. Struct. Dyn.* **2021**, *39*, 6903–6917. [CrossRef]
41. Ullah, H.; Ullah, S.; Li, J.; Yang, F.; Tan, L. An In Silico Design of a Vaccine against All Serotypes of the Dengue Virus Based on Virtual Screening of B-Cell and T-Cell Epitopes. *Biology* **2024**, *13*, 681. [CrossRef] [PubMed]
42. Yuan, L.; Li, X.; Li, M.; Bi, R.; Li, Y.; Song, J.; Li, W.; Yan, M.; Luo, H.; Sun, C. In silico design of a broad-spectrum multiepitope vaccine against influenza virus. *Int. J. Biol. Macromol.* **2024**, *254*, 128071. [CrossRef]
43. Edgar, R.C. MUSCLE: Multiple sequence alignment with high accuracy and high throughput. *Nucleic Acids Res.* **2004**, *32*, 1792–1797. [CrossRef]
44. Thompson, J.D.; Gibson, T.J.; Higgins, D.G. Multiple sequence alignment using ClustalW and ClustalX. *Curr. Protoc. Bioinform.* **2003**, 2.3.1–2.3.22. [CrossRef]
45. Oliver, T.; Schmidt, B.; Nathan, D.; Clemens, R.; Maskell, D. Using reconfigurable hardware to accelerate multiple sequence alignment with ClustalW. *Bioinformatics* **2005**, *21*, 3431–3432. [CrossRef]

46. Jespersen, M.C.; Peters, B.; Nielsen, M.; Marcatili, P. BepiPred-2.0: Improving sequence-based B-cell epitope prediction using conformational epitopes. *Nucleic Acids Res.* **2017**, *45*, W24–W29. [CrossRef]
47. Reynisson, B.; Alvarez, B.; Paul, S.; Peters, B.; Nielsen, M. NetMHCpan-4.1 and NetMHCIIpan-4.0: Improved predictions of MHC antigen presentation by concurrent motif deconvolution and integration of MS MHC eluted ligand data. *Nucleic Acids Res.* **2020**, *48*, W449–W454. [CrossRef]
48. Middleton, D.; Menchaca, L.; Rood, H.; Komerofsky, R. New allele frequency database: http://www.allelefrequencies.net. *Tissue Antigens* **2003**, *61*, 403–407. [CrossRef]
49. Calis, J.J.; Maybeno, M.; Greenbaum, J.A.; Weiskopf, D.; De Silva, A.D.; Sette, A.; Keşmir, C.; Peters, B. Properties of MHC class I presented peptides that enhance immunogenicity. *PLoS Comput. Biol.* **2013**, *9*, e1003266. [CrossRef]
50. Doytchinova, I.A.; Flower, D.R. VaxiJen: A server for prediction of protective antigens, tumour antigens and subunit vaccines. *BMC Bioinform.* **2007**, *8*, 4. [CrossRef]
51. Dimitrov, I.; Bangov, I.; Flower, D.R.; Doytchinova, I. AllerTOP v. 2—A server for in silico prediction of allergens. *J. Mol. Model.* **2014**, *20*, 2278. [CrossRef] [PubMed]
52. Gupta, S.; Kapoor, P.; Chaudhary, K.; Gautam, A.; Kumar, R.; Consortium, O.S.D.D.; Raghava, G.P. In silico approach for predicting toxicity of peptides and proteins. *PLoS ONE* **2013**, *8*, e73957. [CrossRef]
53. Dhanda, S.K.; Vir, P.; Raghava, G.P. Designing of interferon-gamma inducing MHC class-II binders. *Biol. Direct* **2013**, *8*, 30. [CrossRef]
54. Dhanda, S.K.; Gupta, S.; Vir, P.; Raghava, G.P.S. Prediction of IL4 Inducing Peptides. *J. Immunol. Res.* **2013**, *2013*, 263952. [CrossRef]
55. Fadaka, A.O.; Sibuyi, N.R.S.; Martin, D.R.; Goboza, M.; Klein, A.; Madiehe, A.M.; Meyer, M. Immunoinformatics design of a novel epitope-based vaccine candidate against dengue virus. *Sci. Rep.* **2021**, *11*, 19707. [CrossRef]
56. Li, W.; Joshi, M.D.; Singhania, S.; Ramsey, K.H.; Murthy, A.K. Peptide vaccine: Progress and challenges. *Vaccines* **2014**, *2*, 515–536. [CrossRef]
57. Jalal, K.; Khan, K.; Ahmad, D.; Hayat, A.; Basharat, Z.; Abbas, M.N.; Alghamdi, S.; Almehmadi, M.; Sahibzada, M.U.K. Pan-genome reverse vaccinology approach for the design of multi-epitope vaccine construct against Escherichia albertii. *Int. J. Mol. Sci.* **2021**, *22*, 12814. [CrossRef]
58. Dhanushkumar, T.; Kamaraj, B.; Vasudevan, K.; Gopikrishnan, M.; Dasegowda, K.; Rambabu, M. Structural immunoinformatics approach for rational design of a multi-epitope vaccine against triple negative breast cancer. *Int. J. Biol. Macromol.* **2023**, *243*, 125209.
59. Ghaffari-Nazari, H.; Tavakkol-Afshari, J.; Jaafari, M.R.; Tahaghoghi-Hajghorbani, S.; Masoumi, E.; Jalali, S.A. Improving multi-epitope long peptide vaccine potency by using a strategy that enhances CD4+ T help in BALB/c mice. *PLoS ONE* **2015**, *10*, e0142563. [CrossRef]
60. Gasteiger, E.; Hoogland, C.; Gattiker, A.; Duvaud, S.; Wilkins, M.; Appel, R.; Bairoch, A. Protein Identification and Analysis Tools on the ExPASy Server. In *The Proteomics Protocols Handbook*; Humana Press: Totowa, NJ, USA, 2005; pp. 571–607.
61. Magnan, C.N.; Randall, A.; Baldi, P. SOLpro: Accurate sequence-based prediction of protein solubility. *Bioinformatics* **2009**, *25*, 2200–2207. [CrossRef]
62. McGuffin, L.J.; Bryson, K.; Jones, D.T. The PSIPRED protein structure prediction server. *Bioinformatics* **2000**, *16*, 404–405. [CrossRef] [PubMed]
63. Kelley, L.A.; Mezulis, S.; Yates, C.M.; Wass, M.N.; Sternberg, M.J. The Phyre2 web portal for protein modeling, prediction and analysis. *Nat. Protoc.* **2015**, *10*, 845–858. [CrossRef] [PubMed]
64. Heo, L.; Park, H.; Seok, C. GalaxyRefine: Protein structure refinement driven by side-chain repacking. *Nucleic Acids Res.* **2013**, *41*, W384–W388. [CrossRef]
65. Wiederstein, M.; Sippl, M.J. ProSA-web: Interactive web service for the recognition of errors in three-dimensional structures of proteins. *Nucleic Acids Res.* **2007**, *35* (Suppl. S2), W407–W410. [CrossRef]
66. Laskowski, R.A.; MacArthur, M.W.; Moss, D.S.; Thornton, J.M. PROCHECK: A program to check the stereochemical quality of protein structures. *J. Appl. Crystallogr.* **1993**, *26*, 283–291. [CrossRef]
67. Berman, H.M.; Westbrook, J.; Feng, Z.; Gilliland, G.; Bhat, T.N.; Weissig, H.; Shindyalov, I.N.; Bourne, P.E. The Protein Data Bank. *Nucleic Acids Res.* **2000**, *28*, 235–242. [CrossRef]
68. Chen, X.; Li, Y.; Wang, X. Multi-epitope vaccines: A promising strategy against viral diseases in swine. *Front Cell Infect Microbiol.* **2024**, *14*, 1497580. [CrossRef]
69. Yan, Y.; Tao, H.; He, J.; Huang, S.-Y. The HDOCK server for integrated protein–protein docking. *Nat. Protoc.* **2020**, *15*, 1829–1852. [CrossRef]
70. Laskowski, R.A.; Jabłońska, J.; Pravda, L.; Vařeková, R.S.; Thornton, J.M. PDBsum: Structural summaries of PDB entries. *Protein Sci.* **2018**, *27*, 129–134. [CrossRef]

71. Lovell, S.C.; Davis, I.W.; Arendall III, W.B.; De Bakker, P.I.; Word, J.M.; Prisant, M.G.; Richardson, J.S.; Richardson, D.C. Structure validation by Cα geometry: ϕ, ψ and Cβ deviation. *Proteins Struct. Funct. Bioinform.* **2003**, *50*, 437–450. [CrossRef]
72. Case, D.; Babin, V.; Berryman, J.; Betz, R.; Cai, Q.; Cerutti, D.; Cheatham III, T.; Darden, T.; Duke, R.; Gohlke, H. The FF14SB force field. *Amber* **2014**, *14*, 29–31.
73. Maier, J.A.; Martinez, C.; Kasavajhala, K.; Wickstrom, L.; Hauser, K.E.; Simmerling, C. ff14SB: Improving the accuracy of protein side chain and backbone parameters from ff99SB. *J. Chem. Theory Comput.* **2015**, *11*, 3696–3713. [CrossRef]
74. Wang, J.; Wolf, R.M.; Caldwell, J.W.; Kollman, P.A.; Case, D.A. Development and testing of a general amber force field. *J. Comput. Chem.* **2004**, *25*, 1157–1174. [CrossRef]
75. Kräutler, V.; Van Gunsteren, W.F.; Hünenberger, P.H. A fast SHAKE algorithm to solve distance constraint equations for small molecules in molecular dynamics simulations. *J. Comput. Chem.* **2001**, *22*, 501–508. [CrossRef]
76. Allemailem, K.S. A Comprehensive Computer Aided Vaccine Design Approach to Propose a Multi-Epitopes Subunit Vaccine against Genus Klebsiella Using Pan-Genomics, Reverse Vaccinology, and Biophysical Techniques. *Vaccines* **2021**, *9*, 1087. [CrossRef]
77. Roe, D.R.; Cheatham III, T.E. PTRAJ and CPPTRAJ: Software for processing and analysis of molecular dynamics trajectory data. *J. Chem. Theory Comput.* **2013**, *9*, 3084–3095. [CrossRef]
78. DeLano, W.L. Pymol: An open-source molecular graphics tool. *CCP4 Newsl. Protein Crystallogr.* **2002**, *40*, 82–92.
79. Rapin, N.; Lund, O.; Bernaschi, M.; Castiglione, F. Computational immunology meets bioinformatics: The use of prediction tools for molecular binding in the simulation of the immune system. *PLoS ONE* **2010**, *5*, e9862. [CrossRef]
80. Grote, A.; Hiller, K.; Scheer, M.; Münch, R.; Nörtemann, B.; Hempel, D.C.; Jahn, D. JCat: A novel tool to adapt codon usage of a target gene to its potential expression host. *Nucleic Acids Res.* **2005**, *33* (Suppl. S2), W526–W531. [CrossRef]
81. Liu, D.; Chen, Y. Epitope screening and vaccine molecule design of PRRSV GP3 and GP5 protein based on immunoinformatics. *J. Cell Mol. Med.* **2024**, *28*, e18103. [CrossRef]
82. Cai, H.; Zhang, H.; Cheng, H.; Liu, M.; Wen, S.; Ren, J. Progress in PRRSV infection and adaptive immune response mechanisms. *Viruses* **2023**, *15*, 1442. [CrossRef] [PubMed]
83. Yu, Z.; Yi, H.; Ma, J.; Wei, Y.; Cai, M.; Li, Q.; Qin, C.; Chen, Y.-j.; Han, X.; Zhong, R. Ginsenoside Rg1 suppresses type 2 PRRSV infection via NF-κB signaling pathway in vitro, and provides partial protection against HP-PRRSV in piglet. *Viruses* **2019**, *11*, 1045. [CrossRef] [PubMed]
84. Wium, M.; Jonker, H.I.; Olivier, A.J.; Bellstedt, D.U.; Botes, A. DNA vaccines against Mycoplasma elicit humoral immune responses in ostriches. *Front. Immunol.* **2019**, *10*, 1061. [CrossRef] [PubMed]
85. Shabir, N.; Khatun, A.; Nazki, S.; Kim, B.; Choi, E.-J.; Sun, D.; Yoon, K.-J.; Kim, W.-I. Evaluation of the cross-protective efficacy of a chimeric porcine reproductive and respiratory syndrome virus constructed based on two field strains. *Viruses* **2016**, *8*, 240. [CrossRef]
86. Sira, E.M.J.S.; Banico, E.C.; Fajardo, L.E.; Odchimar, N.M.O.; Cruz, K.M.D.; Orosco, F.L. In silico design of multi-epitope vaccine candidate based on structural proteins of porcine reproductive and respiratory syndrome virus. *Vet. Immunol. Immunopathol.* **2025**, *280*, 110881. [CrossRef]
87. Tseng, C.Y.; Murtada, F.; Chou, L.Y. Precision nanoscale patterning of TLR ligands for improved cancer immunotherapy. *Cell Rep. Methods* **2024**, *4*, 100782. [CrossRef]

*veterinary sciences*

MDPI

*Article*

# Etiological Detection, Isolation, and Pathogenicity of Porcine Reproductive and Respiratory Syndrome Virus in China

**Yingbin Du [1,2], Jingyi Chen [1,2], Tianze Ren [1,2], Chunying Xie [1,2], Yiye Zhang [1,2], Liurong Fang [1,2] and Yanrong Zhou [1,2,*]**

1 National Key Laboratory of Agricultural Microbiology, College of Veterinary Medicine, Huazhong Agricultural University, Wuhan 430070, China
2 Key Laboratory of Preventive Veterinary Medicine in Hubei Province, the Cooperative Innovation Center for Sustainable Pig Production, Wuhan 430070, China
* Correspondence: yrzhou@mail.hzau.edu.cn

**Simple Summary:** Porcine reproductive and respiratory syndrome (PRRS) is a global widespread infectious disease that severely impacts the swine industry. The high genetic variability of the PRRS virus (PRRSV) has led to the continuous emergence and spread of novel strains. The surveillance of PRRSV epidemiology and isolation of circulating strains are crucial for developing effective PRRSV control strategies. In this study, we confirmed that NADC30/34-like PRRSV is the dominant epidemic strain in China, followed by highly pathogenic PRRSV (HP-PRRSV). An HP-PRRSV strain (JX03) and a recombinant NADC34-like PRRSV strain (HN08) were isolated, and their whole genome and phylogenetic characterization were analyzed. Animal experiments confirmed that the JX03 strain exhibited the highest pathogenicity, followed by CHN-HB-2018 (a NADC30-like PRRSV strain previously isolated by our lab), while HN08 demonstrated the lowest pathogenicity. These findings provide critical information for PRRSV molecular epidemiology and vaccine strain selection.

**Abstract:** Due to its high genomic variability, the epidemiological landscape of porcine reproductive and respiratory syndrome virus (PRRSV) has become increasingly complex in recent years. From 2022 to 2023, we collected a total of 1044 clinical samples from pigs suspected of PRRSV infection in China and discovered a PRRSV-positive rate of 29.8% (311/1044) using RT-PCR targeting the nsp2 gene. Among these positive samples, NADC30/34-like PRRSV, highly pathogenic PRRSV (HP-PRRSV), and classical PRRSV strains accounted for 60.1%, 37.9%, and 4.5%, respectively. These results indicate that the most prevalent PRRSV strains in China are NADC30/34-like PRRSV, followed by HP-PRRSV. Two PRRSV strains, JX03 and HN08, were isolated, and $TCID_{50}$ assays were performed to determine their titers at different time points post-infection, revealing differences in their proliferation kinetics. Phylogenetic, amino acid sequence, and recombination analyses demonstrated that the JX03 and HN08 strains cluster within lineage 8 (HP-PRRSV) and sublineage 1.5 (NADC34-like PRRSV), respectively. Notably, the HN08 strain was identified as a recombinant between the NADC30-like and NADC34-like strains, while no recombination event was detected in the JX03 strain. Pathogenicity assessments showed that the JX03 strain exhibited higher pathogenicity than the CHN-HB-2018 strain (a NADC30-like PRRSV strain was previously isolated by our lab), as evidenced by differences in clinical signs and mortality rates in piglets. In contrast, HN08 displayed no obvious clinical symptoms or mortality, revealing lower pathogenicity than the CHN-HB-2018 strain. These findings provide valuable information on the epidemiological and genetic characteristics of PRRSV strains in China, laying a foundation for the development of effective strategies against PRRSV.

**Keywords:** porcine reproductive and respiratory syndrome virus; NADC30/34-like strain; genetic variation; phylogenetic analysis; pathogenicity

## 1. Introduction

Porcine reproductive and respiratory syndrome (PRRS) is a highly contagious viral disease that causes significant economic losses in the global swine industry [1,2]. The causative agent, PRRSV, is a single-stranded positive-sense RNA virus belonging to the Arteriviridae family within the order Nidovirales [3]. The PRRSV genome is approximately 15 kb in length and encodes at least 10 open reading frames (ORFs), including ORF1a, ORF1b, ORF2a, ORF2b, ORF3, ORF4, ORF5, ORF5a, ORF6, and ORF7 [4,5]. Notably, ORF1a and ORF1b encode the polyproteins pp1a and pp1ab, which are subsequently cleaved by proteases into at least 16 nonstructural proteins (NSPs) [6]. ORFs 2–7 encode structural proteins, including GP2, envelope protein (E), GP3-GP5, GP5a, membrane protein (M), and nucleocapsid protein (N) [2].

The PRRSV genome exhibits high genetic variability. In 2016, the International Committee on the Taxonomy of Viruses (ICTV) formally classified PRRSV as two distinct viral species, PRRSV-1 and PRRSV-2 [7]. Based on the ORF5 gene sequences, PRRSV-1 is further categorized into four lineages, while PRRSV-2 is classified into eleven lineages [8,9]. In China, PRRSV-2 is the predominant circulating strain, with PRRSV-1 being relatively uncommon [10–13]. Therefore, most epidemiological studies on PRRSV in China primarily focus on PRRSV-2. Currently, PRRSV-2 in China is primarily concentrated in lineage 1 (NADC30/34-like PRRSV), lineage 3 (QYYZ-like PRRSV), lineage 5 (VR-2332-like PRRSV), and lineage 8 (HP-PRRSV), with lineages 1 and 8 being the most prevalent [10,14–17]. Detection rates of NADC30-like and NADC34-like strains have recently increased dramatically, exhibiting widespread distribution across multiple Chinese provinces [18–21].

Various PRRSV strains exhibit distinct pathogenic characteristics [10,22–24]. HP-PRRSV is characterized by severe clinical manifestations, such as high fever and elevated mortality rates in piglets [25]. NADC30-like isolates exhibit marked pathogenicity divergence [13,22,26–28]. Most NADC30 strains caused transient fever without mortality in infected pigs [29], while the NADC30-like PRRSV JL580 strain demonstrated high virulence, causing severe clinical manifestations including persistent pyrexia, respiratory distress, and 100% mortality [28]. Similar pathogenic heterogeneity occurs in NADC34-like strains. Comparative studies have revealed that the majority of NADC34-like strains, such as IA/NADC34, IA/ISU-1, and IA/ISU-5, induced severe interstitial pneumonia with pulmonary edema and hemorrhagic lesions, whereas IA/ISU-2 and HLJDZD32-1901 caused only mild pulmonary pathology or limited interstitial pneumonia [30,31]. Diverse recombination patterns serve as a key determinant of pathogenicity differences among NADC30/34-like variants. For example, the recombinant strain SDLY23-1742 resulting from NADC30-like, NADC34-like, and HP-PRRSV genetic recombination exhibited higher pathogenicity than BJ1805-2 (a NADC34-like strain with no recombination event) [24]. Consequently, continuous monitoring of PRRSV-2 prevalence and clinical pathogenic characteristics remains essential for developing effective prevention and control strategies.

The nsp2 gene is frequently utilized in molecular epidemiological investigations due to its characteristic amino acid deletions and insertions [15,18,20,32]. Compared with the original strain of PRRSV-2 (VR-2332), HP-PRRSV exhibits a discontinuous 30-aa (1 + 29) deletion, while NADC30-like PRRSV displays a more extensive discontinuous deletion of 131-aa (111 + 1 + 19), and the emerging NADC34-like strains feature a contiguous 100-aa deletion [15,33,34]. Similarly, the PRRSV ORF5 gene also exhibits high genetic diversity [16,35]. Classifying PRRSV into lineages or clades through phylogenetic analysis of the ORF5 gene is essential for understanding the genetic relationships among different PRRSV strains [30,36,37]. In the clinical surveillance of PRRSV, simultaneous targeting of both the nsp2 and ORF5 genes for genetic variation analysis significantly enhances detection accuracy and provides comprehensive molecular epidemiological characterization [38].

In this study, we conducted an epidemiological analysis of PRRSV in China from 2022 to 2023. Our findings revealed that the currently prevalent strains are predominantly NADC30/34-like PRRSV, followed by HP-PRRSV. An HP-PRRSV strain (JX03) and a recombinant NADC34-like PRRSV strain (HN08) were isolated. Phylogenetic, amino acid sequence, and recombination analyses demonstrated that the JX03 and HN08 strains cluster within lineages 8 and 1.5, respectively. A comparison of pathogenicity demonstrated variations among different PRRSV lineages, with JX03 exhibiting the highest pathogenicity, followed by CHN-HB-2018, while HN08 had the lowest pathogenicity. These findings provide insights into the molecular characteristics, genetic diversity, pathogenicity, and evolution of PRRSV in China.

## 2. Materials and Methods

### 2.1. Clinical Sample Collection and Treatment

Clinical tissue samples, including lungs, placentas, and aborted fetuses, were collected from pigs suspected of being infected with PRRSV across 27 provinces (including municipalities and autonomous regions) in China. These samples (0.5 g) were homogenized in 2 mL of phosphate-buffered saline (PBS) using a tissue homogenizer for 10 min, followed by centrifugation at 12,000 rpm for 10 min at 4 °C. The supernatants were aliquoted and stored at −80 °C for RNA extraction and virus isolation.

### 2.2. RNA Extraction and RT-PCR Detection

Total RNA was extracted and reverse transcribed into cDNA via the TRIzol (Omega, Georgia, USA) method and an RNA reverse transcription kit (Vazyme, Nanjing, China) following the manufacturer's instructions. PCR was subsequently performed using a pair of specific primers (forward: 5′-CTCCTTTGATTGGAATGTTGTG-3′; reverse: 5′-GATGGCTTGAGCTGAGTATTTTG-3′) to amplify a gene fragment within the nsp2 region, which allows for the differential detection of NADC30/34-like PRRSV (681 bp), HP-PRRSV (984 bp), and classical PRRSV (1072 bp) strains [39].

### 2.3. Virus Isolation and Growth Curve Detection

Porcine alveolar macrophages (PAMs) were cultured in Roswell Park Memorial Institute 1640 medium (RPMI-1640; Sigma, St. Louis, MO, USA) supplemented with 10% fetal bovine serum (FBS). The prepared supernatants of PRRSV-positive samples were filtered through a 0.22 μm filter (Millipore, Burlington, MA, USA). Subsequently, PAMs seeded in 12-well plates were inoculated with the resulting filtrates and incubated for 2 h at 37 °C. Following inoculation, the cells were washed three times with PBS to eliminate the remaining inoculum, supplemented with fresh medium, and incubated for an additional 2 to 3 days prior to harvest. Next, the harvested cell cultures underwent three freeze–thaw cycles to facilitate viral release. The samples were then centrifuged at 12,000 rpm for 10 min, and the supernatant was collected as the F0 generation virus. Three serial passages were conducted following the same procedure. Finally, the isolated viruses were stored at −80 °C.

For viral growth curve detection, PAMs in 12-well plates were infected with PRRSV isolates at a 0.01 multiplicity of infection (MOI) for 2 h (37 °C). After removing the inoculum and three washes with PBS, the plates were supplemented with fresh medium. At 6, 12, 24,

36, 48, 60, and 72 h post-infection (hpi), the supernatant and cells were harvested, and 50% tissue culture infectious dose ($TCID_{50}$) assays were performed to determine viral titers.

*2.4. Indirect Immunofluorescence Assay (IFA)*

PAMs infected or mock-infected with PRRSV isolates were washed with PBS, fixed with 4% paraformaldehyde, and permeabilized with cold methanol. After being blocked with 5% bovine serum albumin, the cells were incubated with a PRRSV-specific monoclonal antibody (mAb) against N protein (previously generated in our lab), followed by a fluorescein isothiocyanate (FITC)-conjugated goat anti-mouse IgG antibody (Invitrogen, Carlsbad, CA, USA). Finally, the nuclei were stained with 0.01% 4′,6-diamidino-2-phenylindole (DAPI), and the fluorescence images were visualized using a fluorescence microscope (Olympus, Tokyo, Japan).

*2.5. Multiple Alignment, Phylogenetic, and Recombination Analyses*

The complete genome sequences of the isolated strains were determined with the paired primers listed in Table S1. Specific primers (forward: 5′-ATGTTGGGGAAATGCTTGACC-3′; reverse: 5′-CTAGAGACGACCCCATTGTTC-3′) targeting the ORF5 gene were used for RT-PCR amplification, followed by sequencing. A total of 51 complete genome sequences and 70 ORF5 gene sequences of PRRSV were retrieved from the NCBI GenBank database https://www.ncbi.nlm.nih.gov/ (accessed on 4 September 2024) to serve as reference sequences. Detailed information about these sequences is provided in Table S2. Phylogenetic analyses were constructed for both whole-genome sequences and ORF5 gene sequences. Sequence alignment was conducted using MAFFT v7.526 [40]. For homology analysis, the MegAlign program within DNASTAR v7.0 was employed [41]. Maximum likelihood (ML) phylogenetic trees were generated using MEGA v12.0. Additionally, the amino acid sequences of the nsp2 and GP5 proteins were analyzed using CLC Main Workbench v6.8. Recombination analyses were performed for the nucleotide sequence of PRRSV isolates using RDP v4.0.

*2.6. Animal Experiments*

Twenty-four 4-week-old piglets were confirmed negative for both PRRSV antigen and antibodies and randomly assigned to four groups, with six piglets in each group. The first group was infected with the JX03 strain ($2 \times 10^5$ $TCID_{50}$ per piglet), while the second and third groups were infected with the HN08 and CHN-HB-2018 strains, respectively ($2 \times 10^{6.5}$ $TCID_{50}$ per piglet for both groups). The fourth group (negative control) was administered an equivalent volume of RPMI-1640. All groups were inoculated intramuscularly (1 mL) and intranasally (1 mL), respectively.

Following viral infection, the piglets were monitored daily for rectal temperature, and body weight was recorded once weekly. Serum and nasal swabs were collected at 0, 3, 5, 7, 10, and 14 days post-infection (dpi). At 5 and 10 dpi, one piglet from each group was randomly euthanized for necropsy. RT-qPCR was used to determine viral loads in serum, nasal swabs, and tissues, as described previously [42]. At necropsy, lung and hilar lymph node samples were collected and subsequently fixed in formalin for histopathological analysis and immunohistochemistry (IHC). IHC was performed with a PRRSV N-specific monoclonal antibody.

*2.7. Statistical Analyses*

Statistical analyses were performed using GraphPad Prism v10.1.0. Multiple comparisons were conducted using a one-way analysis of variance; data are presented as means ± standard deviations.

# 3. Results

## 3.1. RT-PCR Detection and Phylogenetic Analysis of PRRSV in China

From June 2022 to June 2023, a total of 1044 clinical tissue samples, including lung, placenta, and aborted fetuses, were collected from pigs suspected of infection with PRRSV across 27 provinces (including municipalities and autonomous regions) in China. The samples were tested by RT-PCR to amplify a gene fragment within the nsp2 region, allowing for the differential detection of NADC30/34-like PRRSV, HP-PRRSV, and classical PRRSV strains (Figure 1A). As shown in Table 1, the overall positive rate for PRRSV was 29.8% (311/1044). Among the positive samples, NADC30/34-like PRRSV, HP-PRRSV, and classical PRRSV accounted for 60.1% (187/311), 37.9% (118/311), and 4.5% (14/311), respectively. Moreover, co-infections involving NADC30/34-like and HP-PRRSV strains were observed, with an infection rate of 2.6% (8/311).

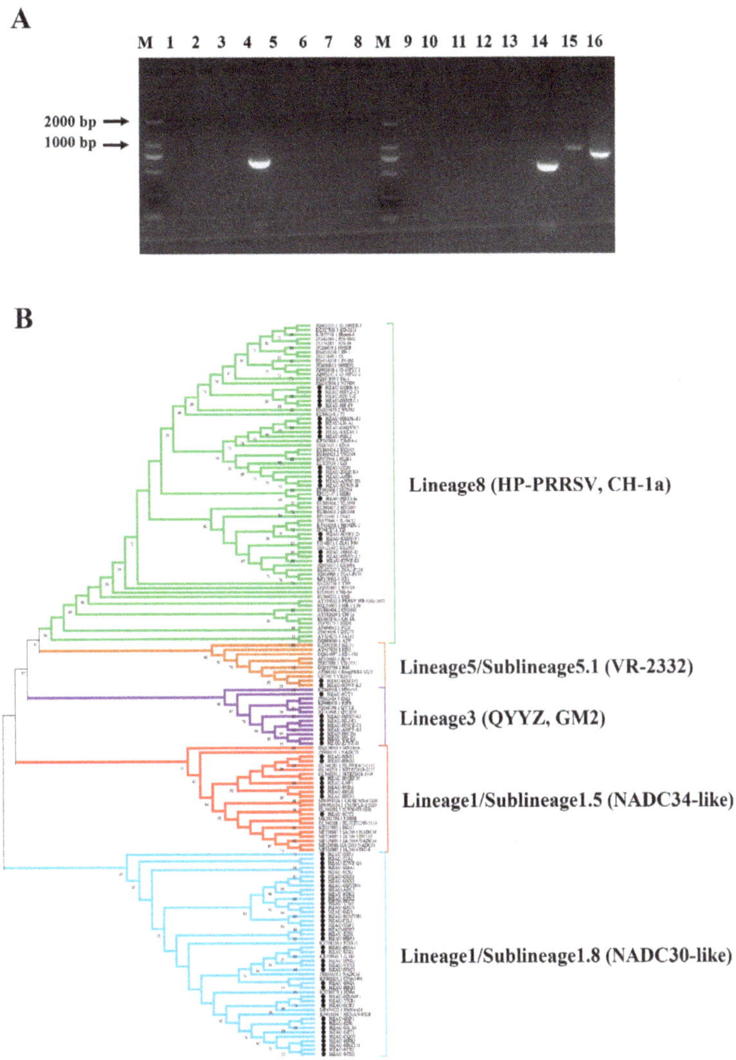


**Figure 1.** PRRSV detection of clinical samples. (**A**) RT-PCR detection (partial) results for PRRSV in China (2022.6–2023.6). M: 2000 bp DNA marker; lane 1–16: clinical samples of PRRSV; lane 4, 14:

NADC30/34-like PRRSV (681 bp); lane 15: classical PRRSV (1072 bp); lane16: HP-PRRSV (984 bp). (**B**) Genetic evolution analysis of PRRSV-positive samples based on nucleotide sequences of ORF5 gene. Phylogenetic trees consist of five subgroups, including lineage 1/sublineage 1.5 (blue), lineage 1/sublineage 1.8 (red), lineage 3 (purple), lineage 5/sublineage 5.1 (orange), and lineage 8 (green). Positive clinical samples are highlighted by black circles.

**Table 1.** The RT-PCR detection results for PRRSV in China (2022–2023).

| Clinical Samples | Positive Numbers | | | Positive Samples | Positive Rate | Percentage [a] | | |
|---|---|---|---|---|---|---|---|---|
| | HP-PRRSV | NADC30/34-like | Classical | | | HP-PRRSV | NADC30/34-like | Classical |
| 1044 | 118 | 187 | 14 | 311 | 29.8% | 37.94% | 60.13% | 4.50% |

[a]: percentage of different PRRSV strains in positive samples.

To further investigate the epidemiological trends of PRRSV strains, ORF5 genes were amplified and sequenced from positive samples, yielding 78 ORF5 gene sequences, which were then used for phylogenetic analysis. The results showed that these sequences clustered into four lineages, including lineage 1, lineage 3, lineage 5, and lineage 8, with proportions of 60.26% (47/78), 2.56% (2/78), 10.26% (8/78), and 26.92% (21/78), respectively (Figure 1B). RT-PCR results based on nsp2 and a phylogenetic analysis based on ORF5 confirmed that the predominant circulating PRRSV strains in China are primarily of lineage 1 (NADC30/34-like PRRSV), followed by lineage 8 (HP-PRRSV).

*3.2. Isolation and Identification of PRRSV Epidemic Strains*

To isolate PRRSV epidemic strains, PAMs were inoculated with the supernatant of tissue homogenates obtained from PRRSV-positive samples. Following a 2 h incubation, the inoculum was removed and replaced with fresh RPMI-1640. Typical CPEs of PRRSV were observed in two samples after about 48 h of inoculation (Figure 2A). IFA was performed using PRRSV N-specific mAb to further confirm the presence of PRRSV. As shown in Figure 2B, green fluorescence signals were observed in virus-infected PAMs, and no green fluorescence signals were observed in the control group.

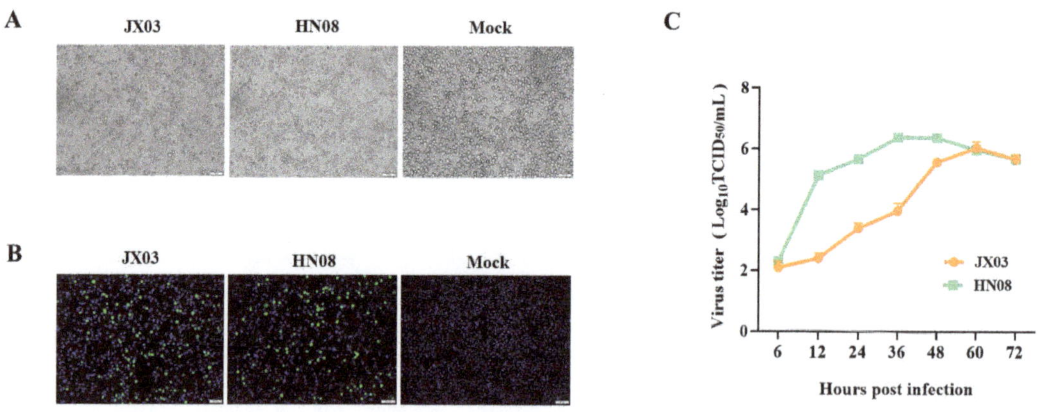


**Figure 2.** Isolation and identification of the PRRSV strains JX03 and HN08. (**A**) Cytopathic effects (CPEs) in PAMs infected with JX03 and HN08 strains. (**B**) Identification of JX03 and HN08 strains in PAMs by indirect immunofluorescence assay with PRRSV N-specific mAb. Fluorescence images were captured via fluorescence microscope (Olympus). (**C**) Growth curve of JX03 strain and HN08 strain in PAMs.

Additionally, we also detected the viral titers of these two strains at different time points post-infection using the $TCID_{50}$ assay and plotted the growth curve. The results showed that these two PRRSV strains exhibited distinct replication kinetics, with peak viral titers occurring at 36 and 60 hpi, respectively, followed by a gradual decline (Figure 2C). These findings collectively confirmed the successful isolation of two PRRSV strains, JX03 and HN08, respectively.

### 3.3. Whole-Genome Characterization and Phylogenetic Analysis of PRRSV Isolates

Twelve pairs of primers were designed to amplify and sequence the whole genomes of the JX03 and HN08 strains, respectively. The sequencing results revealed that the full-length genomes of the JX03 and HN08 strains were 15,024 and 15,325 nucleotides (nt), respectively, excluding the poly(A) tail. Homology analyses of the JX03 and HN08 strains were conducted with seven PRRSV reference strains, including Ch-1a, CH-1R, JXA1, NADC30, NADC34, GM2, and QYYZ. The results are presented in Tables 2 and 3. The JX03 strain exhibited the highest whole-genome nucleotide homology with the HP-PRRSV strain JXA1 at 99.0%. Furthermore, the coding regions (nsp1–12 and ORF2–7) of the JX03 strain showed the highest nucleotide and amino acid homology with the corresponding regions of the JXA1 strain, ranging from 97.6% to 99.8% and from 97.3% to 100.0%, respectively (Table 2). Phylogenetic analysis based on whole-genome nucleotide sequences showed that the JX03 strain clusters within lineage 8 (HP-PRRSV). These homology and phylogenetic analyses further confirm that the JX03 strain belongs to HP-PRRSV.

**Table 2.** Homology analysis of the JX03 strain and reference PRRSV strains.

| JX03 | CH-1a | | CH-1R | | JXA1 | | NADC30 | | NADC34 | | GM2 | | QYYZ | |
|---|---|---|---|---|---|---|---|---|---|---|---|---|---|---|
| | nt | aa | nt | aa | nt | aa | nt | aa | nt | aa | nt | aa | nt | aa |
| Complete Genome | 94.3 | | 94.1 | | 99.0 | | 81.8 | | 81.7 | | 86.3 | | 86.4 | |
| 5′UTR | 96.8 | | 95.3 | | 99.5 | | 92.1 | | 90.0 | | 94.7 | | 94.7 | |
| nsp1 | 94.3 | 92.4 | 94.2 | 92.4 | 99 | 98.7 | 84.8 | 85.4 | 84.4 | 87.2 | 88 | 89.1 | 88.5 | 89.1 |
| nsp2 | 86.5 | 78.4 | 86.7 | 78.6 | 98.9 | 97.3 | 63.7 | 46.3 | 66.3 | 50.3 | 77.3 | 62.8 | 77.5 | 62.9 |
| nsp3 | 94.8 | 98.0 | 94.8 | 97.8 | 99.0 | 98.9 | 83.3 | 90.6 | 82.8 | 90.8 | 81.8 | 88.1 | 82.0 | 88.1 |
| nsp4 | 94.9 | 96.1 | 95.1 | 96.6 | 98.4 | 97.5 | 84.2 | 92.6 | 85.5 | 93.1 | 84.8 | 92.2 | 84.6 | 91.7 |
| nsp5 | 94.5 | 94.1 | 93.9 | 93.5 | 98.9 | 98.8 | 90.0 | 92.9 | 84.3 | 88.8 | 82.7 | 90.0 | 82.4 | 90.0 |
| nsp6 | 94.5 | 94.1 | 93.9 | 93.5 | 98.8 | 98.8 | 90.0 | 92.9 | 84.3 | 88.8 | 82.7 | 90.0 | 82.4 | 90.0 |
| nsp7 | 94.5 | 94.1 | 93.9 | 93.5 | 98.8 | 98.8 | 90.0 | 92.9 | 84.3 | 88.8 | 82.7 | 90.0 | 82.4 | 90.0 |
| nsp8 | 97.1 | 100.0 | 97.1 | 100.0 | 97.8 | 100.0 | 89.1 | 95.7 | 88.4 | 93.5 | 92.8 | 100.0 | 92.0 | 100.0 |
| nsp9 | 96.4 | 98.6 | 96.3 | 98.3 | 98.9 | 99.1 | 87.2 | 97.2 | 87.4 | 96.9 | 98.9 | 99.1 | 91.3 | 96.9 |
| nsp10 | 95.5 | 97.7 | 95.4 | 97.7 | 99.5 | 99.1 | 86.0 | 95.5 | 85.8 | 95.2 | 90.3 | 96.4 | 90.6 | 94.3 |
| nsp11 | 94.9 | 96.9 | 95.2 | 97.8 | 99.0 | 99.1 | 91.2 | 96.4 | 87.2 | 95.5 | 89.0 | 94.6 | 88.8 | 93.8 |
| nsp12 | 95.9 | 97.4 | 95.9 | 97.4 | 98.3 | 99.5 | 89.4 | 96.1 | 82.7 | 91.6 | 87.3 | 96.8 | 86.6 | 96.1 |
| ORF2 | 96.4 | 96.1 | 95.7 | 95.7 | 99.2 | 98.8 | 86.5 | 86.4 | 87.4 | 86.0 | 89.9 | 88.7 | 89.9 | 89.1 |
| ORF3 | 95.3 | 92.5 | 95.0 | 92.9 | 99.0 | 98.0 | 83.7 | 82.0 | 84.2 | 83.5 | 90.6 | 89.8 | 90.8 | 89.4 |
| ORF4 | 96.5 | 96.5 | 96.3 | 96.3 | 97.6 | 97.6 | 87.5 | 87.5 | 86.8 | 86.8 | 95.0 | 95.0 | 94.8 | 94.8 |
| ORF5 | 94.9 | 93.0 | 94.0 | 91.5 | 99.3 | 99.5 | 85.7 | 86.1 | 87.1 | 87.1 | 83.3 | 82.1 | 83.6 | 82.6 |
| ORF6 | 97.3 | 97.1 | 97.0 | 96.6 | 99.8 | 99.4 | 89.1 | 93.7 | 89.3 | 94.3 | 91.6 | 96.6 | 91.0 | 97.1 |
| ORF7 | 95.7 | 94.4 | 95.7 | 95.2 | 99.5 | 98.4 | 90.6 | 90.3 | 89.2 | 90.3 | 87.9 | 88.7 | 89.2 | 91.1 |
| 3′UTR | 96.1 | | 97.2 | | 99.4 | | 90.4 | | 89.9 | | 88.8 | | 89.9 | |

Additionally, the HN08 strain had the highest whole-genome nucleotide homology with the NADC30 strain at 90.0%. Consistent with the homology analysis results, the phylogenetic tree constructed based on the whole-genome nucleotide sequences indicated that the HN08 strain belongs to sublineage 1.8 (NADC30-like PRRSV) (Figure 3A). Consistently, its ORF1a, ORF1b, and ORF6–7 regions also demonstrated the highest homology with the NADC30 strain. Nucleotide homology ranged from 81.5% to 94.1% and amino acid homology ranged from 80.6% to 97.7% (Table 3). Interestingly, the HN08 strain exhibited the highest homology with the NADC34 strain in the ORF2–5 regions. Nucleotide homology ranged from 92.9% to 95.9% and amino acid homology ranged from 92.2% to 97.0%, indicating that it belongs to the NADC34-like PRRSV. In line with nucleotide homology results, the phylogenetic tree constructed from the ORF5 gene nucleotide sequences revealed that the HN08 strain clusters within sublineage 1.5 (NADC34-like PRRSV) (Figure 3B). These results suggest that the HN08 strain may be a recombinant strain derived from NADC30-like PRRSV and NADC34-like PRRSV.

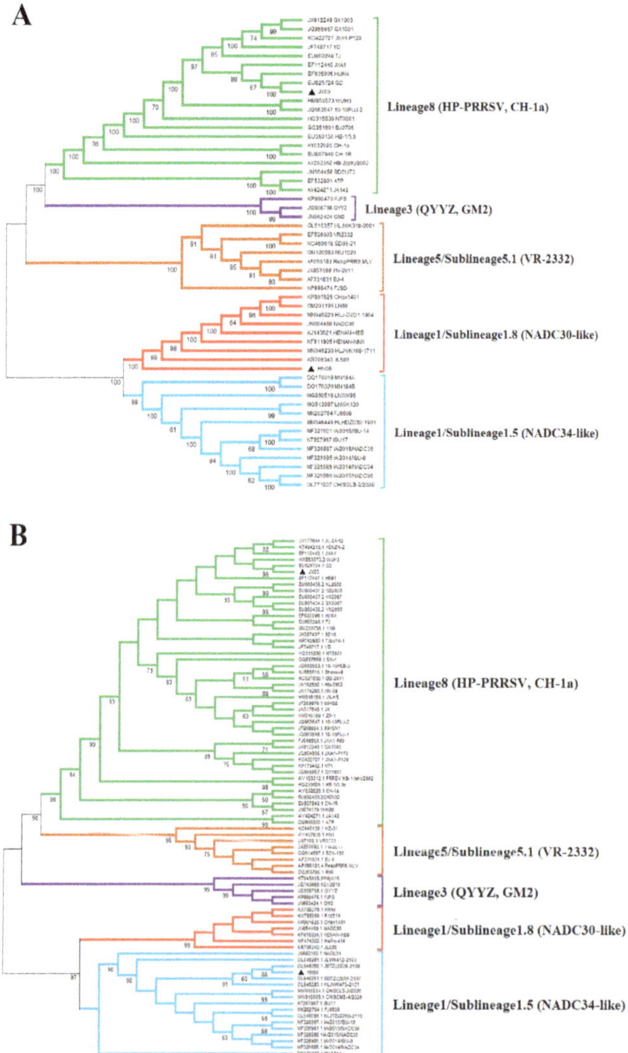


**Figure 3.** Genomic characterization analysis of JX03 and HN08 strains. Phylogenetic trees based on (**A**) complete genomes and (**B**) ORF5 nucleotide sequences of PRRSV. Phylogenetic trees consist of five subgroups, including lineage 1/sublineage 1.5 (blue), lineage 1/sublineage 1.8 (red), lineage 3 (purple), lineage 5/sublineage 5.1 (orange), and lineage 8 (green). Isolated strains are highlighted by black triangles.

**Table 3.** Homology analysis of the HN08 strain and reference PRRSV strains.

| HN08 | CH-1a | | CH-1R | | JXA1 | | NADC30 | | NADC34 | | GM2 | | QYYZ | |
|---|---|---|---|---|---|---|---|---|---|---|---|---|---|---|
| | nt | aa | nt | aa | nt | aa | nt | aa | nt | aa | nt | aa | nt | aa |
| Complete Genome | 82.7 | | 82.8 | | 82.5 | | 90.0 | | 85.3 | | 79.6 | | 79.8 | |
| 5′UTR | 90.8 | | 89.2 | | 90.3 | | 96.4 | | 89.7 | | 88.2 | | 88.2 | |
| nsp1 | 81.2 | 83.6 | 81.3 | 83.6 | 80.5 | 83.9 | 91.0 | 91.1 | 83.2 | 84.9 | 79.2 | 82.8 | 79.7 | 82.6 |
| nsp2 | 60.6 | 42.2 | 60.8 | 42.4 | 61.0 | 42.7 | 90.7 | 80.6 | 74.5 | 55.5 | 59.6 | 40.0 | 59.6 | 40.3 |
| nsp3 | 84.4 | 92.6 | 84.5 | 92.4 | 84.3 | 91.7 | 89.8 | 94.8 | 84.2 | 90.8 | 79.7 | 87.4 | 79.6 | 87.4 |
| nsp4 | 92.3 | 95.1 | 92.5 | 95.6 | 94.3 | 96.6 | 84.2 | 92.6 | 84.0 | 94.6 | 84.0 | 92.2 | 83.8 | 91.7 |
| nsp5 | 90.0 | 92.4 | 89.8 | 91.8 | 92.7 | 93.5 | 87.3 | 92.4 | 81.6 | 88.8 | 81.2 | 90.0 | 80.8 | 90.6 |
| nsp6 | 91.7 | 100.0 | 91.7 | 100.0 | 93.8 | 100.0 | 87.5 | 93.8 | 81.2 | 87.5 | 91.7 | 100.0 | 91.7 | 100.0 |
| nsp7 | 94.5 | 95.0 | 94.1 | 94.6 | 97.4 | 98.1 | 81.5 | 83.8 | 81.2 | 85.3 | 90.3 | 91.1 | 92.7 | 94.2 |
| nsp8 | 97.8 | 100.0 | 97.8 | 100.0 | 97.1 | 100.0 | 89.9 | 95.7 | 89.1 | 93.5 | 93.5 | 100.0 | 92.8 | 100.0 |
| nsp9 | 91.8 | 98.0 | 91.7 | 97.7 | 92.7 | 98.1 | 90.8 | 97.7 | 87.5 | 96.9 | 89.3 | 97.0 | 88.5 | 96.6 |
| nsp10 | 84.2 | 93.2 | 84.2 | 93.2 | 83.8 | 93.4 | 94.1 | 97.5 | 89.8 | 97.0 | 84.8 | 94.8 | 83.6 | 93.6 |
| nsp11 | 89.9 | 95.5 | 90.2 | 96.4 | 87.8 | 96.0 | 93.9 | 97.3 | 85.6 | 94.6 | 85.7 | 93.8 | 85.6 | 92.9 |
| nsp12 | 88.6 | 94.8 | 88.8 | 94.8 | 88.8 | 95.5 | 93.5 | 96.8 | 85.5 | 90.3 | 86.8 | 94.8 | 86.4 | 94.2 |
| ORF2 | 85.1 | 82.9 | 85.1 | 82.5 | 84.8 | 81.3 | 83.7 | 81.7 | 94.8 | 92.2 | 84.4 | 81.3 | 84.6 | 81.7 |
| ORF3 | 83.8 | 80.4 | 84.1 | 80.8 | 83.4 | 81.2 | 84.6 | 83.1 | 92.9 | 92.9 | 83.4 | 81.6 | 83.1 | 82.4 |
| ORF4 | 86.4 | 88.3 | 86.4 | 87.7 | 85.5 | 87.2 | 94.8 | 95.5 | 94.8 | 96.6 | 85.5 | 86.0 | 85.7 | 86.6 |
| ORF5 | 86.4 | 88.1 | 86.1 | 86.6 | 86.2 | 89.6 | 87.2 | 91.0 | 95.9 | 97.0 | 83.6 | 84.1 | 83.6 | 83.6 |
| ORF6 | 86.9 | 92.6 | 86.9 | 92.0 | 87.2 | 93.1 | 94.7 | 96.0 | 94.3 | 95.4 | 89.7 | 93.1 | 88.8 | 93.1 |
| ORF7 | 90.3 | 91.9 | 90.3 | 91.9 | 90.3 | 90.3 | 95.4 | 95.2 | 93.8 | 92.7 | 84.9 | 86.3 | 86.6 | 89.5 |
| 3′UTR | 89.3 | | 88.8 | | 90.4 | | 97.8 | | 95.5 | | 88.8 | | 89.9 | |

*3.4. Gene Recombination Analysis of HN08*

Phylogenetic and homology analyses indicated that the HN08 strain may be a recombinant strain. Therefore, a comprehensive recombination analysis of the HN08 strain was conducted using RDP 4.0 NADC30 PRRSV (GenBank: MH500776) and NADC34 PRRSV (GenBank: MF326985) as reference strains. The results revealed that the HN08 strain is a recombinant virus primarily derived from the NADC30 strain, with the NADC34 strain serving as a secondary parent. The recombinant regions were identified between nt 11,712–12,781 (ORF2–3 region) and nt 13,390–14,030 (ORF5 region). These results are consistent with the homology and phylogenetic analysis findings for the HN08 strain (Figure 4).

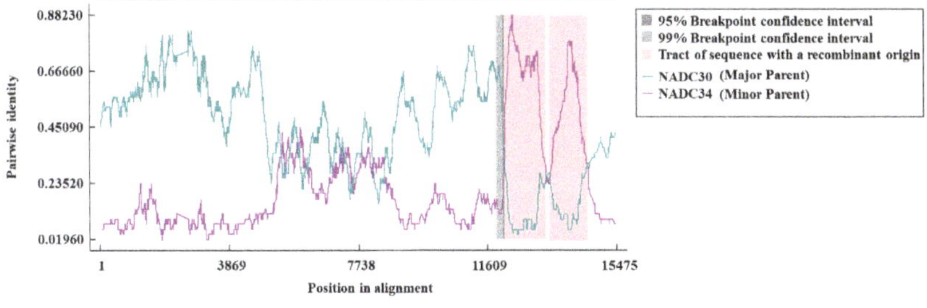


**Figure 4.** The genome recombination analysis of PRRSV HN08 strain. The Y-axis indicates the percentage similarity of the query sequence (HN08) to the NADC30 (blue) and NADC34 (purple) strains. The recombination area is highlighted in pink.

### 3.5. Amino Acid Variations in nsp2 and GP5 of PRRSV Isolates

The nsp2 and GP5 proteins of PRRSV are among the most frequently mutated viral proteins, making them commonly used for phylogenetic analysis [43]. The amino acid sequences of nsp2 and GP5 from the JX03 and HN08 strains were aligned and analyzed against the reference sequences of PRRSV. The results revealed that the JX03 strain exhibits 30-aa deletions (1 + 29) in its nsp2 compared to the PRRSV-2 prototype strain VR-2332, which aligns with the deletion pattern observed in the HP-PRRSV reference strain JXA1. In contrast, the HN08 strain's nsp2 contains 131-aa deletions (111 + 1 + 19), consistent with the deletion pattern of the NADC30 strain (Figure 5A). Moreover, the amino acid sequence alignment of the GP5 protein revealed that the JX03 strain was most homologous to the JXA1 strain, with a single mutation at position 195 (L195R). The HN08 strain showed high homology compared to the NADC34 PRRSV, harboring five mutations, including F25L, N32S, A98T, I161V, and V189I (Figure 5B). These findings agree with the results of the phylogenetic and homology analyses.

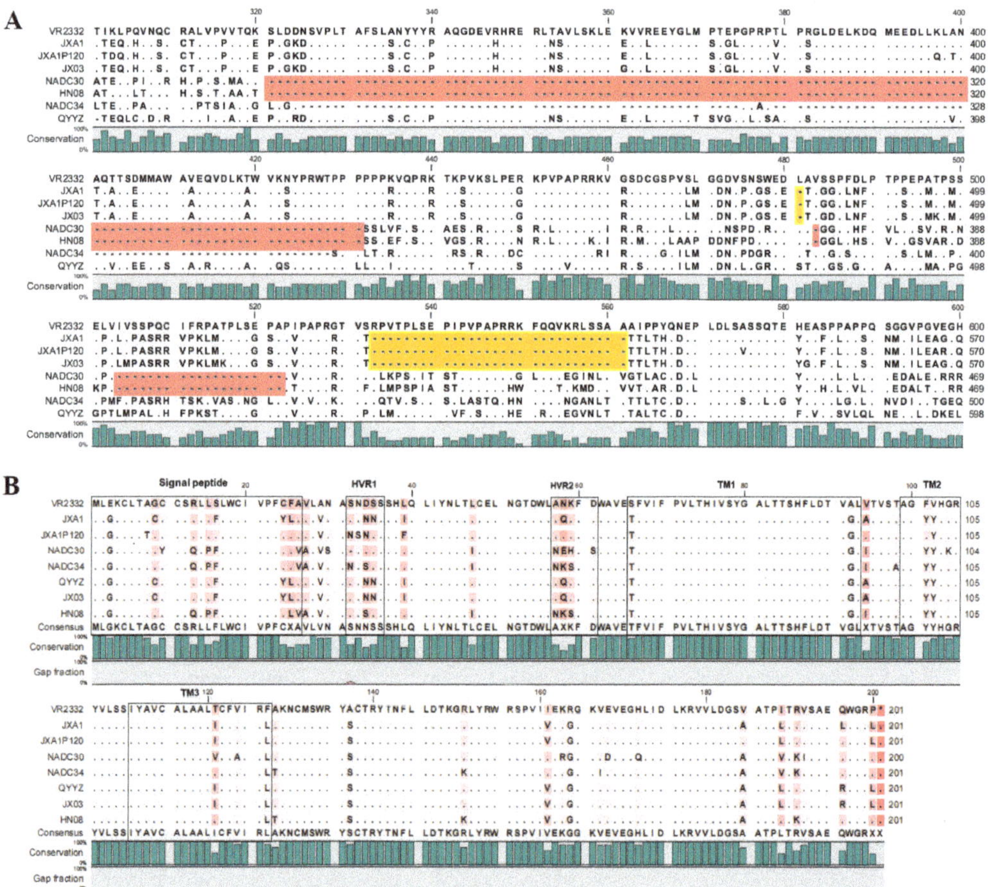

**Figure 5.** Amino acid sequence analyses of JX03 and HN08 strains. (**A**) Amino acid sequence analysis of nsp2 (300–600 aa) in JX03 and HN08 strains. Different colors denote the defined deletion pattern as follows: the "111 + 1 + 19" aa deletion pattern of NADC30-like PRRSV (red); the "1 + 29" aa deletion pattern of HP-PRRSV (yellow). (**B**) An amino acid sequence analysis of GP5 in the PRRSV JX03 and HN08 strains. The signal peptide, hypervariable region, and transmembrane domain are highlighted by black frames.

### 3.6. Analysis of Clinical Symptoms and Viral Loads in Piglets Infected with PRRSV

We evaluated the pathogenicity differences between the JX03 strain (HP-PRRSV, lineage 8) and the HN08 strain (NADC34-like PRRSV, sublineage 1.5) in piglets. The NADC30-like PRRSV (sublineage 1.8) CHN-HB-2018 strain, which was previously isolated by our lab [44], served as a control, alongside a negative control group that did not receive a virus inoculation (mock). The experimental design is illustrated in Figure 6A.

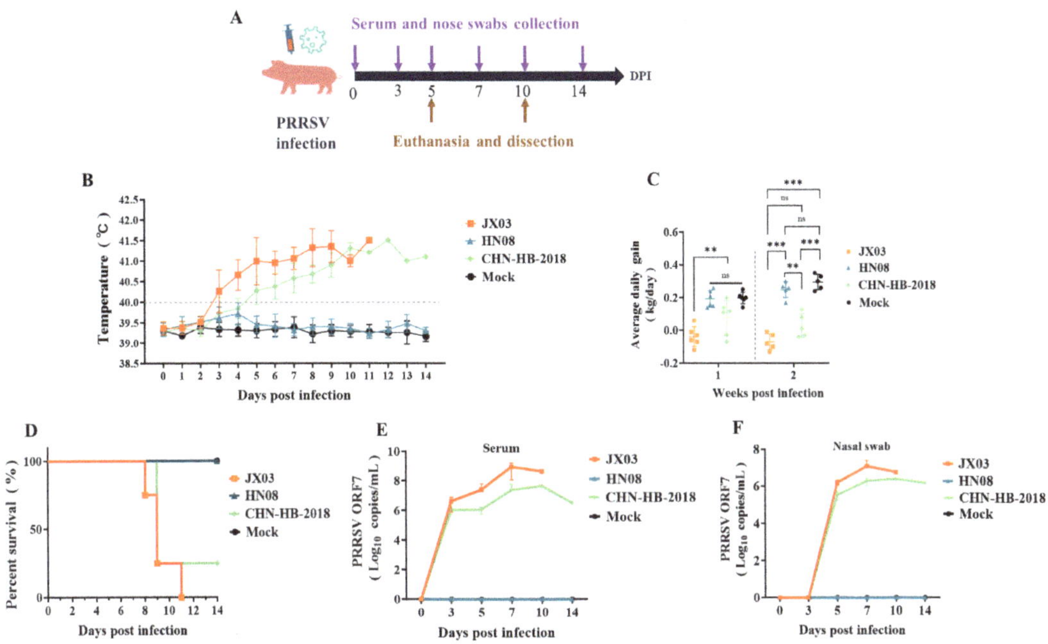


**Figure 6.** Clinical symptoms of different PRRSV strains in piglets. (**A**) Experimental procedure for the pathogenicity assessment of PRRSV in piglets. (**B**) Rectal temperature; (**C**) average weight gain; (**D**) percentage of surviving piglets in each group; (**E**) viral loads in serum; (**F**) viral loads in nose swabs. Error bars indicate the standard error based on at least three independent experimental replicates. Values in are shown as mean ± SD. ** $p < 0.01$; *** $p < 0.001$; ns: no significance.

Rectal temperature monitoring revealed that the JX03 strain infection caused a fever (over 40 °C) starting at 3 dpi and peaking at 42.3 °C, while the CHN-HB-2018 strain infection caused a fever at 5 dpi, peaking at 41.5 °C. In contrast, the HN08 strain group and the mock group maintained normal temperatures throughout the experimental period (Figure 6B). Body weight measurements and average daily gain (ADG) calculations were performed weekly after PRRSV infection. During the first week post-infection (wpi), the piglets infected with the JX03 strain showed a significantly lower ADG than the other three groups. And no significant differences in ADG were observed between the HN08 strain, CHN-HB-2018 strain, and mock groups. In the second wpi, ADG in both the JX03 and CHN-HB-2018 strain groups was significantly lower than those in the HN08 strain and mock groups. However, there were no significant differences between the JX03 and CHN-HB-2018 groups or between the HN08 strain and mock groups (Figure 6C). Additionally, piglets infected with the JX03 and CHN-HB-2018 strains had a survival rate of 0% and 25%, respectively. No mortality was observed in the HN08 strain or mock groups (Figure 6D).

Viral loads in serum and nasal swabs were determined using RT-qPCR for PRRSV ORF7. Viral ORF7 was detectable in both the JX03 and CHN-HB-2018 strain groups starting at 3 dpi in serum and at 5 dpi in nasal swabs, with levels gradually increasing over time and peaking at 7 and 10 dpi, respectively (Figure 6E,F). Notably, the JX03 strain group exhibited higher viral loads compared to the CHN-HB-2018 strain group. In contrast, viral RNA was not detected in piglets infected with the HN08 strain or in mock groups. To investigate differential tissue tropism among PRRSV isolates, viral loads in various tissues were compared at 5 and 10 dpi. The results showed that viral loads at various time points in the HN08 strain group were significantly lower than those observed in the JX03 and CHN-HB-2018 strain groups, consistent with the viral load patterns found in serum and nasal swabs.

Specifically, in the JX03 and HN08 strain groups, the viral loads in the brain tissues and hilar lymph nodes were higher, respectively, at both 5 and 10 dpi (Figure 7A,C). As for the CHN-HB-2018 group, the viral load was highest in brain tissue at 5 dpi (Figure 7A,B) but the highest in lung tissues at 10 dpi (Figure 7C). We also constructed heatmaps based on the RT-qPCR results shown in Figure 7A,C, which more distinctly illustrate the differences in tissue tropism among piglets infected with different PRRSV strains (Figure 7B,D).

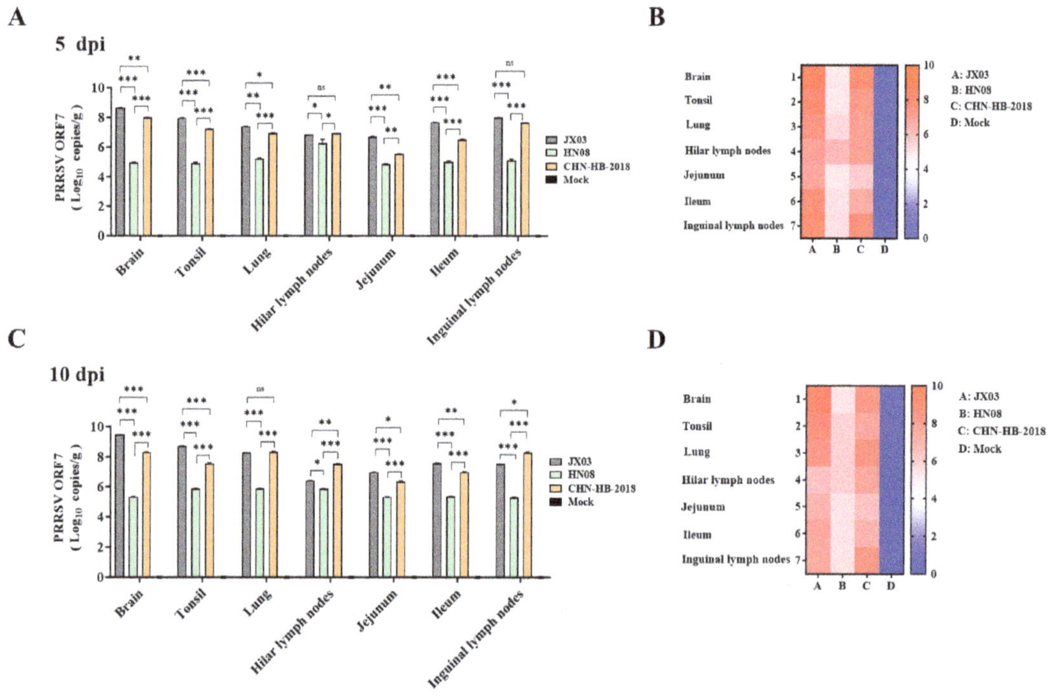


**Figure 7.** Analysis of viral loads in different tissues. (**A**,**B**) Viral loads in various tissues of piglets on 5 dpi. (**C**,**D**) Viral loads in various tissues of piglets on 10 dpi. Error bars indicate the standard error based on at least three independent experimental replicates. Values in are shown as mean ± SD. * $p < 0.05$; ** $p < 0.01$; *** $p < 0.001$; ns: no significance.

### 3.7. Pathological Analysis of Lungs and Hilar Lymph Nodes in Piglets

One piglet from each group was euthanized and necropsied at 5 and 10 dpi. The necropsy at 5 dpi indicated that piglets infected with the JX03 and CHN-HB-2018 strains exhibited focal parenchymal lesions in the lungs. In contrast, the piglets in the HN08 strain displayed interstitial damage, while obvious pathological lesions were observed in the mock group (Figure S1A). The lung lesions in the JX03 and CHN-HB-2018 groups worsened, characterized by widened interstitial lung tissue, pronounced consolidation, and accompanying hemorrhage at 10 dpi. In contrast, the piglets in the HN08 group presented with pulmonary edema. No obvious pathological lesions were observed in the mock group (Figure 8A).

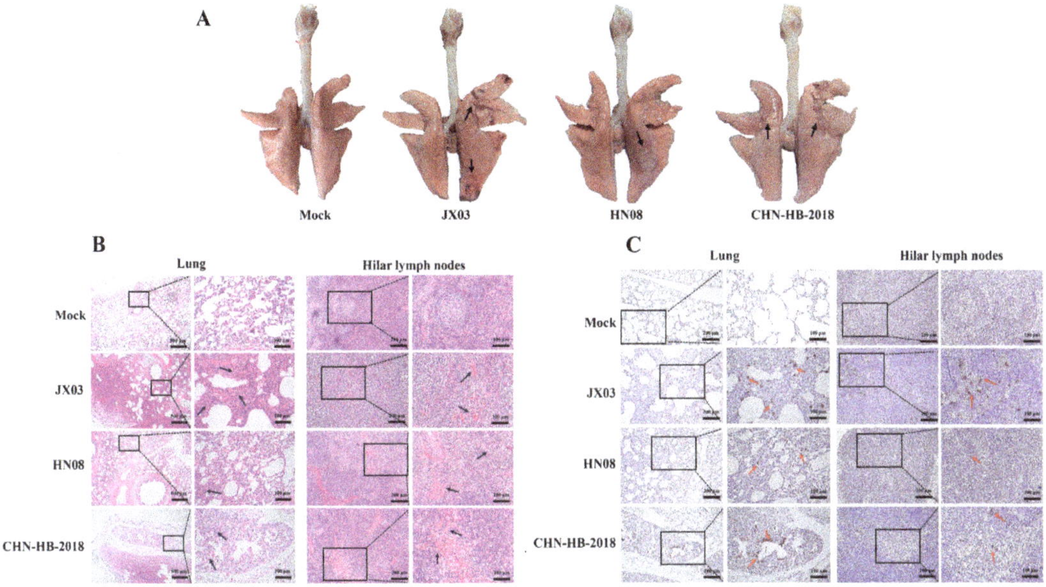


**Figure 8.** Pathological changes in PRRSV-infected piglets on 10 dpi. (**A**) Gross lesions in the lungs. (**B**) Pathological changes in lung and hilar lymph nodes of piglets (HE staining). (**C**) IHC analysis of lungs and hilar lymph nodes. Black arrow: tissue damage of lungs or erythrocyte or neutrophilic infiltration; red arrow: positive signals of PRRSV.

Histopathological examinations using HE staining were conducted on the lungs and hilar lymph nodes. All the PRRSV infection groups exhibited bronchial hyperplasia and thickening of the alveolar walls in the lungs at 5 dpi, while the hilar lymph nodes showed no obvious lesions (Figure S1B). However, the results at 10 dpi revealed that extensive infiltration of the inflammatory cells and pulmonary hemorrhage were severe pathological changes that could be observed in the lung tissues of both the JX03 and CHN-HB-2018 groups (Figure 8B). In addition to these common pathological changes, the alveolar structure was nearly absent in the JX03 infection group, whereas in the CHN-HB-2018 infection group, although many alveoli had collapsed, a few relatively intact alveolar structures remained (Figure 8B). In contrast, most alveolar structures remained intact in the HN08 infection group, showing only mild thickening of the alveolar walls and some bronchial hyperplasia notes (Figure 8B). As for the hilar lymph nodes, the JX03 and CHN-HB-2018 strains caused lymphoid follicle atrophy and the infiltration of red blood cells in the lymphoid medulla and sinuses. Additionally, the hilar lymph nodes of piglets in the HN08 infection group exhibited only sparse lymphocytes in certain areas. No obvious pathological lesions were observed in the lungs and hilar lymph nodes of the mock group (Figure 8B). Collectively, the JX03 and CHN-HB-2018 strains exhibited severe lung and hilar lymph node lesions, whereas the HN08 strain showed mild damage.

IHC analysis was conducted to assess PRRSV antigen levels in the lung tissues and hilar lymph nodes of piglets. PRRSV-positive signals were detected in the lung tissues of piglets infected with both the JX03 and CHN-HB-2018 strains at 5 and 10 dpi, with the positive signal numbers increasing over time (Figure S2). In contrast, a limited number of positive signals were observed in the lung tissues of piglets infected with the HN08 strain at 10 dpi, while no positive signals were observed at 5 dpi (Figures 8C, S1C and S2). These results indicated that the PRRSV antigen levels in piglets infected with the JX03 strain were higher than in the other groups, which was consistent with the findings from RT-qPCR. Furthermore, the PRRSV-positive signals were also detected in the hilar lymph nodes of piglets infected with all three PRRSV infection groups at 5 and 10 dpi, with higher signal numbers observed in the JX03 and CHN-HB-2018 strain groups compared to the HN08 strain group (Figures 8C and S2).

## 4. Discussion

Since the first report of PRRSV in China in 1996, multiple strains have emerged, including classical PRRSV, HP-PRRSV, NADC30-like PRRSV, and NADC34-like PRRSV strains [18,34,45]. Gao et al. analyzed PRRSV strains circulating in China from 1996 to 2016 and found that HP-PRRSV accounted for 85.5% (2079/2430) of cases, while NADC30-like PRRSV represented only 5.2% (126/2430) [46]. However, a more recent epidemiological surveillance survey revealed that NADC30-like PRRSV had a positive rate of 74.18% in clinical samples, followed by an NADC34-like PRRSV detection rate of 11.98% and HP-PRRSV at just 4.04% [34]. These findings suggested that NADC30/34-like PRRSV is displacing HP-PRRSV as the predominant strain in China. Our study indicated that NADC30/34-like PRRSV strains are currently the primary circulating strains in China, accounting for 60.1%, which is in accordance with the findings of Xu et al. [34].

Previous studies have demonstrated that PAMs serve as the primary target cells for PRRSV infection and PAMs can successfully isolate field strains [47,48]. Notably, an increasing number of NADC30/34-like PRRSV strains demonstrate reduced or negligible infectivity in MARC-145 cells, whereas researchers have successfully isolated these strains using PAMs [49]. Consistent with these findings, PRRSV strains isolated by our lab between 2023 and 2025 exhibited similarly poor adaptability to MARC-145 cells (unpublished data). However, PAMs are primary cells that present several challenges, including high procurement costs, complex isolation procedures, and heightened susceptibility to contamination. Consequently, many research teams have employed alternative cell lines such as MARC-145, MA-104, and CL2621 for in vitro PRRSV isolation and propagation [50,51]. Therefore, it is significant to elucidate the mechanisms underlying the differential cytopathogenicity of PRRSV strains. Recently, several researchers have successfully adapted PRRSV to MARC-145 cells by replacing structural proteins or introducing mutations at the key amino acid sites of GP2a [52,53]. These advancements are crucial for the successful isolation and identification of newly emerging field strains of PRRSV.

The frequent recombination of PRRSV is a key factor contributing to the increasingly complex epidemiological landscape of the virus [54]. As a newly evolved branch of lineage 1, NADC34-like PRRSV shares recombination traits with NADC30-like PRRSV. For example, the GD-H1 strain exhibited the same "111 + 1 + 19" aa deletion pattern in nsp2 as NADC30-like strains, while the phylogenetic analysis of its ORF5 gene classified it within NADC34-like sublineage 1.5 [55]. The PRRSV strain SDLY23-1742 was confirmed as a multiclade recombinant of NADC30-like, NADC34-like, and HP-PRRSV strains [24]. In our study, the NADC34-like HN08 strain exhibited similar recombination characteristics to the previously reported GD-H1 strain [55]. Nucleotide homology analysis of the HN08 strain revealed that the nsp2 gene shares the highest sequence identity with the NADC30 strain, while the ORF5 gene is more closely related to the NADC34 strain, so it is defined as an NADC34-like strain. Given the frequent recombination in the nsp2 region of NADC30/34-like PRRSV, incorporating both nsp2 and ORF5 genes as reference markers in clinical PRRSV strain detection could significantly enhance diagnostic accuracy.

Current research consistently demonstrates that infection with HP-PRRSV induces marked fever in piglets, often leading to high mortality, whereas NADC30/34-like PRRSV typically results in milder clinical manifestations [24,31,33,44,53]. Consistent with these findings, our study revealed that the JX03 strain causes high morbidity and mortality rates in piglets. In contrast, piglets infected with the CHN-HB-2018 or HN08 strains exhibited milder clinical symptoms and showed lower or no mortality. However, PRRSV exhibits significant pathogenicity heterogeneity among NADC34-like PRRSV strains [24,56,57]. For instance, the NADC34 strain, first isolated in the U.S. in 2014, demonstrated high pathogenicity in piglets, characterized by high fever among other clinical signs [29]. The FJGD01 strain, isolated in Fujian as a recombinant of NADC30-like, NADC34-like, and JXA1-like PRRSVs, induced high viremia, prolonged fever, severe weight loss, and extensive pulmonary lesions in piglets [56]. In contrast, several other NADC34-like isolates displayed attenuated pathogenicity, causing only persistent fever, moderate respiratory distress, and mild interstitial pneumonia [58]. In addition to differences in pathogenicity, various PRRSV strains may exhibit distinct tissue tropism. Notably, PRRSV has been detected in the intestinal tracts of pigs, suggesting its potential enteric tropism [59]. Furthermore, Zhao et al. demonstrated that PRRSV infection induces intestinal barrier dysfunction in neonatal piglets and confirmed active viral replication in small intestinal tissues [60]. Consistent with these findings, our study revealed that PRRSV strains from different lineages could be detected in the intestinal tract of piglets, although differences in viral loads were observed.

## 5. Conclusions

In this study, we analyzed the epidemiological situation of PRRSV in China and found that the prevalent strains are primarily NADC30/34-like PRRSV, followed by HP-PRRSV. We successfully isolated and characterized an HP-PRRSV strain (JX03) and a recombinant NADC34-like PRRSV strain (HN08). Phylogenetic, amino acid sequence, and recombination analyses demonstrated that the JX03 and HN08 strains cluster within lineage 8 and sublineage 1.5, respectively. Animal experiments indicated that JX03 exhibited the highest pathogenicity, followed by the CHN-HB-2018 strain, while the HN08 strain exhibited the lowest pathogenicity. These findings are expected to aid in PRRSV prevention and control and support vaccine development efforts.

**Supplementary Materials:** The following supporting information can be downloaded at: https://www.mdpi.com/article/10.3390/vetsci12060530/s1, Figure S1: Pathological changes in PRRSV-infected piglets on 5 dpi.; Figure S2: Quantitative comparison of the PRRSV-positive signals in the lungs and hilar lymph nodes of piglets across different groups; Table S1: Primer sequences used for the amplification of the whole-genome of the JX03 and HN08 strains;; Table S2: The reference strains used for PRRSV sequence alignment.

**Author Contributions:** Y.D.: conceptualization, writing—original draft preparation, resources, methodology; Y.D. and J.C.: software, methodology, visualization; L.F. and Y.Z. (Yanrong Zhou): project administration, supervision, methodology; Y.D., J.C. and T.R.: resources, validation; Y.D., J.C., T.R., C.X. and Y.Z. (Yiye Zhang): visualization, formal analysis, validation; Y.Z. (Yanrong Zhou): writing—review and editing, funding acquisition. All authors have read and agreed to the published version of the manuscript.

**Funding:** This work is supported by the National Key Research and Development Program of China (2022YFD1800305), the Fundamental Research Funds for the Central Universities (2662022DKQD003), and the Open Fund Project of the International Joint Research Center of National Animal Immunology, Ministry of Education Key Laboratory for Animal Pathogens and Biosafety, Ministry of Agriculture and Rural Affairs Key Laboratory of Quality and Safety Control of Poultry Products (no. 2025IJRCNAI03).

**Institutional Review Board Statement:** All the piglets were cared for in accordance with the guidelines of the Regulations for the Administration of Affairs Concerning Experimental Animals established by the Ministry of Science and Technology of China. The experimental procedures were conducted following protocols approved by the Scientific Ethics Committee of Huazhong Agricultural University (approval number: HZAUSW-2024-0086).

**Informed Consent Statement:** Not applicable.

**Data Availability Statement:** The data presented in this study are available upon request from the corresponding author.

**Acknowledgments:** The authors would like to thank the National Key Laboratory of Agricultural Microbiology Core Facility and the Cooperative Innovation Center for Sustainable Pig Production for their assistance during the experiments.

**Conflicts of Interest:** The authors declare no conflicts of interest.

## References

1. Nathues, H.; Alarcon, P.; Rushton, J.; Jolie, R.; Fiebig, K.; Jimenez, M.; Geurts, V.; Nathues, C. Cost of porcine reproductive and respiratory syndrome virus at individual farm level—An economic disease model. *Prev. Vet. Med.* **2017**, *142*, 16–29. [CrossRef] [PubMed]
2. Díaz, J.A.C.; Fitzgerald, R.M.; Shalloo, L.; da Costa, M.R.; Niemi, J.; Leonard, F.C.; Kyriazakis, I.; Manzanilla, E.G. Financial analysis of herd status and vaccination practices for porcine reproductive and respiratory syndrome virus, swine influenza virus, and in farrow-to-finish pig farms using a bio-economic simulation model. *Front. Vet. Sci.* **2020**, *7*, 556674.
3. Snijder, E.J.; Kikkert, M.; Fang, Y. Arterivirus molecular biology and pathogenesis. *J. Gen. Virol.* **2013**, *94*, 2141–2163. [CrossRef]
4. Snijder, E.J.; Meulenberg, J.J. The molecular biology of arteriviruses. *J. Gen. Virol.* **1998**, *79*, 961–979. [CrossRef]
5. Wang, L.H.; Hou, J.; Zhang, H.X.; Feng, W.H. Complete genome sequence of a novel highly pathogenic porcine reproductive and respiratory syndrome virus variant. *J. Virol.* **2012**, *86*, 13121. [CrossRef]
6. Steinberger, J.; Kontaxis, G.; Rancan, C.; Skern, T. Comparison of self-processing of foot-and-mouth disease virus leader proteinase and porcine reproductive and respiratory syndrome virus leader proteinase nsp1α. *Virology* **2013**, *443*, 271–277. [CrossRef]
7. Kuhn, J.H.; Lauck, M.; Bailey, A.L.; Shchetinin, A.M.; Vishnevskaya, T.V.; Bào, Y.M.; Ng, T.F.F.; LeBreton, M.; Schneider, B.S.; Gillis, A.; et al. Reorganization and expansion of the nidoviral family. *Arch. Virol.* **2016**, *161*, 755–768. [CrossRef]
8. Yim-Im, W.; Anderson, T.K.; Paploski, I.A.D.; VanderWaal, K.; Gauger, P.; Krueger, K.; Shi, M.; Main, R.; Zhang, J. Refining PRRSV-2 genetic classification based on global ORF5 sequences and investigation of their geographic distributions and temporal changes. *Microbiol. Spectr.* **2023**, *11*, e0291623. [CrossRef]
9. Yim-Im, W.; Anderson, T.K.; Böhmer, J.; Baliellas, J.; Stadejek, T.; Gauger, P.C.; Krueger, K.M.; Vermeulen, C.J.; Buter, R.; Kazlouski, A.; et al. Refining genetic classification of global porcine reproductive and respiratory syndrome virus type 1 (PRRSV-1) and investigating their geographic and temporal distributions. *Vet. Microbiol.* **2025**, *302*, 110413. [CrossRef]
10. Guo, B.Q.; Chen, Z.S.; Liu, W.Q.; Cui, Y.Z. Isolation and identification of porcine reproductive and respiratory syndrome virus (PRRSV) from aborted fetuses suspected of PRRS. *China. J. Prev. Vet. Med.* **1996**, *2*, 1–5.
11. Wu, J.; Li, J.; Tian, F.; Ren, S.; Yu, M.; Chen, J.; Lan, Z.; Zhang, X.; Yoo, D.; Wang, J. Genetic variation and pathogenicity of highly virulent porcine reproductive and respiratory syndrome virus emerging in China. *Arch. Virol.* **2009**, *154*, 1589–1597. [CrossRef] [PubMed]
12. Liu, Y.; Li, J.; Yang, J.; Zeng, H.; Guo, L.; Ren, S.; Sun, W.; Chen, Z.; Cong, X.; Shi, J.; et al. Emergence of different recombinant porcine reproductive and respiratory syndrome viruses, China. *Sci. Rep.* **2018**, *8*, 4118. [CrossRef] [PubMed]
13. Zhu, H.; Wei, L.; Liu, X.; Liu, S.; Chen, H.; Chen, P.; Li, X.; Qian, P. Pathogenicity studies of NADC34-like porcine reproductive and respiratory syndrome virus LNSY-GY and NADC30-like porcine reproductive and respiratory syndrome virus GXGG-8011 in piglets. *Viruses* **2023**, *15*, 2247. [CrossRef]
14. Tong, G.Z.; Zhou, Y.J.; Hao, X.F.; Tian, Z.J.; An, T.Q.; Qiu, H.J. Highly pathogenic porcine reproductive and respiratory syndrome, China. *Emerg Infect Dis.* **2007**, *13*, 1434–1436. [CrossRef]
15. Zhou, L.; Wang, Z.; Ding, Y.; Ge, X.; Guo, X.; Yang, H. NADC30-like strain of porcine reproductive and respiratory syndrome virus, China. *Emerg. Infect. Dis.* **2015**, *21*, 2256–2257. [CrossRef]
16. Zhang, H.L.; Zhang, W.L.; Xiang, L.R.; Leng, C.L.; Tian, Z.J.; Tang, Y.D.; Cai, X.-H. Emergence of novel porcine reproductive and respiratory syndrome viruses (ORF5 RFLP 1-7-4 viruses) in China. *Vet. Microbiol.* **2018**, *222*, 105–108. [CrossRef]
17. Liu, J.; Wei, C.; Lin, Z.; Xia, W.; Ma, Y.; Dai, A.; Yang, X. Full genome sequence analysis of a 1-7-4-like PRRSV strain in Fujian province, China. *PeerJ* **2019**, *7*, e7859. [CrossRef]

18. Li, C.; Zhuang, J.; Wang, J.; Han, L.; Sun, Z.; Xiao, Y.; Ji, G.; Li, Y.; Tan, F.; Li, X.; et al. Outbreak investigation of NADC30-like PRRSV in South-East China. *Transbound. Emerg. Dis.* **2016**, *63*, 474–479. [CrossRef]
19. Zhao, J.; Xu, L.; Xu, Z.; Deng, H.; Li, F.; Sun, X.; Zhou, Y.; Zhu, L. Emergence and spread of NADC34-like PRRSV in Southwest China. *Transbound. Emerg. Dis.* **2022**, *69*, e3416–e3424. [CrossRef]
20. Zhai, W.; Yu, S.; Zhang, P.; Lin, Y.; Ge, S.; Zhang, T.; Zhang, K.; He, S.; Hu, Q.; Tang, X.; et al. Epidemiology and genetic characteristics of porcine reproductive and respiratory syndrome virus in the Hunan and Hebei provinces of China. *Vet. Sci.* **2023**, *10*, 63. [CrossRef]
21. Zhao, H.; Zhou, J.; Zhang, M.; Yang, X.; Xi, Z.; Gao, Q.; Fang, R.; Zhao, P.; Zhao, J. Epidemiological surveillance and economic impact analysis of different porcine reproductive and respiratory syndrome virus infection statuses in 23 breeding pig farms in Hubei, China. *Anim. Dis.* **2024**, *4*, 47. [CrossRef]
22. Guo, J.; Li, C.; Lu, H.; Wang, B.; Zhang, L.; Ding, J.; Jiao, X.; Li, Q.; Zhu, S.; Wang, A.; et al. Reverse genetics construction and pathogenicity of a novel recombinant NADC30-like PRRSV isolated in China. *Front. Vet. Sci.* **2024**, *11*, 1434539. [CrossRef] [PubMed]
23. Chang, H.; Gao, X.; Wu, Y.; Wang, F.; Lai, M.; Zheng, J.; Qiu, Y.; He, Y.; Liang, X.; Yuan, K.; et al. Genomic and pathogenicity analysis of two novel highly pathogenic recombinant NADC30-like PRRSV strains in China, in 2023. *Microbiol. Spectr.* **2024**, *12*, e0036824. [CrossRef]
24. Cui, M.; Qiu, M.; Yang, S.; Qiu, Y.; Qi, W.; Lin, H.; Sun, Z.; Zheng, W.; Zhu, J.; Chen, N. The replication efficacy of NADC34-like porcine reproductive and respiratory syndrome virus 2 is not directly associated with the pathogenicity. *Vet. Microbiol.* **2025**, *301*, 110367. [CrossRef]
25. Tian, K.; Yu, X.; Zhao, T.; Feng, Y.; Cao, Z.; Wang, C.; Hu, Y.; Chen, X.; Hu, D.; Tian, X.; et al. Emergence of fatal PRRSV variants: Unparalleled outbreaks of atypical PRRS in China and molecular dissection of the unique hallmark. *PLoS ONE* **2007**, *2*, e526. [CrossRef]
26. Jiang, P.; Chen, P.Y.; Dong, Y.Y.; Cai, J.L.; Cai, B.X.; Jiang, Z.H. Isolation and genome characterization of porcine reproductive and respiratory syndrome virus in P.R. China. *J. Vet. Diagn. Investig.* **2000**, *12*, 156–158. [CrossRef]
27. Li, L.; Zhao, Q.; Ge, X.; Teng, K.; Kuang, Y.; Chen, Y.; Guo, X.; Yang, H. Chinese highly pathogenic porcine reproductive and respiratory syndrome virus exhibits more extensive tissue tropism for pigs. *Virol. J.* **2012**, *9*, 203. [CrossRef]
28. Zhao, K.; Ye, C.; Chang, X.B.; Jiang, C.G.; Wang, S.J.; Cai, X.H.; Tong, G.-Z.; Tian, Z.-J.; Shi, M.; An, T.-Q. Importation and recombination are responsible for the latest emergence of highly pathogenic porcine reproductive and respiratory syndrome virus in China. *J. Virol.* **2015**, *89*, 10712–10716. [CrossRef]
29. Brockmeier, S.L.; Loving, C.L.; Vorwald, A.C.; Kehrli, M.E., Jr.; Baker, R.B.; Nicholson, T.L.; Lager, K.M.; Miller, L.C.; Faaberg, K.S. Genomic sequence and virulence comparison of four Type 2 porcine reproductive and respiratory syndrome virus strains. *Virus Res.* **2012**, *169*, 212–221. [CrossRef]
30. Van Geelen, A.G.M.; Anderson, T.K.; Lager, K.M.; Das, P.B.; Otis, N.J.; Montiel, N.A.; Miller, L.C.; Kulshreshtha, V.; Buckley, A.C.; Brockmeier, S.L.; et al. Porcine reproductive and respiratory disease virus: Evolution and recombination yields distinct ORF5 RFLP 1-7-4 viruses with individual pathogenicity. *Virology* **2018**, *513*, 168–179. [CrossRef]
31. Song, S.; Xu, H.; Zhao, J.; Leng, C.; Xiang, L.; Li, C.; Fu, J.; Tang, Y.-D.; Peng, J.; Wang, Q.; et al. Pathogenicity of NADC34-like PRRSV HLJDZD32-1901 isolated in China. *Vet. Microbiol.* **2020**, *246*, 108727. [CrossRef] [PubMed]
32. Zhang, H.; Luo, Q.; Zheng, Y.; Sha, H.; Li, G.; Kong, W.; Huang, L.; Zhao, M. Genetic variability and recombination of the NSP2 gene of PRRSV-2 strains in China from 1996 to 2021. *Vet. Sci.* **2023**, *10*, 325. [CrossRef] [PubMed]
33. Li, Z.; He, Y.; Xu, X.; Leng, X.; Li, S.; Wen, Y.; Wang, F.; Xia, M.; Cheng, S.; Wu, H. Pathological and immunological characteristics of piglets infected experimentally with a HP-PRRSV TJ strain. *BMC Vet. Res.* **2016**, *12*, 230. [CrossRef]
34. Xu, H.; Li, C.; Li, W.; Zhao, J.; Gong, B.; Sun, Q.; Tang, Y.D.; Xiang, L.; Leng, C.; Peng, J.; et al. Novel characteristics of Chinese NADC34-like PRRSV during 2020–2021. *Transbound. Emerg. Dis.* **2022**, *69*, e3215–e3224. [CrossRef]
35. Fang, Y.; Snijder, E.J. The PRRSV replicase: Exploring the multifunctionality of an intriguing set of nonstructural proteins. *Virus Res.* **2010**, *154*, 61–76. [CrossRef]
36. Sun, Y.F.; Yu, H.; Jiang, X.; Ma, J.F.; Xu, C.Q.; Yu, X.X.; Li, L.-A. Novel ORF5 deletion of NADC30-like porcine reproductive and respiratory syndrome viruses circulating in northern China from 2016 to 2018. *J. Vet. Diagn. Investig.* **2020**, *32*, 928–932. [CrossRef]
37. Li, P.X.; Shen, Y.S.; Wang, T.L.; Li, J.; Li, Y.; Zhao, Y.R.; Liu, S.; Li, B.; Liu, M.; Meng, F. Epidemiological survey of PRRS and genetic variation analysis of the ORF5 gene in Shandong Province, 2020–2021. *Front. Vet. Sci.* **2022**, *9*, 987667. [CrossRef]
38. Huang, B.; Xu, T.; Luo, Z.; Deng, L.; Jian, Z.; Lai, S.; Ai, Y.; Zhou, Y.; Ge, L.; Xu, Z.; et al. Prevalence and genetic diversity of PRRSV in Sichuan province of China from 2021 to 2023: Evidence of an ongoing epidemic transition. *Virology* **2024**, *600*, 110213. [CrossRef]
39. Song, T. Analysis of PRRSV Genetic Variation and Quantitative Interactomics of Its Nsp2 Protein. Ph.D. Thesis, Huazhong Agricultural University, Wuhan, China, 2016.

40. Katoh, K.; Standley, D.M. MAFFT multiple sequence alignment software version 7: Improvements in performance and usability. *Mol. Biol. Evol.* **2013**, *30*, 772–780. [CrossRef]
41. Burland, T.G. DNASTAR's Lasergene sequence analysis software. *Methods Mol. Biol.* **2000**, *132*, 71–91.
42. Pang, Y.; Li, C.; Wang, Y.; Liu, J.; Su, G.; Duan, C.; Fang, L.; Zhou, Y.; Xiao, S. Porcine reproductive and respiratory syndrome virus infection manipulates central carbon metabolism. *Vet. Microbiol.* **2023**, *279*, 109674. [CrossRef] [PubMed]
43. Yin, B.; Qi, S.; Sha, W.; Qin, H.; Liu, L.; Yun, J.; Zhu, J.; Li, G.; Sun, D. Molecular Characterization of the Nsp2 and ORF5 (ORF5a) Genes of PRRSV Strains in Nine Provinces of China During 2016–2018. *Front. Vet. Sci.* **2021**, *8*, 605832. [CrossRef] [PubMed]
44. Ouyang, Y.; Du, Y.; Zhang, H.; Guo, J.; Sun, Z.; Luo, X.; Mei, X.; Xiao, S.; Fang, L.; Zhou, Y. Genetic characterization and pathogenicity of a recombinant porcine reproductive and respiratory syndrome virus strain in China. *Viruses* **2024**, *16*, 993. [CrossRef] [PubMed]
45. Fang, Y.; Treffers, E.E.; Li, Y.H.; Tas, A.; Sun, Z.; van der Meer, Y.; de Ru, A.H.; van Veelen, P.A.; Atkins, J.F.; Snijder, E.J.; et al. Efficient-2 frameshifting by mammalian ribosomes to synthesize an additional arterivirus protein. *Proc. Natl. Acad. Sci. USA* **2012**, *109*, E2920–E2928. [CrossRef]
46. Gao, J.C.; Xiong, J.Y.; Ye, C.; Chang, X.B.; Guo, J.C.; Jiang, C.G.; Zhang, G.-H.; Tian, Z.-J.; Cai, X.-H.; Tong, G.-Z.; et al. Genotypic and geographical distribution of porcine reproductive and respiratory syndrome viruses in mainland China in 1996–2016. *Vet. Microbiol.* **2017**, *208*, 164–172. [CrossRef]
47. Delputte, P.L.; Nauwynck, H.J. Porcine Arterivirus infection of alveolar macrophages is mediated by sialic acid on the virus. *J. Virol.* **2004**, *78*, 8094–8101. [CrossRef]
48. Xie, J.X.; Vereecke, N.; Theuns, S.; Oh, D.; Vanderheijden, N.; Trus, I.; Sauer, J.; Vyt, P.; Bonckaert, C.; Lalonde, C.; et al. Comparison of primary virus isolation in pulmonary alveolar macrophages and four different continuous cell lines for type 1 and type 2 porcine reproductive and respiratory syndrome virus. *Vaccines* **2021**, *9*, 594. [CrossRef]
49. Zhang, H.L.; Tang, Y.D.; Liu, C.X.; Xiang, L.R.; Zhang, W.L.; Leng, C.L.; Wang, Q.; An, T.-Q.; Peng, J.-M.; Tian, Z.-J.; et al. Adaptions of field PRRSVs in Marc-145 cells were determined by variations in the minor envelope proteins GP2a-GP3. *Vet. Microbiol.* **2018**, *222*, 46–54. [CrossRef]
50. Osemeke, O.H.; Machado, I.; De Conti, E.; Musskopf, M.; Mil-Homens, M.P.; Stutzman, S.; Guo, B.; Petznick, T.; Silva, G.D.-S.; Gauger, P.; et al. Optimizing tongue fluid sampling and testing protocols for enhanced PRRSV isolation from perinatal swine mortalities. *Viruses* **2025**, *17*, 102. [CrossRef]
51. Shi, C.X.; Liu, Y.L.; Ding, Y.Z.; Zhang, Y.G.; Zhang, J. PRRSV receptors and their roles in virus infection. *Arch. Microbiol.* **2015**, *197*, 503–512. [CrossRef]
52. Xie, J.; Trus, I.; Oh, D.; Kvisgaard, L.K.; Rappe, J.C.F. A Triple Amino Acid Substitution at Position 88/94/95 in Glycoprotein GP2a of Type 1 Porcine Reproductive and Respiratory Syndrome Virus (PRRSV1) Is Responsible for Adaptation to MARC-145 Cells. *Viruses* **2019**, *11*, 36. [CrossRef] [PubMed]
53. Ye, Z.; Zhang, Z.; Zhu, Z.; Sun, Z.; Tian, K.; Li, X. A Live-Attenuated Chimeric Vaccine Candidate Against the Emerging NADC34-Like PRRSV. *Vet. Sci.* **2025**, *12*, 290. [CrossRef]
54. Zhou, L.; Han, J.; Yang, H. The evolution and diversity of porcine reproductive and respiratory syndrome virus in China. *Vet. Microbiol.* **2024**, *298*, 110252. [CrossRef] [PubMed]
55. Wu, Y.; Peng, O.; Xu, Q.; Li, Q.; Li, W.; Lin, L.; Zhou, Q.; Cai, X.; Hu, G.; He, Z.; et al. Characterization and Pathogenicity of Two Novel PRRSVs Recombined by NADC30-like and NADC34-like Strains in China. *Viruses* **2022**, *14*, 2174. [CrossRef]
56. Liu, J.; Liu, C.; Xu, Y.; Yang, Y.; Li, J.; Dai, A.; Huang, C.; Luo, M.; Wei, C. Molecular characteristics and pathogenicity of a novel recombinant porcine reproductive and respiratory syndrome virus strain from NADC30-, NADC34-, and JXA1-like strains that emerged in China. *Microbiol. Spectr.* **2022**, *10*, e0266722. [CrossRef]
57. Zhao, H.Z.; Liu, C.Y.; Meng, H.; Sun, C.L.; Yang, H.W.; Wang, H.; Zou, J.; Li, P.; Han, F.Y.; Qi, G.; et al. Evolution characterization and pathogenicity of an NADC34-like PRRSV isolated from Inner Mongolia, China. *Viruses* **2024**, *16*, 683. [CrossRef]
58. Xie, C.Z.; Ha, Z.; Zhang, H.; Zhang, Y.; Xie, Y.B.; Zhang, H.; Nan, F.-L.; Wang, Z.; Zhang, P.; Xu, W.; et al. Pathogenicity of porcine reproductive and respiratory syndrome virus (ORF5 RFLP 1-7-4 viruses) in China. *Transbound. Emerg. Dis.* **2020**, *67*, 2065–2072. [CrossRef]
59. Jelsma, T.; Wijnker, J.J.; van der Poel, W.H.M.; Wisselink, H.J. Intestinal Viral Loads and Inactivation Kinetics of Livestock Viruses Relevant for Natural Casing Production: A Systematic Review and Meta-Analysis. *Pathogens* **2021**, *10*, 173. [CrossRef]
60. Zhao, J.; Wan, S.; Sun, N.; Sun, P.; Sun, Y.; Khan, A.; Guo, J.; Zheng, X.; Fan, K.; Yin, W.; et al. Damage to intestinal barrier integrity in piglets caused by porcine reproductive and respiratory syndrome virus infection. *Vet. Res.* **2021**, *52*, 93. [CrossRef]

veterinary sciences

MDPI

Article

# Isolation and Characterization of Porcine Epidemic Diarrhea Virus G2c Strains Circulating in China from 2021 to 2024

Xi Lu [1,2], Chen Chen [1,2], Zixuan Wang [1,2,3] and Anding Zhang [1,2,4,5,*]

1 State Key Laboratory of Agricultural Microbiology, Hubei Hongshan Laboratory, College of Veterinary Medicine, Huazhong Agricultural University, Wuhan 430070, China; lucylook5@webmail.hzau.edu.cn (X.L.); chen110@webmail.hzau.edu.cn (C.C.); kakawzx@163.com (Z.W.)
2 Key Laboratory of Preventive Veterinary Medicine in Hubei Province, The Cooperative Innovation Center for Sustainable Pig Production, Wuhan 430070, China
3 Shenzhen Kingkey Smart Agriculture Times Co., Ltd., Shenzhen 518000, China
4 Key Laboratory of Development of Veterinary Diagnostic Products, Ministry of Agriculture of the People's Republic of China, Wuhan 430070, China
5 International Research Center for Animal Disease, Ministry of Science and Technology of the People's Republic of China, Wuhan 430070, China
* Correspondence: andye8019@mail.hzau.edu.cn

**Simple Summary:** This study successfully isolated three Porcine epidemic diarrhea virus strains from immune-failure pig farms in China, designated 2021-HBMC, 2024-JXNC, and 2024-JXYX. All three strains were classified as G2c, the most prevalent genotype in China. They exhibit several mutations in the major neutralizing epitopes of the S protein compared to the AJ1102 strain, along with deletions in some N-glycosylation sites, which may explain the reduced neutralizing activity of the vaccine serum against them. These findings highlight the widespread prevalence of the G2c genotype in China and suggest that current vaccines are less effective against it. This underscores the current PEDV epidemic situation and informs future vaccine development and control strategies.

**Abstract:** Porcine epidemic diarrhea virus (PEDV) is a major pathogen responsible for viral diarrhea in pigs, causing particularly high mortality in neonatal piglets. In recent years, genetic variations in PEDV have resulted in alterations in both its virulence and antigenicity, leading to a reduced efficacy of existing vaccines. In this study, diarrheal samples were collected from four commercial pig farms in the Hubei, Guangxi, and Jiangxi provinces, China, which experienced vaccine failure. RT-qPCR confirmed PEDV infection, and three PEDV strains, 2021-HBMC, 2024-JXYX, and 2024-JXNC, were successfully isolated. Sequence analysis and phylogenetic tree construction classified these strains into the G2c genotype, the predominant subtype in China. The neutralization assays revealed a significant reduction in the neutralizing titers of these strains against the immune serum compared with the AJ1102 reference strain. Further amino acid sequence analysis of the spike (S) protein identified several mutations in key neutralizing epitopes compared with the AJ1102 strain, including S27L, E57A, N139D, M214T, and P229L in the S-NTD epitope; A520S, F539L, K566N, D569E, G612V, P634S, E636V/K in the COE epitope; and Y1376H in the 2C10 epitope, along with several deletions at N-glycosylation sites (347NSSD and 510NITV). Additionally, whole-genome sequencing and recombination analysis indicated that the 2021-HBMC strain may have resulted from a recombination event. The findings of this study underscore the challenge posed by the continuous genetic evolution of PEDV to vaccine efficacy and provide valuable insights for future vaccine development and control strategies.

**Keywords:** porcine epidemic diarrhea virus (PEDV); virus isolation; antigenic variation; vaccine efficacy; G2c genotype

## 1. Introduction

Porcine epidemic diarrhea virus (PEDV) is a primary pathogen responsible for viral diarrhea in pigs, causing symptoms such as watery diarrhea, vomiting, and severe dehydration [1,2]. PEDV can infect pigs of various breeds and ages, with the greatest threat to neonatal piglets, where morbidity and mortality rates can reach up to 100%, leading to substantial economic losses to the global swine industry [3–5]. Since 2010, PEDV has caused numerous outbreaks worldwide, particularly in Asia and North America, with a notable increase in epidemic intensity and severity, making it one of the major challenges facing the swine industry [6–8].

PEDV belongs to the *Coronaviridae* family and *Alphacoronavirus* genus and has a single-stranded positive-sense RNA genome, characterized by a high mutation rate and a propensity for recombination [9]. Based on the genetic variability of the highly mutable spike (S) gene, PEDV is classified into two major genotypes, G1 and G2, which are further subdivided into G1a, G1b, G2a, and G2b [10]. PEDV was first discovered in the United Kingdom in 1977 and was introduced into China in the 1980s, initially causing only regional outbreaks [11,12]. However, in 2010, a G2 variant emerged that was distinct from the classic G1 strain, exhibiting increased virulence, which led to higher morbidity and mortality in piglets and caused severe economic losses to the swine industry in Asia [13–15]. In 2013, the highly virulent G2 strain emerged in the United States, rapidly spreading across multiple states and gradually evolving into a global epidemic [16,17]. Subsequently, in 2014, an S-INDEL strain, arising from recombination between G1 and G2 strains, emerged, showing higher homology to the G1 strain in the S protein but relatively lower virulence [16–18].

Vaccination is one of the primary strategies for controlling PEDV; however, the virus' complex and rapid genetic variation in recent years has led to changes in its pathogenicity, tissue tropism, and antigenicity, challenging the effectiveness of current vaccines [19–21]. Between 2010 and 2020, PEDV infections were reported in nearly all provinces of China, with the G2a and G2b strains predominantly circulating between 2010 and 2016 [22,23]. A modified live vaccine developed from the G2b strain AJ1102 was released in late 2017 and widely adopted, helping to curb the spread of PEDV to some extent [24,25]. However, after 2020, despite widespread vaccination in most pig farms, several studies reported PEDV outbreaks, with the strains being identified as a new subtype G2c [26–28]. These studies suggest that the incidence of G2c strains has increased in recent years, and existing vaccines may not effectively prevent G2c PEDV infections. Therefore, ongoing monitoring of PEDV, including strain epidemiology and genetic variation, is essential for the development of more effective vaccines and the formulation of evidence-based control strategies.

In this study, a total of 17 PEDV-positive samples (including feces and intestinal tissue) were collected from commercial pig farms in the Hubei, Guangxi, and Jiangxi provinces of China, where PEDV outbreaks had occurred. RT-qPCR testing confirmed that PEDV was the primary cause of diarrhea in these farms, and three representative PEDV strains were successfully isolated. Sequence analysis and phylogenetic tree construction classified these strains as G2c. Further neutralization tests and analysis of the S protein amino acid sequences identified potential reasons for the vaccination failures. This study examined the genetic evolution of and variation in these G2c strains compared to representative strains from Asia, providing valuable insights into the current PEDV epidemic situation and informing future vaccine development and control strategies.

## 2. Materials and Methods

### 2.1. Sample Collection and Processing

From 2021 to 2024, a total of 17 PEDV-positive samples were conveniently collected from 4 commercial farms with PEDV outbreaks in the Hubei (n = 1), Guangxi (n = 1), and Jiangxi (n = 2) provinces, each housing 2000–3000 sows. Specifically, 1 PEDV-positive sample was collected from a farm in Hubei Province, 6 from a farm in Guangxi Province, and 5 each from two farms in Jiangxi Province. The piglets experienced watery diarrhea and dehydration, leading to mortality. The collected samples were diluted with an equal volume of Dulbecco's Modified Eagle Medium (DMEM) (Gibco, Grand Island, NE, USA) and homogenized using a multi-sample tissue grinder (Lichen-BX, Shanghai, China). The homogenate was then centrifuged at 12,000 rpm for 5 min to separate the supernatant, which was subsequently filtered through a 0.22 μm filter (Sigma-Aldrich, Burlington, MA, USA) to obtain the viral solution.

### 2.2. Viral RNA Isolation and RT-qPCR Detection

Following the manufacturer's instructions, 200 μL of the viral solution was processed for nucleic acid extraction using the Singuway Nucleic Acid Extraction Kit and Extraction Instrument (Singuway, Shenzhen, China). Pathogen detection was performed via a quadruplex RT-qPCR assay developed in our laboratory (unpublished data), which detects the viral loads of PEDV, TGEV, PoRV, and PDCoV in the samples via a probe-based method. Specifically, specific primers, probes for PEDV, TGEV, PoRV, and PDCoV, and a one-step RT-qPCR enzyme (cwbio, Taizhou, China) were prepared in a 15 μL reaction mixture. Then, 5 μL of the sample RNA was added to the mixture, and RT-qPCR was performed using a multi-channel qPCR detection device (Tianlong, Xi'an, China).

### 2.3. PEDV Isolation

The filtered viral solution was supplemented with trypsin (Sigma-Aldrich, Burlington, MA, USA) to achieve a final concentration of 10 μg/mL and then inoculated onto Vero cells, which had been washed three times with PBS. The cells were incubated for 2 h at 37 °C in a 5% $CO_2$ incubator. After incubation, the inoculum was discarded, and the cells were replenished with DMEM containing 10 μg/mL trypsin for further incubation at 37 °C with 5% $CO_2$ for 48 to 72 h. The cells with noticeable cytopathic effects were collected for three freeze–thaw cycles and then passaged into Vero cell monolayers for four consecutive generations. After the fifth passage, viral RNA was extracted, and the viral load was quantified using RT-qPCR.

### 2.4. Immunofluorescence Assay

The 2021-HBMC, 2024-JXNC, and 2024-JXYX strains were inoculated onto Vero cell monolayers and cultured for 24 h. Upon the appearance of significant cytopathic effects, the cells were fixed with a solution of 50% acetone and 50% anhydrous ethanol for 15 min. The cells were then washed three times with PBS. Next, a PEDV N monoclonal antibody (prepared in our laboratory) was added and incubated at 37 °C for 1 h. After three washes with PBS, the cells were incubated with DyLight488-conjugated goat anti-mouse IgG (Abbkine, Wuhan, China) at 37 °C for 1 h. Following three PBS washes, the cells were stained with DAPI (Beyotime, Shanghai, China) for 10 min. After another round of PBS washes, fluorescence was observed under a fluorescence microscope.

### 2.5. Virus $TCID_{50}$ Assay and Serum Neutralization Test

The PEDV strain, capable of inducing cytopathic effects (CPEs) on Vero cells, was used for the $TCID_{50}$ assay. The virus was diluted tenfold in DMEM supplemented with

10 μg/mL trypsin and subsequently added to Vero cell monolayers, which had been washed three times with PBS. $TCID_{50}$ was determined using the Reed–Muench method.

For the serum neutralization test, serum was collected from sows immunized with a commercial vaccine (based on the AJ1102 strain) from a pig farm in Hubei. The serum was inactivated by incubation at 56 °C for 30 min and then diluted twofold in DMEM containing 10 μg/mL trypsin. Equal volumes of the diluted serum were mixed with 200 $TCID_{50}$ of the virus and incubated for 1 h at 37 °C. The serum–virus mixture was then added to Vero cell monolayers, which had been washed twice with PBS, and incubated for 2 h at 37 °C in a 5% $CO_2$ incubator. After discarding the mixture, the cells were washed twice with DMEM containing 10 μg/mL trypsin and cultured for 24 h under the same $CO_2$ incubation conditions. The neutralizing titer of the serum was determined based on the highest serum dilution that inhibited 50% of the CPE.

### 2.6. Sequencing, Multiple-Sequence Alignment, and Phylogenetic Analysis

RNA was extracted from PEDV-positive samples with high viral concentrations. cDNA was synthesized using random primers and reverse transcriptase (Vazyme, Nanjing, China). The S gene was amplified using the three primer pairs listed in Table S1 and Vazyme Phanta Max Super-Fidelity DNA Polymerase (Vazyme, Nanjing, China). The amplified products were sent to Qingke Biotechnology (Wuhan, China) for sequencing and assembly, with the 2021-HBMC sample undergoing whole-genome sequencing. The obtained S gene sequences were aligned with the CV777 and AJ1102 vaccine strains, as well as 2180 other S gene sequences available in the NCBI database (Table S2) for sequence comparison and phylogenetic analysis. The genome sequences of the 2021-HBMC sample and other reference genomes were also aligned and analyzed phylogenetically with 290 reference genomes (Table S3). Sequence alignment was performed using MAFFT 7.526 with default settings. Phylogenetic trees for the S gene and the entire genome were constructed using the Neighbor-Joining method with 1000 bootstrap replicates, implemented in MEGA 7.0 software. The resulting trees were visualized using the Chiplot online platform [29].

### 2.7. Amino Acid Sequence Analysis of the Spike Protein

To examine the genetic characteristics of the PEDV strains, the spike (S) protein sequences of the isolated strains were aligned with representative strains from various genotypes using MEGA 7.0 software. The variations in the neutralizing epitopes of the S protein were analyzed. Sequence editing and refinement were carried out using Jalview 2.11.4.1. Additionally, online tools were employed to predict changes in N-glycosylation sites (https://services.healthtech.dtu.dk/services/NetNGlyc-1.0/) (accessed on 3 February 2025), with N-glycosylation sites scoring "++" or higher considered highly specific [30].

### 2.8. Viral Recombination Analysis

Recombination events in the isolated viruses were assessed using RDP v4.101 software, employing seven methods (RDP, GENECONV, BootScan, MaxChi, Chimaera, SiScan, 3Seq) set to their default configurations. A recombination event was considered reliable if at least six methods provided a positive result with a $p$-value of less than $10^{-6}$.

## 3. Results

### 3.1. Viral Isolation and Identification

Seventeen PEDV-positive samples were collected from farms affected by PEDV in the Hubei, Guangxi, and Jiangxi provinces. These samples were all confirmed to be positive for PEDV using quadruple RT-PCR, while TGEV, PDCoV, and PoRV tested negative. Virus isolation was performed in Vero cells following established protocols. After 4–5 passages, characteristic cytopathic effects of PEDV, primarily syncytia formed through cell fusion, were observed (Figure 1A). An immunofluorescence assay (IFA) was then conducted in Vero cells, revealing specific green fluorescence at the sites of cytopathic changes (Figure 1B). RT-qPCR further confirmed the expression of the PEDV N gene; the cycle threshold (CT) values are 8.21, 15.3, and 14.08, respectively (Figure 1C). Based on the collection sites and dates, the strains were named 2021-HBMC, 2024-JXNC, and 2024-JXYX and uploaded to the GenBank database (Genbank: PV235413-PV235415).

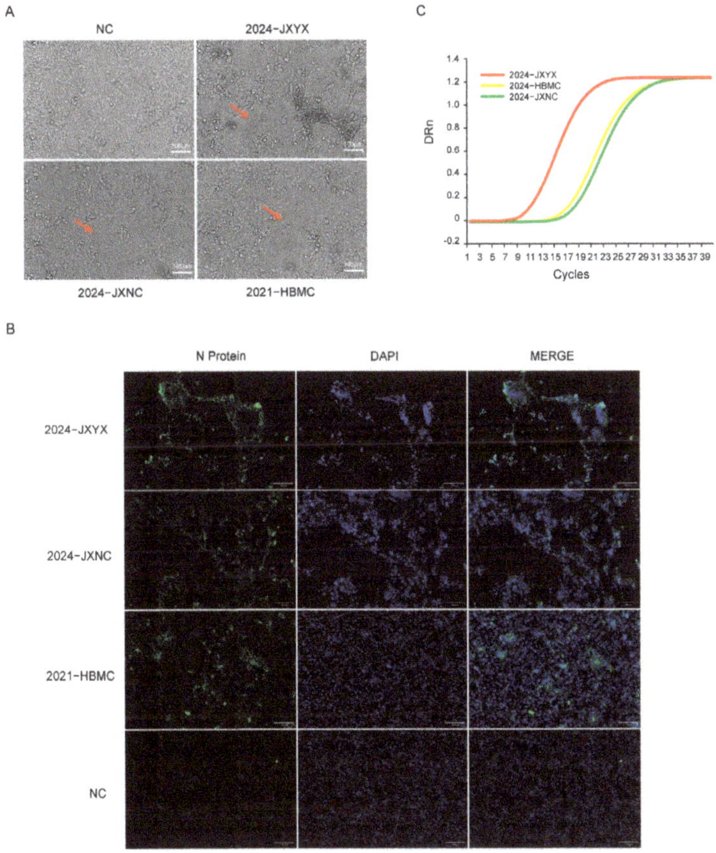


**Figure 1.** Isolation and characterization of the PEDV strains 2021-HBMC, 2024-JXNC and 2024-JXYX. (**A**) Vero cells were infected with the 2021-HBMC, 2024-JXNC, or 2024-JXYX strain for 24 h; the cytopathic effect was observed using a microscope, and the red arrows indicate the syncytium area. (**B**) Vero cells were infected with the 2021-HBMC, 2024-JXNC, or 2024-JXYX strain for 24 h and were detected using an immunofluorescence assay. (**C**) Detection of PEDV-N gene expression levels of generation 5 was carried out using RT-qPCR.

### *3.2. Neutralization Titer of AJ1102 Immune Serum Against Isolated Strains Is Significantly Reduced*

To investigate the cause of immunization failure, we conducted neutralization assays using immune sera from the G2b vaccine against isolated strains. The results revealed that the neutralization titers of the vaccine serum against the 2021-HBMC, 2024-JXNC, and 2024-JXYX strains were significantly lower than the titers observed for the AJ1102 reference strain. These findings suggest that the isolated strains in this study may exhibit antigenic variation (Figure 2).

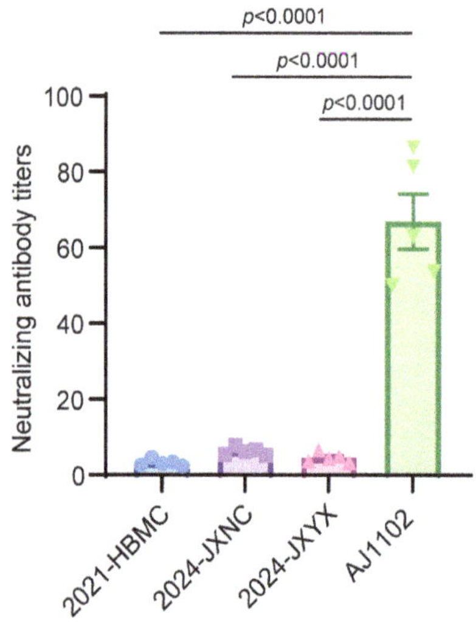


**Figure 2.** The neutralization titer of G2b immune serum against 2021-HBMC, 2024-JXNC, and 2024-JXYX strains. All values are expressed as the mean ± standard error of the mean (SEM). The data were subjected to one-way analysis of variance (ANOVA), and Dunnett's multiple comparisons test was used to compare each group with the AJ1102 group using GraphPad Prism 9.5.1 software.

### 3.3. *Identity and Phylogenetic Analysis of the S Gene*

To identify and analyze the genotype and genetic variation in endemic strains in these disease fields, we amplified and sequenced the S gene and constructed an S gene-based phylogenetic tree using the sequences of 29 representative PEDV strains and complete S gene sequences available in the NCBI database from 2010 to 2024 (Table S2). As shown in Figure 3A, PEDV is primarily classified into three types: G1, G2, and S-INDEL. The G2 type is further subdivided into G2a, G2b, and G2c subtypes. The strains identified in this study all belong to the G2c subtype. Specifically, 2021-HBMC, 2024-GXYL, 2024-JXNC, and 2024-JXYX are located in four distinct branches within G2c. Notably, 2021-HBMC is most closely related to the 2020 strain CH/SDZH/07/2020 (MZ161042), while 2024-JXNC is most closely related to the 2017 strain FY/2017 (MW826593). The strain 2024-GXYL is most closely related to the 2022 strain PEDV/CN/JSYC (PP901936), and 2024-JXYX is most closely related to the 2022 strain FJLY01-2022 (OP171894). These strains showed 92.1% to 93.2% similarity to the vaccine strain CV777 and 96.0% to 97.4% similarity to the vaccine strain AJ1102 (Table 1). Notably, the Chinese strains account for a significant proportion. We calculated the distribution of genotypes in China from 2011 to 2024. As depicted in Figure 3B, the detection rate of G2c strains has significantly increased since 2018, reaching 92.3% in 2024, while the detection rates of G2a and G2b strains have declined steadily. Additionally, all European strains (53/53) were exclusively classified as S-INDEL variants. In contrast, Korean and Japanese isolates predominantly clustered within the G2a genotype (89.03% and 91.78%, respectively), with minor representation of S-INDEL and G2c subtypes. Thai strains demonstrated strong G2b predominance (79.06%), showing closer phylogenetic proximity to the AJ1102 reference strain. Vietnamese strains exhibited the broadest genetic diversity: G2a (25.92%), G2b (51.8%), S-INDEL (11.11%), G2c (9.88%), and sporadic G1 variants. North American isolates primarily comprised the G2a-AH2012 lineage (76.2%), with residual S-INDEL populations (23.8%).

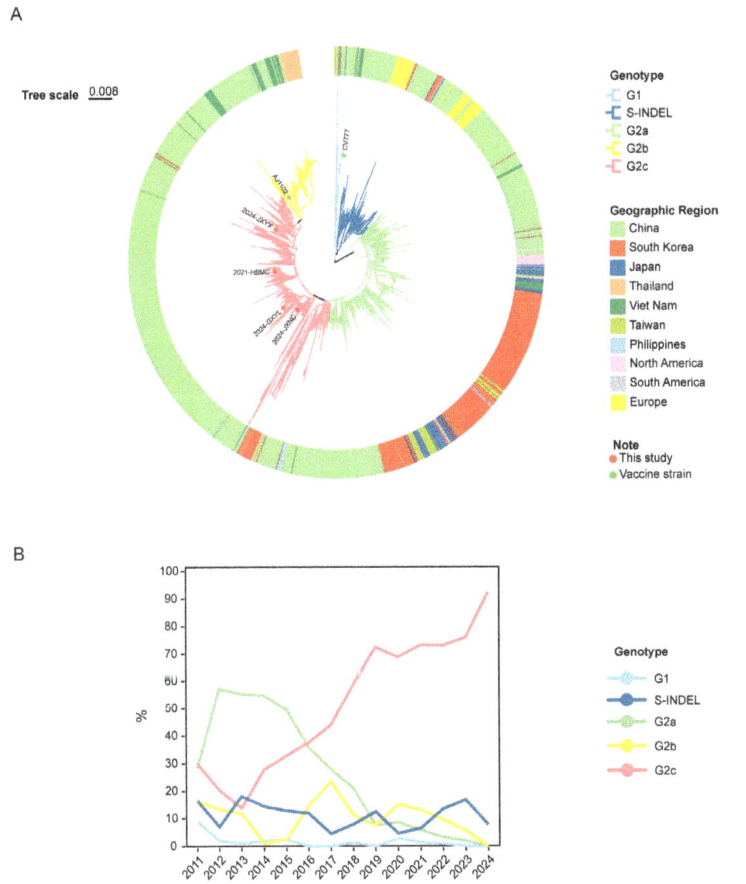


**Figure 3.** Genotyping of the PEDV strains based on S gene sequences from 2010 to 2024. (**A**) The S gene phylogenetic tree was generated by Mega7.0 using the Neighbor-Joining method with 1000 bootstrap replicates. The evolutionary branches of different genotypes are colored, and isolation countries are shown by different color blocks of leaves. (**B**) The isolation rate of each genotype in China from 2011 to 2024 was counted and is represented by line graphs.

**Table 1.** Similarity between G2c strains identified in this study and vaccine strains.

| PEDV Strains | CV777 | AJ1102 |
| --- | --- | --- |
| 2021-HBMC | 93.1% | 97.2% |
| 2024-GXYL | 92.1% | 96.0% |
| 2024-JXNC | 93.2% | 97.2% |
| 2024-JXYX | 93.0% | 97.4% |

### *3.4. Comparative Analysis of Amino Acid Sequences of the Spike Protein*

A comparative analysis of amino acid variations in the spike (S) protein, with a particular focus on the major neutralizing epitopes of PEDV, was conducted to understand the potential causes of vaccine-induced immune failure. As shown in Figure 4A, compared to the vaccine strain CV777, the strains 2021-HBMC, 2024-JXNC, 2024-JXYX, and 2024-GXYL exhibited mutations at multiple sites in the S protein. These strains shared characteristics with other G2-type variants, including consecutive insertions and deletions in the N-terminal domain (NTD) region. Additionally, compared to the vaccine strain AJ1102, mutations such as S27L, E57A, N139D, M214T, and P229L were observed. In the primary neutralizing epitope COE region, eight amino acid variations were identified relative to the AJ1102 strain, including A520S, F539L, K566N, D569E, G612V, P634S, and E636V/K. In the SS6 region, two amino acid mutations (L763S, D765S) were identified compared to CV777, and one mutation (Y1376H) was observed in the 2C10 region relative to AJ1102 (Figure 4B). Furthermore, the N-glycosylation sites of the isolated strains were analyzed. The 2021-HBMC, 2024-JXNC, 2024-JXYX, and 2024-GXYL strains exhibited consistent N-glycosylation sites with the G2c reference strain but lacked several N-glycosylation sites compared to CV777 (112NTSA, 127NKTL, 510NITV, 552NVTN), and compared to AJ1102, they were missing the 510NITV site. Additionally, the 2024-GXYL strain lacked the 347NSSD glycosylation site (Table 2).

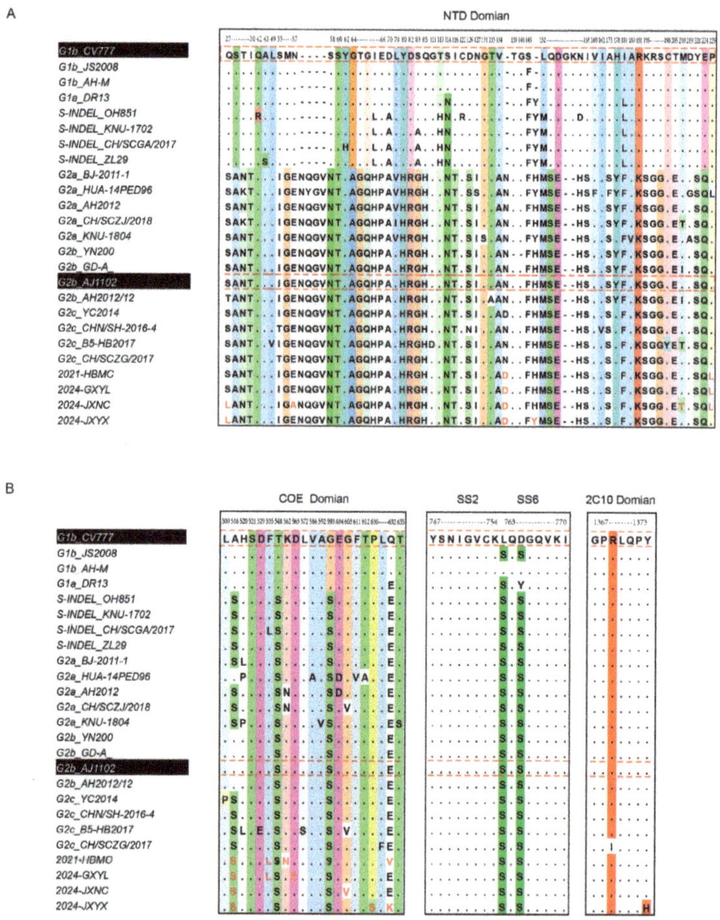


**Figure 4.** Amino acid variation in G2c strains isolated in this study in the S protein. The spike sequences of 2021-HBMC, 2024-GXYL, 2024-JXNC, and 2024-JXYX and the reference strains were aligned with Mega7.0, and the sites with variation were spiced using jalview 2.11.4.1. (**A**) Analysis of amino acid variations in the NTD region. (**B**) Analysis of amino acid variations in the COE, SS2, SS6 and 2C10 domain.

**Table 2.** Variation in highly specific N-glycosylation sites in the spike protein of different genotypes.

| Genotype | Strains | High-Specificity N-Glycosylation Sites | | | | | | | | | | | | | | |
|---|---|---|---|---|---|---|---|---|---|---|---|---|---|---|---|---|
| | | 57 | 62 | 112 | 118 | 127 | 212 | 300 | 320 | 347 | 380 | 510 | 552 | 777 | 1245 | 1258 |
| G1b | KT323979-CV777 | – | – | NTSA | – | NKTL | NVTS | – | – | NSSD | – | NITV | NVTN | NISI | NKTL | – |
| G1b | KC109141-JS2008 | – | – | NTSA | – | NKTL | NVTS | – | – | NSSD | – | NITV | NVTN | NISI | NKTL | – |
| G1a | JQ023161-DR13 | – | – | – | – | NKTL | NVTS | – | NDTS | NSSD | – | NITV | NVTN | NISI | NKTL | NRTG |
| S-INDEL | KU847996-ZL29 | – | – | – | – | NKTL | NVTS | – | NDTS | NSSD | – | – | – | NISI | NKTL | NRTG |
| S-INDEL | KJ399978-OH851 | – | – | – | – | NKTL | NVTS | – | NDTS | NSSD | – | – | – | NISI | NKTL | NRTG |
| S-INDEL | MH243316-KNU-1702 | NSSS | – | – | – | NKTL | NVTS | – | NDTS | NSSD | – | – | – | NISI | NKTL | NRTG |
| G2b | JX188454-AJ1102 | – | NSTW | – | NATA | – | NVTS | – | NDTS | NSSD | – | NITV | – | NISI | NKTL | NRTG |
| G2b | KU646831-AH2012_12 | – | NSTW | – | NATA | – | NVTS | – | NDTS | NSSD | – | NITV | – | NISI | NKTL | NRTG |
| G2a | JN825712-BJ-2011-1 | – | NSTW | – | NATA | – | NVTS | – | NDTS | NSSD | – | – | – | NISI | NKTL | NRTG |
| G2a | KC210145-AH2012 | – | NSTW | – | NATA | – | NVTS | – | NDTS | NSSN | – | – | – | NISI | NKTL | NRTG |
| G2a | MH061342-CH_SCZJ_2018 | – | NSTW | – | NATA | – | NVTS | NKTI | NDTS | NSSD | – | – | – | NISI | NKTL | NRTG |
| G2c | KU252649-YC2014 | – | NSTW | – | NATA | – | NVTS | – | NDTS | NSSD | – | – | – | NISI | NKTL | NRTG |
| G2c | MH161337-CH_SCZG_2017 | – | NSTW | – | NATA | – | NVTS | – | NDTS | NSSN | NSTV | – | – | NISI | NKTL | NRTG |
| G2c | 2021-HBMC | – | NSTW | – | NATA | – | NVTS | – | NDTS | NSSD | – | – | – | NISI | NKTL | NRTG |
| G2c | 2024-GXYL | – | NSTW | – | NATA | – | NVTS | – | NDTS | – | – | – | – | NISI | NKTL | NRTV |
| G2c | 2024-JXNC | – | NSTW | – | NATA | – | NVTS | – | NDTS | NSSN | – | – | – | NISI | NKTL | NRTG |

### *3.5. Comparative Analysis of the Complete Genome of the 2021-HBMC Strain*

Whole-genome sequencing of the 2021-HBMC strain was performed, followed by genetic evolution and recombination analysis with 290 reference strains (Table S3). The results indicated that 2021-HBMC belongs to the G2c genotype, sharing the closest genetic distance with PEDV TRS2021 (OL762461). Both strains are placed in the same small branch, with a genetic similarity of 99.4% (Figure 5A). Recombination analysis revealed that the genomic regions ORF1b (4975–7962) and the S gene (1–914) of 2021-HBMC are likely the result of recombination between an unknown primary parent and the secondary parent FJzz1 (Figure 5B). This finding was supported by seven detection algorithms, all of which yielded $p$-values < $10^{-6}$.

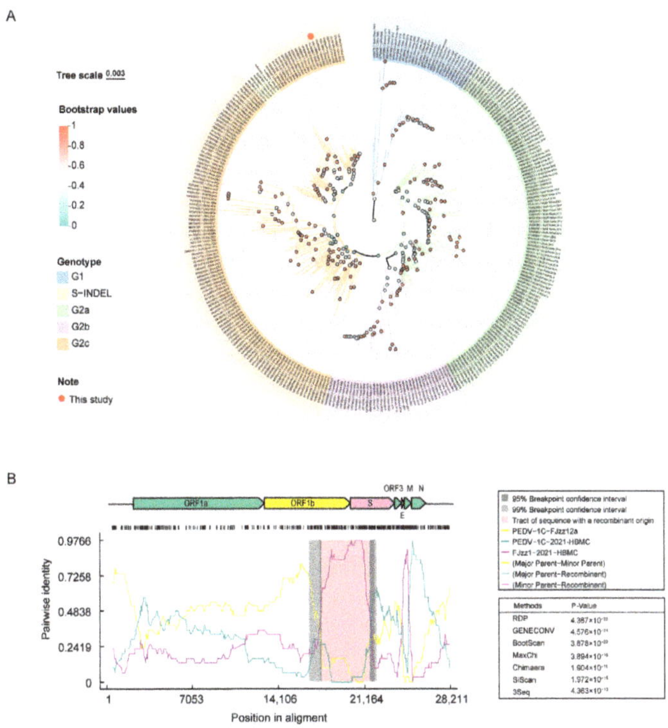


**Figure 5.** Phylogenetic and recombination analyses of full-genome sequences of 2021-HBMC and 290 reference strains. (**A**) The whole-genome phylogenetic tree was generated by Mega 7.0 using the

maximum likelihood method with 1000 bootstrap replicates. The evolutionary genotypes are marked with different color blocks of leaves. And the information for the reference strains, such as accession number, name, isolation year, and countries, is marked on the leaves' names. (**B**) Genome recombination analysis of 2021-HBMC was conducted using RDP4 with seven methods under default settings; recombination events were considered reliable if at least six methods identified them as positive and the $p$-value was below $10^{-6}$. The recombination region is marked with a pink box. The identity between the major parent and 2021-HBMC is represented by a green line, the identity between the minor parent and 2021-HBMC is represented by a purple line, and the identity between the major parent and minor parent is represented by a yellow line.

## 4. Discussion

The porcine epidemic diarrhea (PED) epidemic has been widely reported in Asia, Europe, and North America since 2010, significantly impacting global swine production and attracting considerable attention [17]. According to statistics, PEDV infection was reported in almost all provinces of China from 2010 to 2020 [13]. Meanwhile, the high variability of PEDV has led to genetic diversity, posing challenges to vaccine efficacy. To better control PEDV infections, enhanced surveillance of prevalent PEDV strains is necessary, with next-generation sequencing (NGS) technology playing a critical role. However, the current PEDV genotyping standards are not yet standardized. These discrepancies primarily arise in the classification of G2 variant strains, mainly due to the diversity of the variant strains and the selection of reference strains [31]. Additionally, there is a significant issue regarding the diverse nomenclature for recombinant S-indel strains [16,26]. Additionally, the G2c strain is a recently defined subtype in several studies, previously included under G2a or G2b, but gradually evolving into a distinct branch, primarily composed of Chinese strains, and these strains have been responsible for PEDV infections in multiple regions of China [26–28]. Specifically, G2c is regarded as the major circulating subtype in East, North, South, and Central China between 2017 and 2021 [26]. From 2020 to 2022, the prevalence of G2c in Sichuan Province, China, reached as high as 87.28% [32]. From 2023 to 2024, G2c has continued to be the primary circulating subtype in Southwest China [33]. In conclusion, G2c has become the predominant circulating subtype in various regions of China. The spread of PEDV in China may be facilitated by the inter-provincial live pig trade, with Guangdong and Henan identified as key transmission hubs, and Hubei Province potentially serving as a third hub facilitating the spread to the southwest region [8]. However, no reports of the G2c strain have yet been documented in other countries.

To further investigate the genetic evolution of PEDV, this study conducted a genetic evolution analysis of PEDV S gene sequences from the NCBI database starting from 2010, using genotyping methods similar to Wan Liang et al., Yingshuo Sun et. al., and Cheng Yang et al. [27,28,34]. And the genotype distribution of Chinese strains was subsequently analyzed and summarized. The results showed that G2a strains dominated from 2010 to 2013, but their prevalence has steadily declined in the following years. In contrast, the number of G2c strains has increased consistently since 2018, reaching 92.3% by 2024. This trend indicates that G2c has become the predominant circulating strain in China, which aligns with recent research reports [26,27]. Therefore, it can be inferred that while current vaccines effectively control infections by some G2a and G2b strains, their efficacy against G2c PEDV may be limited. Additionally, although G2c strains were not previously reported in other countries, our analysis revealed that certain Korean strains (such as ON263454, OR125556) and Vietnamese strains (such as MK435380.1, MK435392.1) were also classified as G2c in the phylogenetic tree of the S gene. This suggests that the spread of G2c may not be limited to China alone, and the risk of its cross-border dissemination warrants further monitoring.

To investigate the cause of vaccine failure, this study collected piglet clinical samples from farms in Hubei, Jiangxi, and Guangxi provinces with immune failures (the pigs had been vaccinated with a commercial vaccine based on the G2b strain). Four PEDV strains were identified, and three epidemic strains were successfully isolated. Genetic evolution analysis revealed that all identified and isolated strains belonged to the G2c genotype. Notably, the neutralizing activity of the G2b vaccine immune serum was significantly reduced against these three G2c strains, suggesting that the neutralizing epitopes of the G2c strains may have undergone substantial variation. The S gene is the most mutation-prone region in the PEDV genome, and mutations in this region may alter the antigenicity, pathogenicity, and neutralizing properties of the S protein [7,23,35]. This study found that the S protein of the four G2c strains exhibited only 92.1–93.2% similarity to the CV777 strain and 96.0–97.4% similarity to the AJ1102 vaccine strain. Further analysis revealed several mutations in the key neutralizing epitopes (such as S-NTD, COE, SS6, and 2C10) of these strains compared to the AJ1102 vaccine strain, including the S-NTD epitopes S27L, E57A, N139D, M214T, and P229L; the COE epitopes A520S, F539L, K566N, D569E, G612V, P634S, and E636V/K; and the 2C10 epitope Y1376H. Among them, N139D, P229L, and A520S may affect the phosphorylation level of the PEDV S protein [36]. These mutations may significantly alter the virus's antigenicity and neutralizing characteristics, thus leading to vaccine failure. But further in vitro and in vivo experiments are required to validate the specific effect of these mutations on viral virulence and immune evasion.

In addition, changes in the N-glycosylation sites of the coronavirus S protein are regarded as critical factors influencing viral infectivity, pathogenicity, and immunogenicity [37–40]. Our analysis revealed that the four G2c strains lacked five glycosylation sites compared to the CV777 strain and lacked the 510NITV site compared to the AJ1102 strain. Notably, the 2024-GXYL strain additionally lacked the 347NSSD site. The loss of these glycosylation sites may further alter the virus' antigenicity and neutralizing characteristics, thereby exacerbating the risk of vaccine failure.

Recombination events play an essential role in the evolution of PEDV, contributing to its genetic diversity and allowing newly emerged viruses to rapidly adapt to environmental changes, such as evading host immune responses or enhancing viral virulence [31,41,42]. In this study, whole-genome sequencing of the 2021-HBMC strain was performed, followed by genetic evolution analysis using 290 reference strains. The results showed that the 2021-HBMC strain belonged to the G2c genotype. Notably, some reference strains exhibited discordance in their genotypes when comparing the whole-genome phylogenetic tree with the S gene phylogenetic tree. For instance, the S-INDEL strain OH851 was included in the G2c branch of the whole-genome phylogenetic tree, which could be due to viral recombination events. Recombination analysis using RDP4 indicated that the 2021-HBMC strain might have originated from a recombination event, with the minor parent identified as FJzz1, although the major parent remains undetermined. Further genome-wide comparisons are required to clarify its origin. In addition, to address the rapid evolution of PEDV, future efforts should focus on enhancing surveillance of circulating strains, as well as the development of novel vaccines based on these newly identified variants and the optimization of immunization strategies to improve PEDV control and mitigate its impact on the swine industry.

## 5. Conclusions

In conclusion, this study successfully isolated three PEDV strains from Chinese farms experiencing vaccine failures, all of which belong to the G2c genotype. These strains exhibited multiple mutations in key neutralizing epitopes of the S protein, leading to a significant reduction in neutralizing titers of vaccine-induced immune sera, thus contributing to vaccine failure. Moreover, the loss of glycosylation sites and potential recombination events may further exacerbate the antigenic variation of the virus. The findings of this study highlight the challenge posed by ongoing genetic variations of PEDV on vaccine efficacy and provide valuable insights for future vaccine development and control strategies.

**Supplementary Materials:** The following supporting information can be downloaded at https://www.mdpi.com/article/10.3390/vetsci12050444/s1: Table S1: Primers used for S gene amplification; Table S2: PEDV strains used for phylogenetic analysis of the S gene; Table S3: PEDV strains used for phylogenetic analysis of the genome.

**Author Contributions:** Conceptualization, A.Z. and X.L.; methodology and formal analysis, X.L. and C.C.; investigation, X.L., C.C. and Z.W.; writing—original draft preparation, X.L.; writing—review and editing, X.L. and A.Z.; data curation, X.L. and Z.W.; visualization, X.L.; resources, supervision, project administration, and funding acquisition, A.Z. All authors have read and agreed to the published version of the manuscript.

**Funding:** This study was supported by grants from Science and Technology Development Plan Project of the Silk Road Economic Belt Innovation-Driven Development Pilot Zone and the Urumqi-Changji-Shihezi National Innovation Demonstration Zone, grant number 2022LQ01003.

**Institutional Review Board Statement:** The animal study protocol was approved by the Scientific Ethics Committee of Huazhong Agricultural University (HZAUSW-2025-0050, 17 April 2025).

**Informed Consent Statement:** Not applicable.

**Data Availability Statement:** All data analyzed during this study are available from the corresponding author on reasonable request.

**Conflicts of Interest:** Zixuan wang is an employee in Company Shenzhen Kingkey Smart Agriculture Times. The remaining authors have no conflicts of interest to declare.

## References

1. Yang, S.; Liu, G.; Savelkoul, H.F.J.; Jansen, C.A.; Li, B. Mini-review: Microbiota have potential to prevent PEDV infection by improved intestinal barrier. *Front. Immunol.* **2023**, *14*, 1230937. [CrossRef] [PubMed]
2. Stevenson, G.W.; Hoang, H.; Schwartz, K.J.; Burrough, E.R.; Sun, D.; Madson, D.; Cooper, V.L.; Pillatzki, A.; Gauger, P.; Schmitt, B.J.; et al. Emergence of Porcine epidemic diarrhea virus in the United States: Clinical signs, lesions, and viral genomic sequences. *J. Vet. Diagn. Investig.* **2013**, *25*, 649–654. [CrossRef]
3. Li, Q.; Xu, Z.; Wu, T.; Peng, O.; Huang, L.; Zhang, Y.; Xue, C.; Wen, Z.; Zhou, Q.; Cao, Y. A flagellin-adjuvanted PED subunit vaccine improved protective efficiency against PEDV variant challenge in pigs. *Vaccine* **2018**, *36*, 4228–4235. [CrossRef]
4. Shibata, I.; Tsuda, T.; Mori, M.; Ono, M.; Sueyoshi, M.; Uruno, K. Isolation of porcine epidemic diarrhea virus in porcine cell cultures and experimental infection of pigs of different ages. *Vet. Microbiol.* **2000**, *72*, 173–182. [CrossRef] [PubMed]
5. Wang, Q.; Vlasova, A.N.; Kenney, S.P.; Saif, L.J. Emerging and re-emerging coronaviruses in pigs. *Curr. Opin. Virol.* **2019**, *34*, 39–49. [CrossRef]
6. Song, D.; Zhou, X.; Peng, Q.; Chen, Y.; Zhang, F.; Huang, T.; Zhang, T.; Li, A.; Huang, D.; Wu, Q.; et al. Newly Emerged Porcine Deltacoronavirus Associated With Diarrhoea in Swine in China: Identification, Prevalence and Full-Length Genome Sequence Analysis. *Transbound. Emerg. Dis.* **2015**, *62*, 575–580. [CrossRef]
7. Lin, C.M.; Saif, L.J.; Marthaler, D.; Wang, Q. Evolution, antigenicity and pathogenicity of global porcine epidemic diarrhea virus strains. *Virus Res.* **2016**, *226*, 20–39. [CrossRef] [PubMed]
8. He, W.-T.; Bollen, N.; Xu, Y.; Zhao, J.; Dellicour, S.; Yan, Z.; Gong, W.; Zhang, C.; Zhang, L.; Lu, M.; et al. Phylogeography Reveals Association between Swine Trade and the Spread of Porcine Epidemic Diarrhea Virus in China and across the World. *Mol. Biol. Evol.* **2022**, *39*, 364. [CrossRef]

9. Kocherhans, R.; Bridgen, A.; Ackermann, M.; Tobler, K. Completion of the porcine epidemic diarrhoea coronavirus (PEDV) genome sequence. *Virus Genes* **2001**, *23*, 137–144. [CrossRef]
10. Huang, Y.-W.; Dickerman, A.W.; Piñeyro, P.; Li, L.; Fang, L.; Kiehne, R.; Opriessnig, T.; Meng, X.-J. Origin, evolution, and genotyping of emergent porcine epidemic diarrhea virus strains in the United States. *mBio* **2013**, *4*, e00737-13. [CrossRef]
11. Wood, E.N. An apparently new syndrome of porcine epidemic diarrhoea. *Vet. Rec.* **1977**, *100*, 243–244. [CrossRef]
12. Wang, D.; Fang, L.; Xiao, S. Porcine epidemic diarrhea in China. *Virus Res.* **2016**, *226*, 7–13. [CrossRef] [PubMed]
13. Park, J.E. Porcine Epidemic Diarrhea: Insights and Progress on Vaccines. *Vaccines* **2024**, *12*, 212. [CrossRef] [PubMed]
14. Li, W.; Li, H.; Liu, Y.; Pan, Y.; Deng, F.; Song, Y.; Tang, X.; He, Q. New variants of porcine epidemic diarrhea virus, China, 2011. *Emerg. Infect Dis.* **2012**, *18*, 1350–1353. [CrossRef] [PubMed]
15. Chen, J.; Liu, X.; Shi, D.; Shi, H.; Zhang, X.; Li, C.; Chi, Y.; Feng, L. Detection and molecular diversity of spike gene of porcine epidemic diarrhea virus in China. *Viruses* **2013**, *5*, 2601–2613. [CrossRef]
16. Guo, J.; Fang, L.; Ye, X.; Chen, J.; Xu, S.; Zhu, X.; Miao, Y.; Wang, D.; Xiao, S. Evolutionary and genotypic analyses of global porcine epidemic diarrhea virus strains. *Transbound. Emerg. Dis.* **2019**, *66*, 111–118. [CrossRef]
17. Lee, C. Porcine epidemic diarrhea virus: An emerging and re-emerging epizootic swine virus. *Virol. J.* **2015**, *12*, 193. [CrossRef]
18. Wang, L.; Byrum, B.; Zhang, Y. New variant of porcine epidemic diarrhea virus, United States, 2014. *Emerg. Infect. Dis.* **2014**, *20*, 917–919. [CrossRef]
19. Li, W.; van Kuppeveld, F.J.M.; He, Q.; Rottier, P.J.M.; Bosch, B.J. Cellular entry of the porcine epidemic diarrhea virus. *Virus Res.* **2016**, *226*, 117–127. [CrossRef]
20. Hou, Y.; Lin, C.-M.; Yokoyama, M.; Yount, B.L.; Marthaler, D.; Douglas, A.L.; Ghimire, S.; Qin, Y.; Baric, R.S.; Saif, L.J.; et al. Deletion of a 197-Amino-Acid Region in the N-Terminal Domain of Spike Protein Attenuates Porcine Epidemic Diarrhea Virus in Piglets. *J. Virol.* **2017**, *91*, 227. [CrossRef]
21. Li, Z.; Ma, Z.; Dong, L.; Yang, T.; Li, Y.; Jiao, D.; Han, W.; Zheng, H.; Xiao, S. Molecular Mechanism of Porcine Epidemic Diarrhea Virus Cell Tropism. *mBio* **2022**, *13*, e0373921. [CrossRef] [PubMed]
22. Wen, Z.; Li, J.; Zhang, Y.; Zhou, Q.; Gong, L.; Xue, C.; Cao, Y. Genetic epidemiology of porcine epidemic diarrhoea virus circulating in China in 2012-2017 based on spike gene. *Transbound. Emerg. Dis.* **2018**, *65*, 883–889. [CrossRef]
23. Su, M.; Li, C.; Qi, S.; Yang, D.; Jiang, N.; Yin, B.; Guo, D.; Kong, F.; Yuan, D.; Feng, L.; et al. A molecular epidemiological investigation of PEDV in China: Characterization of co-infection and genetic diversity of S1-based genes. *Transbound. Emerg. Dis.* **2020**, *67*, 1129–1140. [CrossRef] [PubMed]
24. Bi, J.; Zeng, S.; Xiao, S.; Chen, H.; Fang, L. Complete genome sequence of porcine epidemic diarrhea virus strain AJ1102 isolated from a suckling piglet with acute diarrhea in China. *J. Virol.* **2012**, *86*, 10910–10911. [CrossRef]
25. Won, H.; Lim, J.; Noh, Y.H.; Yoon, I.; Yoo, H.S. Efficacy of Porcine Epidemic Diarrhea Vaccines: A Systematic Review and Meta-Analysis. *Vaccines* **2020**, *8*, 642. [CrossRef]
26. Li, X.; Li, Y.; Huang, J.; Yao, Y.; Zhao, W.; Zhang, Y.; Qing, J.; Ren, J.; Yan, Z.; Wang, Z.; et al. Isolation and oral immunogenicity assessment of porcine epidemic diarrhea virus NH-TA2020 strain: One of the predominant strains circulating in China from 2017 to 2021. *Virol. Sin.* **2022**, *37*, 646–655. [CrossRef]
27. Yang, C.; Sun, J.-Y.; Li, X.L.; Cheng, N.; Wang, K.-Y.; Li, L.-Q.; Cheng, X.-J.; Sun, Y.-F. Emerging and re-emerging genotype 2c porcine epidemic diarrhoea virus with high pathogenicity in China. *J. Infect.* **2024**, *89*, 106192. [CrossRef] [PubMed]
28. Sun, Y.; Gong, T.; Wu, D.; Feng, Y.; Gao, Q.; Xing, J.; Zheng, X.; Song, Z.; Liu, X.; Chen, X.; et al. Isolation, identification, and pathogenicity of porcine epidemic diarrhea virus. *Front. Microbiol.* **2023**, *14*, 1273589. [CrossRef]
29. Xie, J.; Chen, Y.; Cai, G.; Cai, R.; Hu, Z.; Wang, H. Tree Visualization By One Table (tvBOT): A web application for visualizing, modifying and annotating phylogenetic trees. *Nucleic Acids Res.* **2023**, *51*, W587–W592. [CrossRef]
30. Yu, J.; Chai, X.; Cheng, Y.; Xing, G.; Liao, A.; Du, L.; Wang, Y.; Lei, J.; Gu, J.; Zhou, J. Molecular characteristics of the spike gene of porcine epidemic diarrhoea virus strains in Eastern China in 2016. *Virus Res.* **2018**, *247*, 47–54. [CrossRef]
31. Tian, Y.; Yang, X.; Li, H.; Ma, B.; Guan, R.; Yang, J.; Chen, D.; Han, X.; Zhou, L.; Song, Z.; et al. Molecular characterization of porcine epidemic diarrhea virus associated with outbreaks in southwest China during 2014–2018. *Transbound. Emerg. Dis.* **2021**, *68*, 3482–3497. [CrossRef] [PubMed]
32. Peng, Q.; Fu, P.; Zhou, Y.; Lang, Y.; Zhao, S.; Wen, Y.; Wang, Y.; Wu, R.; Zhao, Q.; Du, S.; et al. Phylogenetic Analysis of Porcine Epidemic Diarrhea Virus (PEDV) during 2020-2022 and Isolation of a Variant Recombinant PEDV Strain. *Int. J. Mol. Sci.* **2024**, *25*, 10878. [CrossRef] [PubMed]
33. Xie, B.; Yan, W.; Yang, X.; Fan, H. Molecular characterization of porcine epidemic diarrhea virus in Sichuan from 2023 to 2024. *Microb. Pathog.* **2025**, *203*, 107486. [CrossRef] [PubMed]
34. Liang, W.; Zhou, D.; Geng, C.; Yang, K.; Duan, Z.; Guo, R.; Liu, W.; Yuan, F.; Liu, Z.; Gao, T.; et al. Isolation and evolutionary analyses of porcine epidemic diarrhea virus in Asia. *PeerJ* **2020**, *8*, e10114. [CrossRef]

35. Su, M.; Wang, Y.; Yan, J.; Xu, X.; Zheng, H.; Cheng, J.; Du, X.; Liu, Y.; Ying, J.; Zhao, Y.; et al. Isolation and characterization of a novel S1-gene insertion porcine epidemic diarrhea virus with low pathogenicity in newborn piglets. *Virulence* **2024**, *15*, 2397512. [CrossRef]
36. Sun, J.; Cheng, J.; Shi, D.; Xu, X.; Liu, Y.; Ying, J.; Zhao, Y.; Zheng, H.; Yan, J.; Sun, D.; et al. Genetic Epidemiology of Porcine Epidemic Diarrhea Virus Circulating in China From 2010 to 2024, Characterization of Phylogenetic and Genetic Diversity of S1-Based Genes. *J. Med. Virol.* **2025**, *97*, e70198. [CrossRef]
37. Le, B.T.; Gallage, H.C.; Kim, M.-H.; Park, J.-E. Molecular Characterization of Porcine Epidemic Diarrhea Virus from Field Samples in South Korea. *Viruses* **2023**, *15*, 2428. [CrossRef]
38. Zhu, H.; Lou, J.; Yang, Z.; Bai, J.; Jiang, P.; Wang, X.; Liu, X. STT3B promotes porcine epidemic diarrhea virus replication by regulating N-glycosylation of PEDV S protein. *J. Virol.* **2025**, *99*, e0001825. [CrossRef]
39. Huang, H.-C.; Lai, Y.-J.; Liao, C.-C.; Wang, F.-Y.; Huang, K.-B.; Lee, I.-J.; Chou, W.-C.; Wang, S.-H.; Wang, L.-H.; Hsu, J.-M.; et al. Targeting conserved N-glycosylation blocks SARS-CoV-2 variant infection in vitro. *EBioMedicine* **2021**, *74*, 103712. [CrossRef]
40. Aloor, A.; Aradhya, R.; Venugopal, P.; Gopalakrishnan Nair, B.; Suravajhala, R. Glycosylation in SARS-CoV-2 variants: A path to infection and recovery. *Biochem. Pharmacol.* **2022**, *206*, 115335. [CrossRef]
41. Wang, Y.; Liu, D.; Shi, W.; Lu, R.; Wang, W.; Zhao, Y.; Deng, Y.; Zhou, W.; Ren, H.; Wu, J.; et al. Origin and Possible Genetic Recombination of the Middle East Respiratory Syndrome Coronavirus from the First Imported Case in China: Phylogenetics and Coalescence Analysis. *mBio* **2015**, *6*, e01280-15. [CrossRef] [PubMed]
42. Liu, H.; Yin, X.; Tian, H.; Qiu, Y.; Wang, Z.; Chen, J.; Ma, D.; Zhao, B.; Du, Q.; Tong, D.; et al. The S protein of a novel recombinant PEDV strain promotes the infectivity and pathogenicity of PEDV in mid-west China. *Transbound. Emerg. Dis.* **2022**, *69*, 3704–3723. [CrossRef] [PubMed]

veterinary sciences

MDPI

Article

# Development of a Paper-Based Microfluidic Chip for Point-of-Care Detection of PEDV

Renfeng Li [1], Xiangqin Tian [2], Wenyan Cao [1], Jiaxin Jiang [1], Jiakang Yuan [1], Linyue Li [1], Yonghe You [3], Yanlin Zhou [3], Ziliang Wang [1,*] and Fangyu Wang [4,*]

1 College of Animal Science and Veterinary Medicine, Henan Institute of Science and Technology, Xinxiang 453003, China; lirenfeng@sina.com (R.L.); caowenyan@stu.hist.edu.cn (W.C.); jiangjiaxin@stu.hist.edu.cn (J.J.); yjkhist2024@163.com (J.Y.); m18236356245@163.com (L.L.)
2 Henan Key Laboratory of Medical Tissue Regeneration, Xinxiang Medical University, Xinxiang 453003, China; tianxiangqin@xxmu.edu.cn
3 Sanquan College, Xinxiang Medical University, Xinxiang 453003, China; sqjcyyh2013@126.com (Y.Y.); zhouyanlin@xxmu.edu.cn (Y.Z.)
4 Key Laboratory of Animal Immunology, Institute for Animal Health, Henan Academy of Agricultural Sciences, Zhengzhou 450002, China
* Correspondence: wziliang2000@163.com (Z.W.); sprinkle.w@126.com (F.W.)

**Simple Summary:** This study introduces an innovative cut-and-stack 3D microfluidic device for point-of-care detection of the porcine epidemic diarrhea virus (PEDV), utilizing a foldable paper chip coupled with loop-mediated isothermal amplification (LAMP) and a pH-responsive indicator. By integrating viral nucleic acid enrichment, amplification, and visual quantitative detection within a single device, this assay effectively addresses the limitations of conventional PEDV diagnostic techniques for field applications. Furthermore, it demonstrates significant potential for broader application in the detection of other pathogens, particularly in resource-limited settings.

**Abstract:** PEDV poses a significant threat to the global swine industry, necessitating rapid and accurate diagnostic methods for effective disease management. In this study, we developed a foldable, easy-to-use paper-based microfluidic analytical device ($\mu$PAD) for on-site detection of PEDV. The device seamlessly integrates paper-based nucleic acid enrichment, LAMP reaction, and visual lateral flow detection into a single platform. Key parameters, including nucleic acid extraction protocols, chromatographic channel configurations, colorimetric indicators, and reaction temperature and duration, were systematically optimized. The resulting LAMP-$\mu$PAD assay detects PEDV within 30 min at 60 °C, achieving a limit of detection of $4.82 \times 10^2$ copies/µL with no cross-reactivity against other viruses. When evaluated against RT-PCR using clinical specimens, the assay demonstrated a specificity of 100%, a sensitivity of 95.3%, and an overall concordance of 98.5%. This paper-based sensor offers a promising alternative for the rapid, on-site detection of PEDV and other highly transmissible pathogens.

**Keywords:** porcine epidemic diarrhea virus; loop-mediated isothermal amplification; paper-based microfluidic analytical device; point-of-care detection

## 1. Introduction

PEDV causes an acute, highly contagious enteric disease in swine, characterized by vomiting, watery diarrhea, and dehydration [1]. While PEDV infects swine of all age groups, it exhibits particularly severe mortality rates approaching 100% in neonatal piglets less than seven days old [2]. Since its first identification in the UK in 1971, PEDV has spread globally, causing significant economic losses to the pork industry [3,4]. The rapid transmission and devastating impact of PEDV underscore the critical need for effective on-site diagnostic tools.

Currently employed methods for PEDV detection primarily relies on virus isolation, nucleic acid amplification, and serological techniques [5–7]. Virus isolation, though reliable, is time-consuming and impractical for field use. Nucleic acid detection methods, including reverse transcription-polymerase chain reaction (RT-PCR) and real-time quantitative RT-PCR (qPCR), offer high sensitivity but require specialized equipment and a large amount of reagents and expertise. Serological tests, such as neutralization assays and enzyme-linked immunosorbent assay (ELISA), generally exhibit lower specificity and sensitivity compared to nucleic acid detection methods. These limitations highlight the pressing necessity to develop rapid, straightforward, and user-friendly diagnostic tools that are appropriate for field implementation, thereby improving the detection and management of PEDV.

Recent advances in nucleic acid isothermal amplification techniques have led to the emergence of LAMP as an innovative alternative to traditional thermal cycling methods. Utilizing Bst DNA polymerase and 4–6 specially designed primers, LAMP enables rapid target sequence amplification under constant temperature conditions [8,9]. and the results can be visualized through turbidity, dye-based methods, or immunochromatographic strips [10–12]. LAMP demonstrates high specificity due to its targeting of six distinct regions and offers rapid amplification using simple, cost-effective equipment, making it a promising tool for on-site diagnostics in resource-limited settings [13].

Microfluidic chips represent an innovative diagnostic platform that enables rapid sample analysis through the utilization of microliter-scale reaction volumes. $\mu$PADs constitute an advanced evolution of conventional microfluidic chips, utilizing paper as the foundational substrate material. These devices integrate precisely engineered microchannels fabricated on diverse paper types, leveraging capillary action to autonomously drive fluid flow without external energy input [14,15]. This approach offers a more cost-effective and simplified solution for analytical purposes. To date, $\mu$PADs have demonstrated efficacy in the detection of diverse human and animal pathogens, including porcine reproductive and respiratory syndrome virus (PRRSV) [16], pseudorabies virus (PRV) [17], human papillomavirus [18], severe acute respiratory syndrome coronavirus 2 (SARS-CoV-2) [19,20], Zika virus (ZIKV) [21], Avian influenza virus (AIV) [22], and mycobacterium tuberculosis [23], among others. In this study, we introduce the integration of LAMP with paper-based microfluidic technology to develop an innovative LAMP-$\mu$PAD assay for the detection of PEDV. This approach demonstrates high accuracy, specificity, and sensitivity, while maintaining cost-effectiveness and ease of use.

## 2. Materials and Methods

### 2.1. Materials, Reagents and Instruments

PEDV, porcine deltacoronavirus (PDCoV), porcine circovirus (PCV2), PRRSV, porcine transmissible gastroenteritis virus (TGEV), and standard plasmid harboring the PEDV ORF1a/b gene were obtained from the Key Laboratory of Animal Viral Diseases Prevention and Control at Henan Institute of Science and Technology. WarmStart® RTx reverse transcriptase and Bst 2.0 WarmStart® DNA polymerase were purchased from NEB (Beijing, China). Methyl red and neutral red indicator dyes were purchased from Macklin Co., Ltd. (Shanghai, China). Whatman No. 1 filter paper and GF/F glass fiber membrane were purchased from GE Healthcare Life Sciences (Pittsburgh, PA, USA). Nitrocellulose (NC) membrane (HiFlow Plus 180) was purchased from Merck Millipore Ltd. (Burlington, VT, USA);

The absorbent paper pad (Grade 222) was purchased from Ahlstrom (Helsinki, Finland); Portrait 3 electronic cutting machine was purchased from Silhouette (Lindon, UT, USA).

*2.2. Optimization of RT-LAMP Reaction System*

The reaction mixture comprised 2.5 μL of 10× isothermal amplification buffer, 1.5 μL of $MgSO_4$ (100 mM), 3.5 μL of dNTPs Mix (10 mM), 1 μL of Bst 2.0 WarmStart DNA polymerase (8000 U/mL), 1 μL each of F3/B3 primers (5 μM), 1 μL each of FIP/BIP primers (40 μM), 1 μL of standard plasmid as the template, and nuclease-free water to a final volume of 25 μL. Utilizing the LAMP reaction system, a comprehensive screening of 11 primer sets was conducted to select the most suitable candidates for subsequent visual RT-LAMP assay (Table S1). Concurrently, we undertook an optimization process for various parameters implicated in the LAMP reaction, including reaction time, reaction temperature, concentrations of inner and outer primers, Bst 2.0 enzyme concentration, $Mg^{2+}$ concentration, dNTPs concentration, and buffer concentration. The optimal conditions were determined based on agarose gel electrophoresis results.

*2.3. Fabrication of the μPAD*

The paper chip architecture was designed using AutoCAD 2016 software (Autodesk, San Francisco, CA, USA), incorporating multiple functional zones. The design was exported to Silhouette Studio software V4.4.259.001 for fabrication using a Portrait 3 cutting machine. As illustrated in Figure 1, the paper-based chip comprises five 4 × 6 cm functional modules. Module A is a waste reservoir made of absorbent paper that collects impurities eluted from the enrichment zone. Module B contains a glass fiber membrane disk for enriching viral nucleic acids. Module C includes four filter paper detection zones for performing LAMP amplification and subsequent colorimetric detection. Module D features a nucleic acid chromatography channel for transferring nucleic acids from the enrichment zone to the amplification zone. Module E is a sealing tape that prevents evaporation and cross-contamination during the LAMP amplification reaction. All components were precision-cut and assembled using adhesive tape, resulting in an integrated design that enables efficient nucleic acid enrichment, amplification, and detection within a single paper-based device.

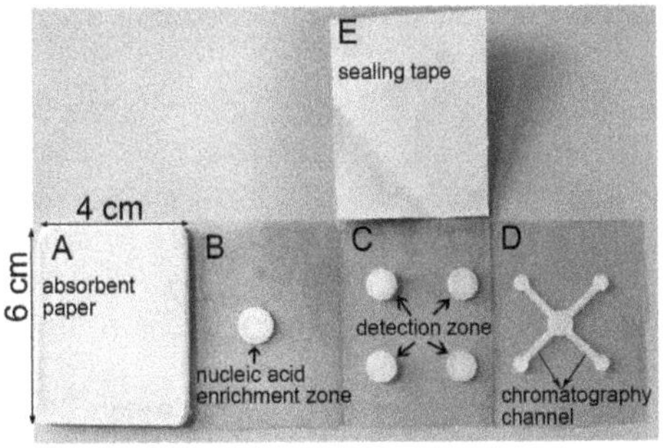


**Figure 1.** Photograph of the LAMP-μPAD. (**A**) Waste liquid zone. (**B**) Nucleic acid enrichment zone. (**C**) Nucleic acid amplification zone. (**D**) Nucleic acid chromatography zone. (**E**) Film sealing tape.

*2.4. Selection of Paper Substrate*

Circular discs of Whatman No. 1 filter paper and NC membrane were precision-cut using a Portrait 3 cutter, and oriented horizontally in 1.5 mL microcentrifuge tubes, respectively. LAMP reaction mixture and DNA template (PEDV cDNA) were dispensed onto each disc, followed by incubation at 60 °C for 40 min. The amplification products were separated from the paper matrix by centrifugation at 13,000× *g* for 1 min. After removing the paper, concentrated products were analyzed by 1% agarose gel electrophoresis. Based on electrophoretic results, the substrate showing optimal amplification efficiency was selected as the base material for the nucleic acid amplification zone in the paper-based chip design.

### 2.5. Optimization of Viral Nucleic Acid Lysis Buffer

The extraction and release of viral nucleic acids play a crucial role in establishing paper-based chip detection methods. To enhance the efficiency of this process, an optimization of the virus lysis buffer was conducted prior to LAMP reaction. The PEDV propagated in Vero cell culture was equally divided into three aliquots, each subjected to lysis using one of three distinct lysis buffers: buffer I (4.5 M guanidine thiocyanate, 50 mM Tris-HCl, 30% Triton X-100, and 1 mg/mL proteinase K), buffer II (6 M guanidine hydrochloride, 1% Tween, and 1% SDS), and buffer III (3 M guanidine hydrochloride and 4 M urea). Subsequently, nucleic acid enrichment was performed using glass fiber membranes.

To evaluate the efficacy of each lysis buffer, the extracted nucleic acids underwent a series of analytical procedures, including RT-PCR, qPCR, and RNA agarose gel electrophoresis. The results obtained from these assays were comprehensively analyzed to determine the optimal virus lysis buffer for the paper-based chip detection system.

### 2.6. Optimization of the Chromatography Channel and Nucleic Acid Amplification Zone Geometry

The optimization of the nucleic acid chromatography channels on the paper chip involved adjusting both their lengths and widths. Initially, various channel lengths (12, 16, 20, 24, 28, and 32 mm) were tested by adding 30 μL of blue ink. The time taken for the ink to fill the channel and the residual ink in the sample well were observed and recorded to determine the most suitable length. Using this optimal length, different channel widths (2, 3, 4, 5, and 6 mm) were then tested with 30 μL of blue ink. The time required for the ink to fill the channel and the residual ink in the well were again observed and recorded to identify the optimal width.

To determine the optimal size of the reaction well, a series of experiments were conducted using wells with varying diameters ranging from 7 mm to 18 mm. In each trial, 30 μL of blue ink was deposited onto the reaction well. The extent of coverage and the amount of residual ink were recorded and analyzed. This systematic approach allowed for the identification of the most suitable reaction well size, balancing complete reagent distribution with minimal waste.

### 2.7. Optimization of Indicator

The selection of an appropriate pH-sensitive indicator is crucial for the visual detection of nucleic acid amplification in paper-based microfluidic chips. Two candidate dyes were evaluated: cresol red, which exhibits a color transition from purple–red to yellow over a pH range of 7.2 to 8.8, and neutral red, which changes from yellow–orange to magenta within a pH range of 6.4 to 8.0. To determine the most suitable indicator, both cresol red and neutral red were independently employed in LAMP amplification reactions, conducted both in solution and on paper-based microfluidic chips. The efficacy of each dye was assessed based on two criteria: (1) the distinctiveness of the color change, and (2) the correlation between the visual results and the amplification products as verified by 1% agarose gel electrophoresis. This comprehensive evaluation enabled the identification of the optimal colorimetric indicator for the LAMP-$\mu$PAD assay.

### 2.8. Detection Procedure of the LAMP-$\mu$PAD

The operational procedure is delineated as follows: (1) Clinical fecal specimens are deposited into 1.5 mL centrifuge tubes, to which 500 μL of 0.9% NaCl ($w/v$) is added. The mixture is centrifuged at 5180$\times$ $g$ for 10 min, after which the supernatant is transferred to a new collection tube. (2) A 200 μL aliquot of the supernatant is transferred to a 1.5 mL centrifuge tube, followed by the addition of 200 μL of lysis buffer. This mixture is incubated at room temperature for 10 min. (3) Module A of the paper-based device is folded beneath Module B. The lysis mixture is then slowly applied to the glass fiber membrane in Module B (nucleic acid enrichment zone) and allowed to incubate for 2–3 min. (4) 200 μL of washing solution A (20 mM Tris-HCl, 40% (*v*/*v*) ethanol, pH 6.6) is added onto nucleic acid enrichment zone in Module B to remove extraneous proteins, with Module A absorbing the waste liquid. (5) 200 μL of washing solution B (50% (*v*/*v*) ethanol, 20 mM NaCl, 2 mM Tris-HCl, pH 7.5) is applied to eliminate high salt concentrations, with Module A again absorbing the waste liquid. (6) Step (5) is repeated, followed by the excision of Module A. (7) RT-LAMP reaction mixture, including reverse transcriptase, Bst polymerase, primers, dNTPs, $Mg^{2+}$ and indicator, are pre-loaded to detection zone in Module C. Following the indicated direction, Module D is folded onto Module C, and Module B is then folded onto Module D. A total of 200 μL of isothermal amplification buffer is applied centrally to elute nucleic acids. (8) Once the reagents have fully permeated the detection zone in Module C, Module B and D are excised, and Module E is folded downward to seal the device. (9) The paper-based microfluidic device is then incubated in a water bath or incubator in preparation for the amplification and subsequent detection. (10) The color change is observed, and images are taken with a smartphone for further analysis.

### 2.9. Image Processing of the LAMP-μPAD Assay

A smartphone camera was utilized to capture amplification results from paper chips, followed by analysis using ImageJ software V1.54i. Images were decomposed into red (R), green (G), and blue (B) channels, with color intensity data extracted from designated detection zones. The colorimetric shift from yellow–orange to magenta was characterized by a significant increase in green intensity and a slight rise in red intensity. Consequently, green intensity served as the primary metric for quantifying amplification progression, with red intensity as a reference. The normalized G/R intensity was adopted as the standard measure throughout the study. All kinetic measurements were performed in quintuplicate ($n$ = 5) under strictly controlled identical conditions to ensure reproducibility, with results presented as mean ± SD.

### 2.10. Determination of Diagnostic Cut-Off Values and Performance Evaluation of the LAMP-μPAD Assay

The threshold cut-off value was evaluated by receiver operator characteristic (ROC) analysis (MedCalc software V23.0.9) using 20 PEDV-positive and 20 negative-PEDV specimens, with RT-PCR as the reference method to classify all specimens.

To determine the optimal reaction kinetics of the LAMP-μPAD assay, amplification efficiency was evaluated using PEDV cDNA as template. The time-dependent analysis was performed at 15-min intervals ranging from 15 to 40 min under isothermal conditions. The optimal reaction time was determined through analysis of colorimetric changes on the devices and 1% agarose gel electrophoresis results. The quantitative analysis of device images using ImageJ software established a correlation between reaction time and the normalized green-to-red intensity ratio (G/R). This relationship was graphically represented, elucidating the kinetics of the RT-LAMP reaction on the paper-based platform.

The optimized assay underwent rigorous specificity and sensitivity assessments to validate its potential as a reliable PEDV diagnostic tool. Specificity was evaluated using cDNA templates from diverse porcine pathogens, including PEDV, PDCoV, PCV2, PRRSV, and TGEV, with nuclease-free water as a negative control. Sensitivity was determined using 10-fold serial dilutions of PEDV cDNA ($4.82 \times 10^7$ to $4.82 \times 10^1$ copies/μL). Both assessments employed a multi-faceted approach, combining visual colorimetric analysis, 1% agarose gel electrophoresis, and quantitative image analysis.

### 2.11. Clinical Implementation of the LAMP-μPAD Assay for PEDV Detection

To validate the diagnostic utility of the LAMP-μPAD assay in clinical settings, we conducted a comparative analysis using 132 archived fecal specimens [24]. These specimens were simultaneously evaluated using both the developed LAMP-μPAD and a reference RT-PCR method (see details in Supplementary Materials). Diagnostic accuracy parameters (sensitivity and specificity) were calculated with their corresponding exact binomial 95% confidence intervals (CIs). The diagnostic agreement between LAMP-μPAD and RT-PCR was assessed using Cohen's kappa coefficient [25].

### 2.12. Statistical Analysis

Statistical analyses were conducted using R version 4.3.3 (https://www.r-project.org/, accessed on 29 February 2024). The correlation between reaction time and G/R intensity were analyzed by using linear regression. Group comparisons were analyzed by a two-sided Student' s *t*-test, and the difference was considered significant when $p$ value < 0.05. Numbers of asterisks mean significant difference (* $p < 0.05$; ** $p < 0.01$; *** $p < 0.001$; **** $p < 0.0001$).

## 3. Results

### 3.1. Development of RT-LAMP Reaction System

Electrophoresis analysis of 11 RT-LAMP primer sets for PEDV detection was performed, revealing that primer set #2 (boxed in red, Figure 2) exhibited optimal performance. This primer set was selected for downstream experiments based on its strong target-specific amplification (lanes 3–4). The primer sequences of primer set #2 were as follows: F3 (forward outer primer): 5′-TCAACAAGTTTCACAAAAACC-3′, B3 (backward outer primer): 5′-ATGGCTTCAAGCAATGCA-3′; FIP (forward inner primer): 5′-CTTCAGGGTTGCACTCATAGAAATTTGCCTAATTTTGAACCTTTCA-3′, BIP (backward inner primer): 5′-TAGGTGCTGACAAGCTGGTGGGTCACATTGTTAAGACACTT-3′ (Figure S1). The optimized reaction time and temperature were determined to be 30 min and 60 °C, respectively. The inner primer concentration was found to be optimal at 32 μM, and the outer primer concentration was determined to be 5 μM for subsequent experiments. The Bst enzyme concentration was optimized at 10,000 U/mL, at which point the band intensity reached its maximum. Furthermore, the optimal concentrations for $Mg^{2+}$, dNTPs, and buffer were established at 80 mM, 10 mM, and 0.8-times dilution, respectively (Figure S1). Based on these findings, a final RT-LAMP reaction system was developed (Figure 3A).

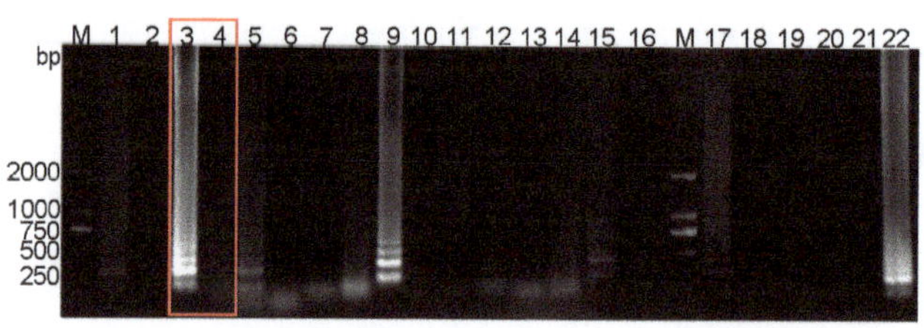


**Figure 2.** Screening of PEDV RT-LAMP primer sets. M for 2000 bp DNA marker, lanes 1 and 2 represent the positive and negative amplification results of primer set 1, respectively; lanes 3 and 4 for primer set 2; lanes 5 and 6 for primer set 3; lanes 7 and 8 for primer set 4; lanes 9 and 10 for primer set 5; lanes 11 and 12 for primer set 6; lanes 13 and 14 for primer set 7; lanes 15 and 16 for primer set 8; lanes 17 and 18 for primer set 9; lanes 19 and 20 for primer set 10; lanes 21 and 22 for primer set 11.

Red rectangular frames highlight primer set 2, which was selected as the optimal primer pair based on specificity and amplification efficiency analyses.

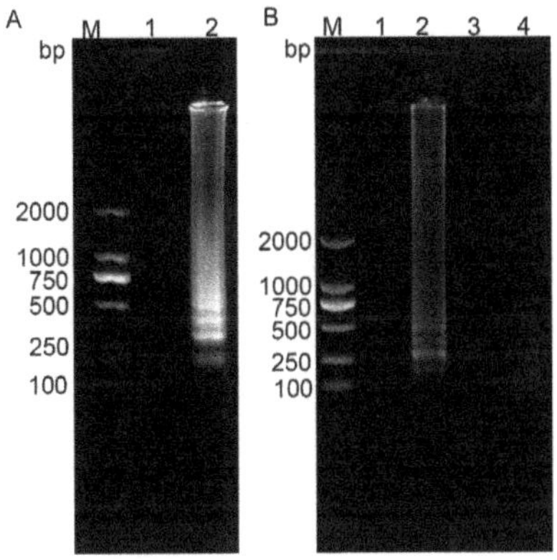


**Figure 3.** Development of RT-LAMP assay and application on different types of paper. (**A**) RT-LAMP for PEDV testing. M: DL2000 DNA marker; 1: negative control ($ddH_2O$); 2: PEDV positive. (**B**) Comparative analysis of paper types for nucleic acid amplification zone. M: DL2000 DNA marker; 1 and 2 represent PEDV negative and positive amplification using RT-LAMP on Whatman No. 1 filter paper, respectively; 3 and 4 represent PEDV negative and positive amplification using RT-LAMP on NC membrane, respectively. ($n$ = 5).

### 3.2. Selection of Paper Substrates

Comparative analyses of LAMP were conducted on two distinct substrates: Whatman No. 1 filter paper and NC membranes. The results demonstrated superior amplification efficiency on Whatman No. 1 filter paper relative to NC membranes, as evidenced by enhanced signal intensity and reaction uniformity (Figure 3B). Based on these findings, Whatman No. 1 filter paper was selected as the optimal substrate for the nucleic acid amplification zone, owing to its balanced hydrophilicity, porosity, and compatibility with LAMP reagent immobilization.

### 3.3. Screening of Nucleic Acid Lysis Buffer

The PEDV solution was divided into three aliquots, each treated with lysis buffer I, II, and III for viral nucleic acid lysis, respectively, followed by enrichment using a glass fiber membrane. The extracted nucleic acids underwent comprehensive analysis via RT-PCR, qPCR, and RNA agarose gel electrophoresis. The results revealed that lysis buffer I produced the highest viral RNA concentration among the three buffers tested, as demonstrated by the most pronounced target bands observed after RT-PCR amplification (Figure 4A,B). Comparative qPCR analysis revealed Ct values of approximately 19, 35, 23, and 18 for viral nucleic acids extracted using lysis buffers I, II, III, and commercial RNA extraction kit, respectively (Figure 4C). RNA gel electrophoresis demonstrated that lysis buffers I and III achieved RNA integrity comparable to the commercial kit, as indicated by distinct 28S and 18S bands (Figure 4D). Despite this, buffer I was selected due to its superior nucleic acid recovery (lower Ct values: ~19 for buffers I vs. ~23 for buffers III), enhanced compatibility with LAMP through optimized detergent/proteinase K composition, and greater reproducibility (smaller inter-replicate variability). Furthermore, lysis buffer I's cost-effectiveness (USD 0.12/sample vs. USD 2.50–5.00 for commercial kits), non-proprietary formulation, and alignment with WHO ASSURED criteria make it particularly suitable for point-of-care applications in resource-limited settings.

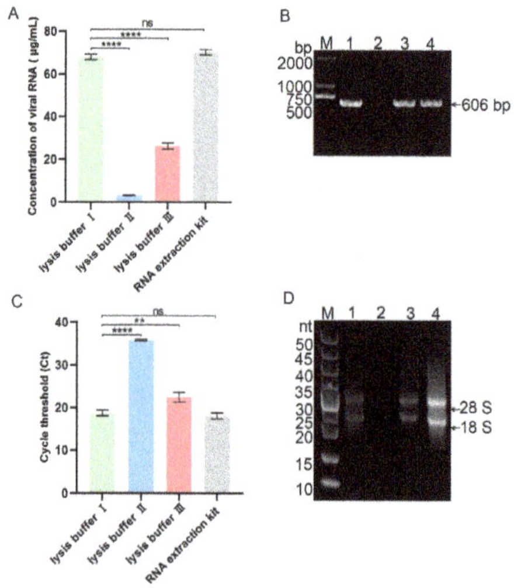


**Figure 4.** Screening of nucleic acid lysis buffers. (**A**) Concentration measurements using ultra-micro nucleic acid protein analyzer. (**B**) RT-PCR results. M: DL2000 DNA marker; Lane 1: Lysis buffer I; Lane 2: Lysis buffer II; Lane 3: Lysis buffer III; Lane 4: RNA extraction kit. (**C**) qPCR results. (**D**) RNA agarose gel electrophoresis results. Lane 1: Lysis buffer I; Lane 2: Lysis buffer II; Lane 3: Lysis buffer III; Lane 4: RNA extraction kit. (* indicates comparison between groups, ns denotes non-significant differences, ** $p < 0.01$, **** $p < 0.0001$, $n = 3$).

### 3.4. Optimization of the Channel and Nucleic Acid Amplification Zone Geometry

For channel length, tests on 12 mm to 32 mm lengths revealed that 16 mm achieved complete fluid penetration in 10 s with minimal residual ink, making it optimal (Figure 5A). Channel width optimization (2 mm to 6 mm) identified 4 mm as ideal, enabling full penetration in 12 s with the least residual ink (Figure 5B). For the dimensions of nucleic acid amplification zone, the results indicated that a chamber diameter of 13 mm provided optimal performance. At this dimension, the blue ink precisely filled the chamber with minimal residual volume (Figure 5C). Consequently, a reaction chamber with a diameter of 13 mm was selected as the optimal configuration for the nucleic acid amplification zone in the paper-based chip design. These optimizations enhanced fluid flow efficiency, reduced mixing, and improved detection sensitivity and accuracy.

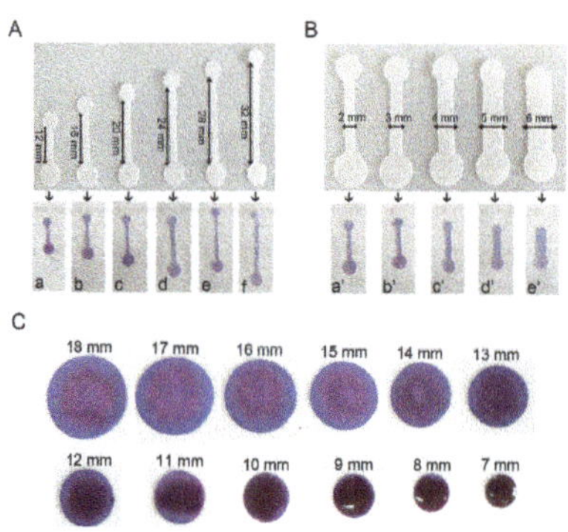


**Figure 5.** Optimization of nucleic acid chromatography channel geometry. (**A**) Channel length optimization (**a**–**f**: 12, 16, 20, 24, 28, and 32 mm, respectively). (**B**) Channel width optimization (**a'**–**e'**: 2, 3, 4, 5, and 6 mm, respectively). (**C**) Dimensional optimization of amplification reaction chambers for nucleic acid detection. ($n = 5$).

### 3.5. Selection of Colorimetric Indicators for the LAMP-μPAD Assay

To enhance the colorimetric detection system, the standard isothermal amplification buffer was replaced with a weak buffer (pH 8.0). Cresol red and neutral red were evaluated as potential indicators for both conventional and paper-based RT-LAMP assays. Assessment criteria included colorimetric transitions and amplicon detection via agarose gel electrophoresis. Both dyes exhibited distinct color changes in solution-based RT-LAMP (Figure 6A,B), with gel electrophoresis confirming specific amplification (Figure 6C). In paper-based assays, while both dyes demonstrated color shifts, cresol red's water solubility led to substrate diffusion, compromising quantitative analysis (Figure 5B). Conversely, neutral red's oil solubility minimized diffusion while maintaining clear color transitions (Figure 5A). Consequently, neutral red was selected as the optimal indicator for paper-based RT-LAMP, offering superior stability and visual clarity.

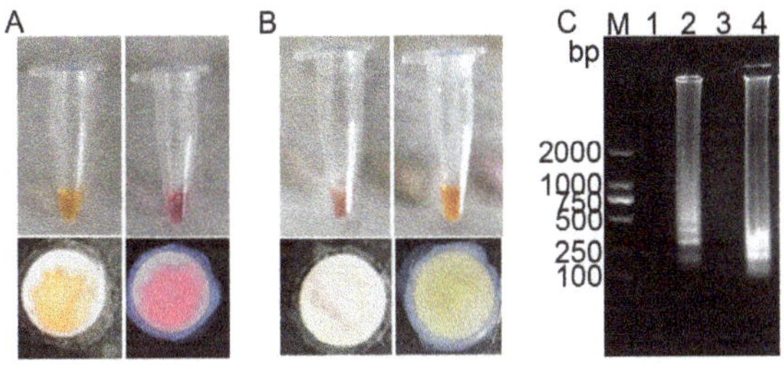


**Figure 6.** Comparison of colorimetric indicators. (**A**) Neutral red and (**B**) cresol red color transitions before and after LAMP amplification. (**C**) Agarose gel electrophoresis analysis. Lane M: DL2000 DNA marker; Lanes 1 and 2: electrophoresis results of neutral red before and after LAMP amplification, respectively; Lanes 3 and 4: electrophoresis results of cresol red before and after LAMP amplification, respectively ($n = 5$).

### 3.6. Cut-Off Value and Diagnostic Performance Evaluation of the LAMP-μPAD Assay

A total of 20 clinically validated PEDV-positive and 20 PEDV-negative fecal specimens were subjected to both LAMP-μPAD and RT-PCR assays. The results demonstrated a significant difference in the G/R intensity between PEDV-positive and negative specimens ($p < 0.001$). Based on the ROC curve analysis of the LAMP-μPAD, the G/R intensity of the 20 PEDV-positive specimens varied from a minimum of 0.666 to a maximum of 0.811, and from 0.796 to 0.908 for the 20 PEDV-negative specimens. The G/R intensity of 0.795 was selected as the optimal cut-off value, with an area under the curve (AUC) of 0.995 ± 0.00607 and 95% confidence interval (CI) ranging from 0.902 to 1.000 (Figure 7).

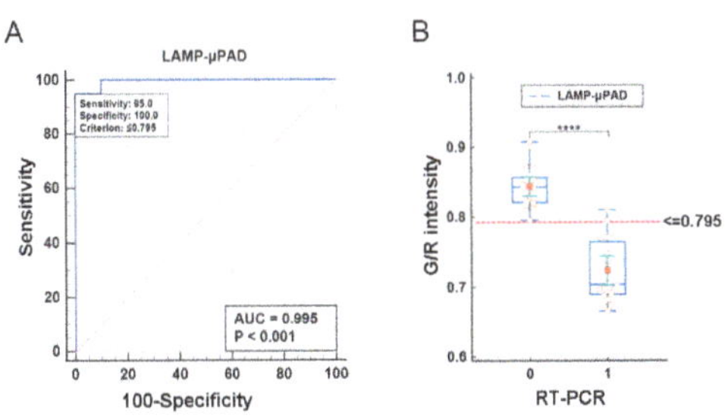


**Figure 7.** Cut-off value and diagnostic performance evaluation of the LAMP-μPAD assay. (**A**) Determination of Cut-off value. (**B**) ROC curve for diagnostic performance evaluation. (**** $p < 0.0001$).

### 3.7. Reaction Time for the LAMP-μPAD Assay

Following the previously optimized conditions, we sought to determine the optimal reaction time for the LAMP-μPAD assay. As illustrated in Figure 8A, a distinct color transition from yellow–orange to magenta was observed in the positive specimens at the 30-min mark, while the negative controls remained unchanged. Corroborating these visual results, gel electrophoresis analysis revealed clearly discernible bands at 30 min (Figure 8B). Furthermore, a colorimetric response intensity analysis using ImageJ software demonstrated a significant disparity in the G/R intensity between positive and negative specimens at 30 min. Notably, the G/R intensity of positive specimens exhibited a linear correlation with reaction time (Figure 8C). These findings indicate that amplification commenced at approximately 25 min, with a pronounced color change manifesting at 30 min. Consequently, the optimal reaction time for the paper-based assay was determined to be 30 min.

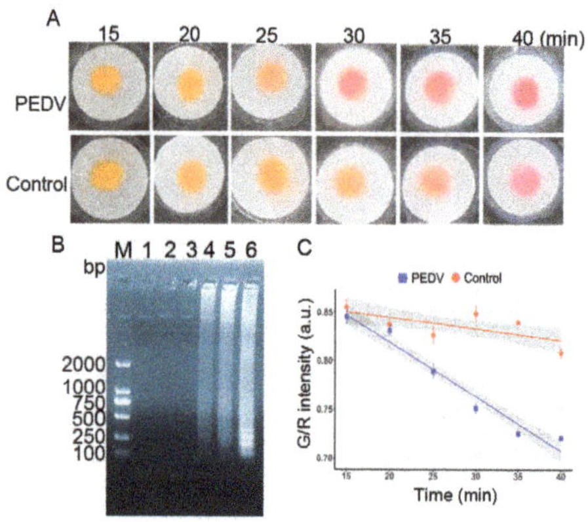


**Figure 8.** Time-dependent analysis of the LAMP-μPAD assay. (**A**) Colorimetric results on paper chips for PEDV and negative control. (**B**) Time course analysis by agarose gel electrophoresis. M: DL2000 DNA marker; Lanes 1–6 represent 15, 20, 25, 30, 35, and 40 min, respectively. (**C**) Regression analysis of G/R intensity versus reaction time (error bars represent standard deviation from five replicates for each time point) ($n$ = 5).

### 3.8. Specificity and Sensitivity of the LAMP-μPAD Assay

As illustrated in Figure 9A, a distinct color transition from yellow–orange to magenta was observed exclusively for the PEDV-positive specimen. The paper chips for other viruses and the negative control maintained their original yellow–orange hue. Corroborating these findings, the agarose gel electrophoresis results revealed characteristic ladder-like bands only in the PEDV lane, while other lanes remained devoid of amplification products (Figure 9B). Further quantitative analysis of the G/R intensity using ImageJ software demonstrated that the PEDV-positive specimen had a significantly lower G/R intensity compared to other viruses and the negative control (Figure 9C). This marked difference in colorimetric intensity further substantiates the specificity of the assay. Collectively, these results provide compelling evidence for the high specificity of the developed LAMP-μPAD method for PEDV detection.

The sensitivity of the LAMP-μPAD assay was evaluated through colorimetric analysis and gel electrophoresis. As illustrated in Figure 10A, a distinct chromatic shift from yellow–orange to magenta was observed at PEDV cDNA concentration threshold of $4.82 \times 10^2$ copies/μL. This color transition ceased at $4.82 \times 10^1$ copies/μL, a finding that was corroborated by agarose gel electrophoresis results (Figure 10B). These findings indicate that the limit of detection for the paper-based assay was $4.82 \times 10^2$ copies/μL. Further quantitative analysis using ImageJ software unveiled an inverse relationship between the G/R intensity and template concentration. A progressive decrease in the G/R intensity was observed with increasing cDNA concentrations, demonstrating a linear correlation (Figure 10C).

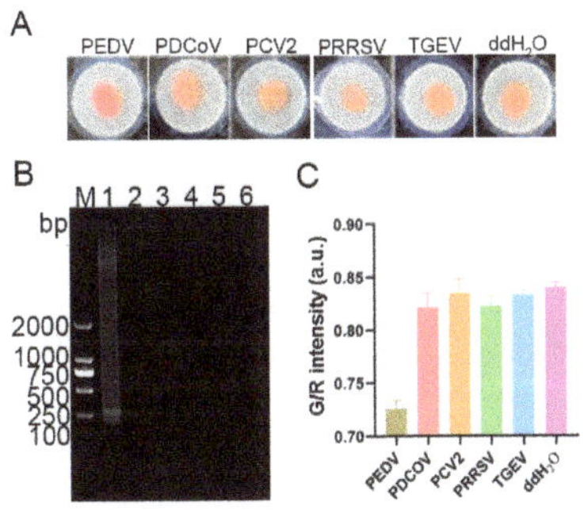


**Figure 9.** Specificity analysis of the LAMP-$\mu$PAD assay. (**A**) Visual colorimetric detection results on paper chips. (**B**) Agarose gel electrophoresis analysis. M: DL2000 DNA marker; Lanes 1–6 represents PEDV, PDCoV, PCV2, PRRSV, TGEV and $ddH_2O$, respectively. (**C**) Quantitative analysis of colorimetric intensity (G/R).

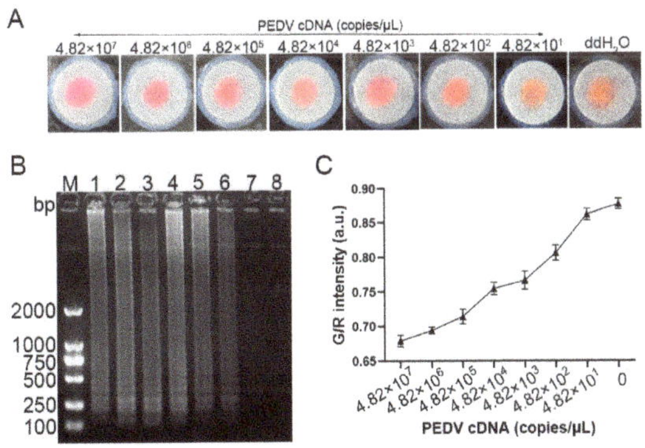


**Figure 10.** Sensitivity evaluation of the LAMP-$\mu$PAD assay. (**A**) Visual detection results on the paper chips at different template concentrations. (**B**) Agarose gel electrophoresis analysis. M: DL2000 DNA marker; Lanes 1–7: Serial dilutions of template from $4.82 \times 10^7$ to $4.82 \times 10^1$ copies/µL; Lane 8: negative control ($ddH_2O$). (**C**) G/R ratio changes across plasmid concentration gradients. ($n = 5$).

### *3.9. Application of the LAMP-µPAD in Clinical Validation*

A total of 132 clinical specimens were analyzed using both RT-PCR and the LAMP-$\mu$PAD assay. RT-PCR analysis identified 43 PEDV-positive and 89 PEDV-negative specimens, corresponding to a positivity rate of 32.6% (43/132). The LAMP-$\mu$PAD detected 41 PEDV-positive and 91 PEDV-negative specimens, yielding a positivity rate of 31.1% (41/132). Among these results, 41 specimens were concordantly positive by both methods. However, two specimens that positive tested by RT-PCR yielded negative results with the LAMP-$\mu$PAD. All 89 specimens that tested negative by RT-PCR were also negative by LAMP-$\mu$PAD. Compared with RT-PCR, the specificity and sensitivity of the LAMP-$\mu$PAD were 100% (95% CI: 91.8–100%) and 95.3% (95% CI: 86.2–98.8%), respectively, with a concordance rate of 98.5% and a kappa value of 0.965 (Table 1). These findings indicate that the LAMP-$\mu$PAD assay developed in this study possesses good clinical detection capabilities and can be employed for PEDV clinical testing.

**Table 1.** Comparison between the LAMP-$\mu$PAD and RT-PCR for detecting PEDV.

| | | RT-PCR | | Total |
|---|---|---|---|---|
| | | **Positive** | **Negative** | |
| LAMP-$\mu$PAD | Positive | 41 | 0 | 41 |
| | Negative | 2 | 89 | 91 |
| | Total | 43 | 89 | 132 |

## 4. Discussion

PEDV has emerged as a significant viral pathogen causing severe diarrheal disease in neonatal piglets. The global dissemination of PEDV in swine populations has inflicted considerable economic losses, fundamentally disrupting the swine industry's development [26,27]. Although vaccination strategies have been implemented, recurring PED outbreaks underscore the urgent demand for rapid, sensitive diagnostic platforms. Recent advances in isothermal amplification techniques, lateral flow assays, and microfluidic platforms have revolutionized PEDV detection strategies [28,29]. These innovative diagnostic approaches, characterized by rapid turnaround times, high analytical sensitivity, exceptional specificity, and visual readout capabilities, establish a compelling framework for advancing clinical PEDV diagnostics.

Paper-based microfluidics offers advantages over traditional microfluidic systems by addressing challenges like sample processing and fluid control. Paper serves as an ideal material for transitioning from lab to consumer use, enabling liquid transport through capillary action without pumps. This technology aligns with WHO's ASSURED criteria for point-of-care diagnostics by being affordable, sensitive, specific, user-friendly, rapid, equipment-free, and deliverable to end-users [30]. The selection of paper substrates for $\mu$PAD critically influences their performance in terms of fluid flow dynamics, reagent compatibility, and detection sensitivity. Optimal paper choice requires balancing key parameters such as porosity, thickness, and material composition—for instance, cellulose for cost-effectiveness or nitrocellulose for enhanced protein-binding capacity. Compatibility with chemical reagents and fabrication techniques (e.g., wax patterning, precision cutting, or laser ablation) must be carefully evaluated. The inherent hydrophilicity of cellulose-based materials facilitates capillary-driven flow, while surface functionalization through hydroxyl groups enables covalent reagent immobilization. Cost-effective options like Whatman No. 1 filter paper are preferred for general applications, whereas nitrocellulose membranes provide high analytical sensitivity, and glass fiber substrates enable rapid fluid transport [31,32]. In this study, we engineered a 3D foldable paper-based microfluidic chip that leverages glass fiber to concentrate viral nucleic acids. We utilized Whatman No. 1 filter paper as both the chromatographic channel and the substrate for the LAMP amplification reaction, achieving excellent detection performance for PEDV. The extraction and enrichment of viral nucleic acid are pivotal steps in the design of $\mu$PAD that enhance the efficiency, sensitivity, and reliability of pathogen detection in various applications, particularly in low-resource or acute settings where rapid results are essential [33,34]. To enable streamlined sample pretreatment in point-of-care testing (POCT) applications, we developed a novel paper-based lysis and enrichment method that facilitates rapid nucleic acid extraction directly from fecal specimens. Through systematic optimization of nucleic acid lysis buffer formulations, we enhanced both RNA yield and extraction stability. The optimized lysis buffer I (4.5 M guanidine thiocyanate, 50 mM Tris-HCl, 30% Triton X-100, and 1 mg/mL proteinase K) demonstrated comparable performance to commercial RNA extraction kits in processing viral targets while maintaining lower cost and non-toxic composition—a critical advantage for decentralized POC diagnostic implementations. This integrated pretreatment method demonstrated exceptional sensitivity ($4.82 \times 10^2$ copies/µL) in downstream LAMP-$\mu$PAD assays, matching the detection limits of conventional column-based purification systems.

The geometric configuration of $\mu$PAD chromatographic channels critically determines detection performance. Narrower channels impede fluid flow due to inadequate capillary forces, whereas wider channels enhance flow dynamics at the expense of analyte mixing [35]. Channel length directly modulates detection time, with extended lengths delaying signal acquisition [28]. Advanced fabrication techniques (e.g., precision cutting, hot-pressing) enable optimized detection zone architectures that improve sensitivity and reproducibility. Mitigating evaporation through reduced exposed surface area and enhancing structural integrity collectively ensure operational reliability [36]. In the present study, we systematically optimized the channel length, width, and reaction chamber size of paper-based microfluidic chips to improve detection performance. These optimizations enhanced fluid flow efficiency, reduced mixing, and improved detection sensitivity and accuracy. When compared with a recently reported 3D-printed RT-LAMP microfluidic chip for multiplex detection of PEDV, TGEV and PDCoV [29], our LAMP-$\mu$PAD platform shows lower sensitivity ($4.82 \times 10^2$ copies/µL vs. $10^1$ copies/reaction for PEDV), but it offers significant practical advantages over photopolymer resin-based 3D-printed chips, particularly in terms of simplified fabrication processes and substantially lower production costs.

One limitation of our study is that we did not extensively investigate the long-term stability and durability of the LAMP-$\mu$PAD assay under varying environmental conditions and various biological samples (tap water, feed, urine, and blood plasma samples, etc.), which could affect its practical utility. Future research should address this limitation to further enhance the robustness and applicability of this diagnostic tool. Nevertheless, to the best knowledge, this investigation represents the first reported application in the literature of a paper-based sensor for PEDV test. It provides an innovative technological framework for rapid and effective on-site diagnosis, thereby enhancing timely disease management and control. Notably, distinct banding patterns were observed across different substrates (Whatman No. 1, NC membrane, and standard RT-LAMP system), suggesting substrate-dependent effects on amplification efficiency. These variations likely stem from a complex interplay of: (1) substrate-specific nucleic acid retention characteristics that modulate template availability and local concentration gradients during amplification; (2) differential non-covalent interactions between paper matrices and DNA fragments (particularly during elution) that may sterically hinder polymerase access or promote alternative priming; (3) competitive binding of reaction components to cellulose fibers that alter effective primer/template ratios; and (4) substrate-dependent formation of primer-dimers or other amplification byproducts due to microenvironmental effects on reaction kinetics, as evidenced by the characteristic ~100 bp band appearing only in specific substrate conditions. These phenomena collectively demonstrate how physical substrate properties can fundamentally influence LAMP reaction fidelity and efficiency through both direct chemical interactions and spatial organization of reaction components. In the future, systematic optimization of substrate material properties (e.g., pore size, surface charge, and chemical modifications) should be conducted to mitigate these matrix effects while maintaining field-deployable simplicity. Additionally, real-time monitoring of amplification kinetics on different substrates could help elucidate the precise mechanisms underlying these substrate-dependent performance variations, enabling rational design of next-generation µPAD platforms.

## 5. Conclusions

In summary, this study establishes a novel paper-based LAMP platform for rapid and precise detection of PEDV. The platform synergistically integrates nucleic acid enrichment, amplification, and visual lateral flow detection, identifying PEDV in under 30 min. This cost-effective and user-friendly μPAD addresses the urgent need for PEDV on-site diagnostics in resource-limited settings, offering a practical alternative to traditional methods. The design's flexibility also hints at detecting other pathogens, expanding its value in veterinary and public health diagnostics.

**Supplementary Materials:** The following supporting information can be downloaded at: https://www.mdpi.com/article/10.3390/vetsci12050427/s1, Figure S1: optimization of the PEDV RT-LAMP reaction system. (**A**) Reaction time (lanes 1–8 correspond to 10, 20, 30, 40, 50, 60, 70, and 80 min, respectively); (**B**) reaction temperature (lanes 1–8 correspond to 54, 56, 58, 60, 62, 64, 66, and 68 °C, respectively); (**C**) $Mg^{2+}$ concentration (lanes 1–8 correspond to 0, 20, 40, 60, 80, 100, 120, and 140 mM, respectively); (**D**) inner primer concentration (lanes 1–8 correspond to 8, 16, 24, 32, 40, 48, 56, and 64 μM, respectively); (**E**) outer primer concentration (lanes 1–8 correspond to 1, 2, 3, 4, 5, 6, 7, and 8 μM, respectively); (**F**) dNTPs concentration (lanes 1–8 correspond to 4, 5.5, 7, 8.5, 10, 11.5, 13, and 14.5 mM, respectively); (**G**) Bst enzyme concentration (lanes 1–8 correspond to 1000, 2000, 4000, 6000, 8000, 10,000, 12,000, and 14,000 U/mL, respectively); (**H**) buffer dilution (lanes 1–8 correspond to 0.4, 0.6, 0.8, 1.0, 1.2, 1.4, 1.6, and 1.8-times dilution, respectively); Table S1: the information of the primers used for RT-LAMP.

**Author Contributions:** Conceptualization, R.L., Z.W. and F.W.; methodology, X.T., W.C., J.Y. and L.L.; software, L.L. and J.J.; validation, Y.Y.; formal analysis, Y.Z.; investigation, R.L. and J.Y.; data curation, X.T.; writing—original draft preparation, R.L.; writing—review and editing, R.L. and F.W.; supervision, Z.W. and F.W.; funding acquisition, Z.W. All authors have read and agreed to the published version of the manuscript.

**Funding:** This research was funded by the Key Science and Technology Program of Henan Province (Grant Nos. 242102111010 and 232102310011).

**Institutional Review Board Statement:** Not applicable.

**Informed Consent Statement:** Not applicable.

**Data Availability Statement:** Data are contained within the article and Supplementary Materials.

**Acknowledgments:** The authors would like to thank Qin Baoliang from Animal Disease Prevention and Control Center of Xinxiang City for providing swine fecal specimens.

**Conflicts of Interest:** The authors declare no conflicts of interest.

## References

1. Pensaert Mb Fau-de Bouck, P.; de Bouck, P. A New Coronavirus-like Particle Associated with Diarrhea in Swine. *Arch. Virol.* **1978**, *58*, 243–247. [CrossRef] [PubMed]
2. Luo, H.; Liang, Z.; Lin, J.; Wang, Y.; Liu, Y.; Mei, K.; Zhao, M.; Huang, S. Research progress of porcine epidemic diarrhea virus S protein. *Front. Microbiol.* **2024**, *15*, 1396894. [CrossRef]
3. Wood, E.N. An apparently new syndrome of porcine epidemic diarrhoea. *Vet. Rec.* **1977**, *100*, 243–244. [CrossRef]
4. Zhang, Y.; Chen, Y.; Zhou, J.; Wang, X.; Ma, L.; Li, J.; Yang, L.; Yuan, H.; Pang, D.; Ouyang, H. Porcine Epidemic Diarrhea Virus: An Updated Overview of Virus Epidemiology, Virulence Variation Patterns and Virus–Host Interactions. *Viruses* **2022**, *14*, 2434. [CrossRef] [PubMed]
5. Fu, Y.; Li, B.; Liu, G. Rapid and efficient detection methods of pathogenic swine enteric coronaviruses. *Appl. Microbiol. Biotechnol.* **2020**, *104*, 6091–6100. [CrossRef]
6. Sun, Y.; Gong, T.; Wu, D.; Feng, Y.; Gao, Q.; Xing, J.; Zheng, X.; Song, Z.; Liu, X.; Chen, X.; et al. Isolation, Identification, and Pathogenicity of Porcine Epidemic Diarrhea Virus. *Front. Microbiol.* **2023**, *14*, 1273589. [CrossRef]
7. Olech, M. Current State of Molecular and Serological Methods for Detection of Porcine Epidemic Diarrhea Virus. *Pathogens* **2022**, *11*, 1074. [CrossRef]
8. Padzil, F.; Mariatulqabtiah, A.R.; Tan, W.S.; Ho, K.L.; Isa, N.M.; Lau, H.Y.; Abu, J.; Chuang, K.-P. Loop-Mediated Isothermal Amplification (LAMP) as a Promising Point-of-Care Diagnostic Strategy in Avian Virus Research. *Animals* **2021**, *12*, 76. [CrossRef] [PubMed]

9. Fang, M.; Wang, Y.; Yang, T.A.-O.; Zhang, J.; Yu, H.; Luo, Z.A.-O.; Su, B.A.-O.; Lin, X.A.-O. Nucleic acid plate culture: Label-free and naked-eye-based digital loop-mediated isothermal amplification in hydrogel with machine learning. *ACS Sens.* **2024**, *9*, 2010–2019. [CrossRef]
10. Seyrig, G.; Stedtfeld, R.D.; Tourlousse, D.M.; Ahmad, F.; Towery, K.; Cupples, A.M.; Tiedje, J.M.; Hashsham, S.A. Selection of fluorescent DNA dyes for real-time LAMP with portable and simple optics. *J. Microbiol. Methods* **2015**, *119*, 223–227. [CrossRef]
11. Xiong, J.A.-O.; Huang, B.; Xu, J.S.; Huang, W.S. A closed-tube loop-mediated isothermal amplification assay for the visual detection of staphylococcus aureus. *Appl. Biochem. Biotechnol.* **2020**, *191*, 201–211. [CrossRef]
12. Allgöwer, S.M.; Hartmann, C.A.; Lipinski, C.; Mahler, V.; Randow, S.; Völker, E.; Holzhauser, T.A.-O. LAMP-LFD Based on Isothermal Amplification of Multicopy Gene Orf160b: Applicability for Highly Sensitive Low-Tech Screening of Allergenic Soybean (Glycine Max) in Food. *Foods* **2020**, *9*, 1741. [CrossRef] [PubMed]
13. Quyen, T.L.; Vinayaka, A.A.-O.; Golabi, M.; Nguyen, T.; Ngoc, H.V.; Bang, D.D.; Wolff, A.A.-O. Multiplexed detection of pathogens using solid-phase loop-mediated isothermal amplification on a supercritical angle fluorescence array for point-of-care applications. *ACS Sens.* **2022**, *7*, 3343–3351. [CrossRef]
14. Choopara, I.; Suea-Ngam, A.A.-O.; Teethaisong, Y.; Howes, P.A.-O.; Schmelcher, M.; Leelahavanichkul, A.; Thunyaharn, S.; Wongsawaeng, D.; deMello, A.J.; Dean, D.; et al. Fluorometric paper-based, loop-mediated isothermal amplification devices for quantitative point-of-care detection of methicillin-resistant staphylococcus aureus (MRSA). *ACS Sens.* **2021**, *6*, 742–751. [CrossRef] [PubMed]
15. Tong, X.; Lin, X.A.-O.; Duan, N.; Wang, Z.; Wu, S.A.-O. Laser-printed paper-based microfluidic chip based on a multicolor fluorescence carbon dot biosensor for visual determination of multiantibiotics in aquatic products. *ACS Sens.* **2022**, *7*, 3947–3955. [CrossRef]
16. Chaumpluk, P.; Suriyasomboon, A. A Simple Paper-Based Lab-on-a-Chip for the Detection of a Highly Pathogenic Strain of Porcine Reproductive and Respiratory Syndrome Virus. *Aust. J. Chem.* **2014**, *67*, 1434–1440. [CrossRef]
17. Huang, L.; Xiao, W.; Xu, T.; Chen, H.; Jin, Z.; Zhang, Z.; Song, Q.; Tang, Y. Miniaturized paper-based smartphone biosensor for differential diagnosis of wild-type pseudorabies virus infection versus vaccination immunization. *Sens. Actuators B-Chem.* **2021**, *327*, 128893. [CrossRef]
18. Rodriguez, N.M.; Wong, W.S.; Liu, L.; Dewar, R.; Klapperich, C.M. A Fully Integrated Paperfluidic Molecular Diagnostic Chip for the Extraction, Amplification, and Detection of Nucleic Acids from Clinical Samples. *Lab Chip* **2016**, *16*, 753–763. [CrossRef] [PubMed]
19. Kim, S.; Akarapipad, P.; Nguyen, B.T.; Breshears, L.E.; Sosnowski, K.; Baker, J.; Uhrlaub, J.L.; Nikolich-Žugich, J.; Yoon, J.-Y. Direct Capture and Smartphone Quantification of Airborne SARS-CoV-2 on a Paper Microfluidic Chip. *Biosens. Bioelectron.* **2022**, *200*, 113912. [CrossRef]
20. Yin, K.; Ding, X.; Li, Z.; Sfeir, M.M.; Ballesteros, E.; Liu, C. Autonomous Lab-on-Paper for Multiplexed, CRISPR-Based Diagnostics of SARS-CoV-2. *Lab Chip* **2021**, *21*, 2730–2737. [CrossRef]
21. Kaarj, K.; Akarapipad, P.; Yoon, J.-Y. Simpler, faster, and sensitive Zika virus assay using smartphone detection of loop-mediated isothermal amplification on paper microfluidic chips. *Sci. Rep.* **2018**, *8*, 12438. [CrossRef] [PubMed]
22. Kamel, M.; Davidson, J.L.; Schober, J.M.; Fraley, G.S.; Verma, M.S. A Paper-Based Loop-Mediated Isothermal Amplification Assay for Highly Pathogenic Avian Influenza. *Sci. Rep.* **2025**, *15*, 12110. [CrossRef]
23. Dong, T.; Qin, L.; Wang, Z.; Fan, C.; Shen, C.; Feng, P.; Kong, Q.; Ke, B.; Ying, B.; Li, F. Point-of-Care Diagnosis of Tuberculosis Using a Portable Nucleic Acid Test with Distance-Based Readout. *Anal. Chem.* **2024**, *96*, 20204–20212. [CrossRef]
24. Li, R.; Cao, W.; Yuan, J.; Li, L.; Zhou, Y.; Wang, F.; Wang, Z.; Tian, X. Development of a Visual Detection Method of Porcine Deltacoronavirus Using Loop-Mediated Isothermal Amplification. *Front. Microbiol.* **2024**, *15*, 1465923. [CrossRef] [PubMed]
25. Landis, J.R.; Koch, G.G. The Measurement of Observer Agreement for Categorical Data. *Biometrics* **1977**, *33*, 159–174. [CrossRef] [PubMed]
26. Zhang, F.; Luo, Y.; Lin, C.; Tan, M.; Wan, P.; Xie, B.; Xiong, L.; Ji, H. Epidemiological Monitoring and Genetic Variation Analysis of Pathogens Associated with Porcine Viral Diarrhea in Southern China from 2021 to 2023. *Front. Microbiol.* **2024**, *15*, 1303915. [CrossRef]
27. Niederwerder, M.C.; Hesse, R.A. Swine Enteric Coronavirus Disease: A Review of 4 Years with Porcine Epidemic Diarrhoea Virus and Porcine Deltacoronavirus in the United States and Canada. *Transbound. Emerg. Dis.* **2018**, *65*, 660–675. [CrossRef]
28. Ma, L.; Lian, K.; Zhu, M.; Tang, Y.; Zhang, M. Visual Detection of Porcine Epidemic Diarrhea Virus by Recombinase Polymerase Amplification Combined with Lateral Flow Dipstrip. *BMC Vet. Res.* **2022**, *18*, 140. [CrossRef]
29. El-Tholoth, M.; Bai, H.; Mauk, M.G.; Saif, L.; Bau, H.H. A Portable, 3D Printed, Microfluidic Device for Multiplexed, Real Time, Molecular Detection of the Porcine Epidemic Diarrhea Virus, Transmissible Gastroenteritis Virus, and Porcine Deltacoronavirus at the Point of Need. *Lab Chip* **2021**, *21*, 1118–1130. [CrossRef]
30. Anushka; Bandopadhyay, A.; Das, P.K. Paper Based Microfluidic Devices: A Review of Fabrication Techniques and Applications. *Eur. Phys. J. Spec. Top.* **2023**, *232*, 781–815. [CrossRef]

31. Tong, X.; Ga, L.; Zhao, R.; Ai, J. Research Progress on the Applications of Paper Chips. *RSC Adv.* **2021**, *11*, 8793–8820. [CrossRef] [PubMed]
32. Noviana, E.; Ozer, T.; Carrell, C.S.; Link, J.S.; McMahon, C.; Jang, I.; Henry, C.S. Microfluidic Paper-Based Analytical Devices: From Design to Applications. *Chem. Rev.* **2021**, *121*, 11835–11885. [CrossRef]
33. Reynolds, J.; Loeffler, R.S.; Leigh, P.J.; Lopez, H.A.; Yoon, J.-Y. Recent Uses of Paper Microfluidics in Isothermal Nucleic Acid Amplification Tests. *Biosensors* **2023**, *13*, 885. [CrossRef] [PubMed]
34. Mei, J.; Wang, D.; Zhang, Y.; Wu, D.; Cui, J.; Gan, M.; Liu, P. Portable Paper-based Nucleic Acid Enrichment for Field Testing. *Adv. Sci.* **2023**, *10*, 2205217. [CrossRef] [PubMed]
35. Thuo, M.M.; Martinez, R.V.; Lan, W.-J.; Liu, X.; Barber, J.; Atkinson, M.B.J.; Bandarage, D.; Bloch, J.-F.; Whitesides, G.M. Fabrication of Low-Cost Paper-Based Microfluidic Devices by Embossing or Cut-and-Stack Methods. *Chem. Mater.* **2014**, *26*, 4230–4237. [CrossRef]
36. Kumawat, N.; Soman, S.S.; Vijayavenkataraman, S.; Kumar, S. Rapid and Inexpensive Process to Fabricate Paper Based Microfluidic Devices Using a Cut and Heat Plastic Lamination Process. *Lab Chip* **2022**, *22*, 3377–3389. [CrossRef]

*veterinary sciences*

MDPI

*Article*

# Molecular Epidemic Characteristics and Genetic Evolution of Porcine Circovirus Type 2 in Henan, China

**Zhifeng Peng [1], Huifang Lv [1], Han Zhang [1], Li Zhao [1], Huawei Li [2], Yanyu He [3], Kangdi Zhao [1,4], Hongxing Qiao [1], Yuzhen Song [1,*] and Chuanzhou Bian [1,*]**

[1] College of Veterinary Medicine, Henan University of Animal Husbandry and Economy, Zhengzhou 450046, China; zfp2017@hnuahe.edu.cn (Z.P.); 80980@hnuahe.edu.cn (H.L.); zhanghan.zz@hotmail.com (H.Z.); 80289@hnuahe.edu.cn (L.Z.); kd18638621887@163.com (K.Z.); 80414@hnuahe.edu.cn (H.Q.)

[2] Institute of Animal Product Quality and Safety Technology, Henan University of Animal Husbandry and Economy, Zhengzhou 450046, China; 80013@hnuahe.edu.cn

[3] The School of Food Technology and Nature Science, Massey University, Palmerston North 4410, New Zealand; y.he@massey.ac.nz

[4] College of Marine Sciences, South China Agriculture University, Guangzhou 510642, China

* Correspondence: 80152@hnuahe.edu.cn (Y.S.); 80305@hnuahe.edu.cn (C.B.)

**Simple Summary:** This study provides an updated overview of the porcine circovirus 2 (PCV2) situation in China following the African swine fever outbreak. The molecular detection rate at population level was 71.17%. The molecular detection rates in 2020, 2021, 2022 and 2023 were 81.16%, 72.41%, 62.50% and 53.49%, respectively. Phylogenetic clustering patterns based on the origin year of the samples were observed, suggesting that nucleotide mutations occur over time. The PCV2d genotype is the dominant genotype in China. The substitution rate for the PCV2 ORF2 sequence in this study was estimated at $1.098 \times 10^{-3}$ substitutions/site/year (ssy), aligning with the high substitution rate expected for PCV2 strains. These findings provide valuable insights into the emergence and evolution of PCV2, and genetic diversity of PCV2, enhancing our understanding of its evolutionary dynamics.

**Abstract:** The widespread distribution and genetic diversity of porcine circovirus type 2 (PCV2) seriously threatens the swine industry worldwide. This study investigates the molecular epidemiology of PCV2 in Henan Province (2020–2023) through PCR screening (385 samples) and whole-genome sequencing (34 strains). The overall detection rate was 71.17% (274/385), with annual rates of 81.16% (112/138) in 2020, 72.41% (84/116) in 2021, 62.50% (55/88) in 2022, and 53.49% (23/43) in 2023, indicating a declining trend. Phylogenetic analysis revealed the dominance of the PCV2d genotype, comprising 82.4% (28/34) of sequenced strains. Evolutionary analysis identified strong negative selection pressure on ORF2, with an elevated substitution rate of $1.098 \times 10^{-3}$ ssy. These findings provide critical insights into the predominance and adaptive evolution of PCV2d, and significantly improve our understanding of its genetic diversity and evolutionary dynamics.

**Keywords:** pigs; porcine circovirus type 2; epidemiology; complete genome; genetic diversity

## 1. Introduction

Porcine circovirus-associated diseases (PCVADs) are one of major threats to the swine industry worldwide, causing huge economic losses each year [1–7]. PCVADs are characterized by immunosuppression and diverse clinical manifestations, including postweaning multisystemic wasting syndrome (PMWS), porcine respiratory disease complex (PRDC), porcine dermatitis and nephropathy syndrome (PDNS), and reproductive disorders in sows [8,9]. These diseases pose serious challenges to the sustainable development of the swine industry worldwide. Key risk factors driving PCV2 transmission include high-density farming practices, suboptimal biosecurity protocols, frequent cross-regional pig movements, and co-infections with pathogens such as PRRSV and *Mycoplasma hyopneumoniae* [10–12]. Additionally, incomplete vaccination coverage and viral mutations in immunodominant epitopes further exacerbate viral persistence and spread [11].

The etiologic agent PCV2 is a member of family *Circoviridae*, genus *Circovirus* [13]. To date, nine PCV2 genotypes (PCV2a–PCV2i) have been identified [6]. Epidemiological investigations indicate that prior to 2008, circulating PCV2 strains were predominantly restricted to three genotypes (PCV2a–PCV2c). Among these, PCV2c was initially isolated from asymptomatic porcine sera in Denmark, and later in Brazil, China, and the Netherlands [14–16]. PCV2d, identified in 2010, has dominated globally since 2012 [17]. Recent studies further characterize novel genotypes (PCV2e–PCV2i), underscoring ongoing viral diversification [6,18,19]. The genome of PCV2 is predicted to contain 11 open reading frames (ORFs), with ORF1 and ORF2 being the primary functional ORFs [20]. The ORF1 encodes replication-related proteins [21], while the ORF2 encodes the capsid protein (Cap), which facilitates the viral entry and induces the production of neutralizing antibodies [22–24]. ORF2 is highly variable and is commonly used in phylogenetic analyses of PCV2 [2,25,26].

PCVADs result in 4–20% mortality in affected herds and reduce average daily weight gain by 15–30% in pigs. Furthermore, in breeding herds, PCV2 infection increases stillbirth rates ($\leq$40%) and reduces litter sizes, with annual losses of GBP 5–10 per sow in the EU [27]. Since its first report in China, PCV2 has become an important pathogen affecting the country's swine industry [28]. Notably, as the world's largest pork producer, China incurs annual losses of CNY 12–15 billion (USD 1.7–2.1 billion) due to PCV2-related morbidity. Several studies have investigated the epidemiological and genetic characteristics of PCV2 in different regions of China [3,26,29–32]. Similar to the global frequency situation, PCV2a, PCV2b, and PCV2d are also the predominate genotypes of PCV2 in China [3,15,33]. Moreover, PCV2 has undergone rapid evolution, with a genotype shift from PCV2b to PCV2d [1,34]. Previous studies have investigated the genetic evolution and genetic characteristics of PCV2 in Henan Province [26,31]. During 2020 and 2021, high pig prices led to the transportation of a large number of piglets to Henan Province. Nevertheless, due to the ongoing epidemic of African swine fever (ASF), most small-scale pig farms were shut down in 2022 and 2023, significantly reducing the density of pig herds and farms. These changes undoubtedly affected the prevalence of PCV2. Notably, while ASF caused acute mortality and immediate production collapse [35], PCV2 imposed chronic economic burdens through subclinical infections and secondary diseases [27]. However, recent data on PCV2 epidemiology remain limited. Given the continuing economic impact of PCV2 on the global swine industry, updating the genetic characteristics of current PCV2 strains is crucial for optimizing PCVAD management and control. In this study, the epidemiological characteristics of PCV2 in Henan Province during 2020 and 2023 were investigated. Additionally, the complete genome sequencing and analysis of 34 PCV2 strains were performed to enhance our understanding of its genetic evolution and diversity.

## 2. Materials and Methods

### 2.1. Sample Collection and Processing

A total of 385 tissue samples were collected from pig farms across 18 cities in Henan Province between February 2020 and May 2023 (clinical case backgrounds summarized in

Supplementary Table S1). During necropsy, tissue samples including spleen and inguinal lymph nodes were systematically collected from pigs that had died from illness with different clinical symptoms. A subset of samples submitted to the laboratory for analysis also included lung tissue and blood. Samples in this study were from Xinyang ($n = 16$), Zhenghzou ($n = 11$), Shangqiu ($n = 22$), Zhoukou ($n = 29$), Xuchang ($n = 19$), Luoyang ($n = 18$), Kaifeng ($n = 27$), Sanmenxia ($n = 18$), Pingdingshan ($n = 21$), Luohe ($n = 15$), Jiaozuo ($n = 24$), Nanyang ($n = 26$), Zhumadian ($n = 32$), Xinxiang ($n = 36$), Hebi ($n = 8$), Jiyuan ($n = 9$), Puyang ($n = 16$), and Anyang ($n = 38$). Samples were processed following standard operating procedures. The tissue samples from the same pig were diluted with three volumes of phosphate-buffered saline (PBS, pH 7.2), freeze-thawed three times, and homogenized with TissueLyser II (QIAGEN, Germany). Subsequently, the samples were centrifuged at 8000× $g$ for 10 min at 4 °C. The supernatants were transferred to a 1.5 mL tube and stored at −80 °C until use. Information on the clinical samples is presented in Figure 1.

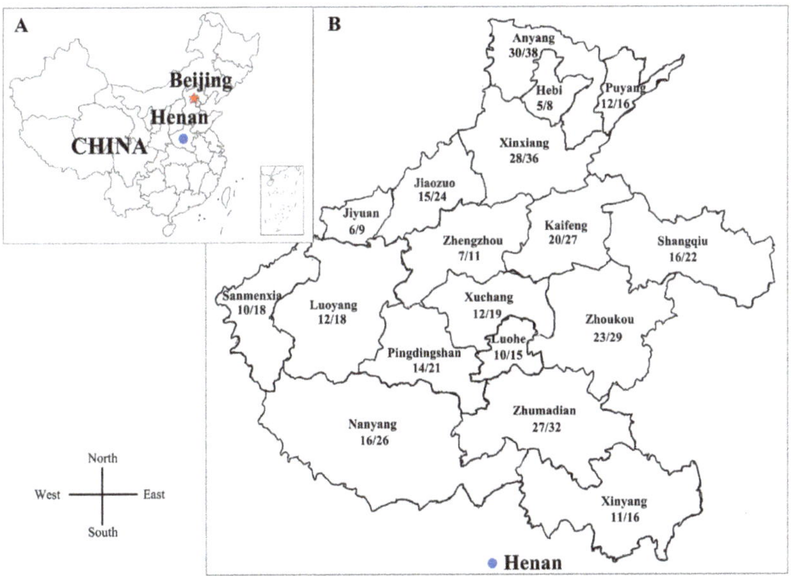


**Figure 1.** The geographical locations in Henan Province of the samples collected in this study. (**A**) Geographic location of Henan Province in China. (**B**) The geographical locations in Henan Province of China where 385 samples were collected. The numbers indicate the detection rate of PCV2 in different cities. Red star: Beijing (China's capital); Blue dot: Henan Province.

### *2.2. DNA Extraction and PCR Detection of PCV2*

Viral genomic DNA was extracted from the supernatant with the Magnetic Universal Genomic DNA Kit (TianGen, Beijing, China) following the manufacturer's instructions. All extracted genomic DNA (310–650 ng/μL) met the minimum threshold for PCR detection and was stored at −80 °C until use. The viral DNA extracted from the tissue samples was used to detect PCV2. To detect PCV2 in clinical samples, a PCR test was performed with the primers (Table 1) as described previously [36]. In brief, the 20 μL reaction mixture consisted of 10 μL of 2× Taq PCR Mix (TianGen, Beijing, China), 0.5 μL (25 μM) of each primer, 1 μL of DNA template, and 8 μL of $ddH_2O$. Amplification was performed with an initial denaturation at 94 °C for 10 min, followed by 40 cycles of 94 °C for 30 s, 58 °C for 45 s, 72 °C for 1 min, and a final extension at 72 °C for 10 min. Nucleic acid electrophoresis of PCR products was analyzed by nucleic acid electrophoresis on a 1.5% agarose gel stained with StarStain Red Plus Nucleic Acid Dye (GenStar, Beijing, China) and visualized using ChemiDoc XRS+ Imager (Bio-Rad, Hercules, CA, USA). Samples were considered PCV2-positive if a 486 bp amplicon was detected on the agarose gels using the molecular marker DL2000 DNA ladder (100–2000 bp range; veterinary diagnostic laboratory, Henan University of Animal Husbandry and Economy).

**Table 1.** Primers used in this study.

| Primer Name | Sequence (5′–3′) | Length | Annealing Temperature | Purpose |
|---|---|---|---|---|
| PCV2-F | CTGTTTTCGAACGCAGTGCC | 486 bp | 56 °C | Detection of PCV2 [36] |
| PCV2-R | GCATCTTCAACACCCGCCT | | | |
| PCV2-C-F | CGGGGTACCACTGAGTCTTTTTTATCACTTCG | 1767–1768 bp | 58 °C | PCV2 whole-genome amplification [37] |
| PCV2-C-R | CCCAAGCTTAAGACTCAGTAATTTATTTCATATGG | | | |

### 2.3. *Amplification and Sequencing of Complete Genome of PCV2 Strains*

To characterize the PCV2 strains circulating in Henan Province between 2020 and 2023, 1 PCV2-positive sample was randomly selected from each batch of 3–6 samples. A total of 34 complete PCV2 genomes were amplified from the positive samples by PCR with the primers listed in Table 1, as previously described [37]. PCR amplification was performed as follows: initial denaturation at 94 °C for 5 min, followed by 30 cycles of 94 °C for 30 s, 58 °C for 30 s, and 72 °C for 60 s, with a final extension at 72 °C for 10 min. The amplified PCR products were purified using the TIANgel Purification Kit (TianGen, Beijing, China) and subsequently cloned into the vector pMD-18 using the original TA Cloning kit (TaKaRa, Dalian, China). The recombinant plasmids were introduced into *E. coli* DH5$\alpha$ cells via electroporation (2.5 kV, 200 $\Omega$, 25 $\mu$F) and cultured on LB agar containing 100 $\mu$g/mL ampicillin. To avoid misleading results caused by PCR artefacts, three random clones were sequenced for each of the PCV2 strains. In total, the 102 clones containing PCV2 genomic DNAs were sequenced by Sangon Biotech (Shanghai, China). The full-length PCV2 genomes were assembled using the CExpress software (v6.2, NovelBio Inc., Winnipeg, Canada) with a redundant base coverage strategy (minimum 10 $\times$ depth per position). The PCV2 sequences obtained in this study have been deposited in GenBank.

### 2.4. *Sequence Comparison and Phylogenetic Analysis*

The complete genomes of PCV2 representative strains, available PCV2 commercial vaccine strains, and all PCV2 strains obtained from Henan Province were downloaded from NCBI database (https://www.ncbi.nlm.nih.gov/) (accessed on 28 August 2023) for genetic analysis. The GenBank accession numbers of PCV2 reference strains are listed in Supplementary Table S2. Multiple sequence alignments of the PCV2 stains obtained in this study, along with reference strains, were performed using the Clustal W function of the MegAlign V7.1.0 program (Lasergene, DNAStar, Madison, WI, USA). The deduced amino acids of ORF2 were analyzed in a similar manner. Furthermore, phylogenetic reconstruction of the complete PCV2 nucleic acid sequence and ORF2 deduced amino acid sequences was conducted in MEGA 6.0 using the neighbor-joining method under a p-distance model, respectively. The robustness of the topology was evaluated by 1000 bootstrap replications.

### 2.5. *Rates of Substitution*

The substitution rates of the PCV2 ORF2 nucleotide sequences were estimated using the Bayesian Markov chain Monte Carlo (MCMC) algorithm implemented in BEAST v1.10.4 to [38]. Three independent MCMC computations were performed on 200 PCV2 ORF2 sequences under the relaxed molecular clock model, the Hasegawa–Kishono–Yano

(HKY) substitution model, and the $\gamma$-site heterogeneity model. To account for dynamic population changes over time, the Bayesian Skygrid tree was selected, configured with 52 parameters and a single transition point at the end [39]. MCMC chains were run for $2 \times 10^8$ iterations, with posterior probability distributions sampled every 1000 steps. Convergence was assessed using Tracer v1.7.2 via visual inspection of trace plots and ensuring an effective sample size exceeded 200 after a 10% burn-in. Substitution rates from three independent runs were obtained using LogCombiner v1.10.4 (part of the BEAST v1.10.4 package), with 95% confidence intervals calculated. Subsequently, the MCMC runs were independently repeated on subsets of the PCV2a, PCV2b, and PCV2d cap gene nucleotide sequence data using the same methodology.

## 3. Results

### *3.1. The Frequency of PCV2 in Different Regions During 2020 and 2023*

From 2020 and 2023, a total of 385 clinical samples were collected, with 274 testing positive for PCV2, yielding an overall positivity rate of 71.17% (274/385) (Table 2). Among the 18 regions in Henan Province, the top three detection rates of PCV2 were 84.38% in Zhumadian, 79.31% in Zhoukou, and 78.95% in Anyang. The lowest detection rate of PCV2 was 55.56% in Sanmenxia (Figure 1). A temporal analysis of PCV2 detection rates, conducted using SPSS v28.0 with sampling duration-adjusted weights, revealed a significant decreasing trend over time: 81.16% (112/138) in 2020, 72.41% (84/116) in 2021, 62.50% (55/88) in 2022, and 53.49% (23/43) in 2023 ($\chi^2 = 9.8$, df = 3, $p = 0.021$). Post hoc comparisons with Bonferroni correction confirmed a marked decline between 2020 and 2023 ($p = 0.003$). However, key limitations should be noted. The small sample size ($n = 43$) may have reduced statistical power, and potential residual bias may exist due to the truncated 2023 sampling period (five months). Additionally, weighting adjustments for partial-year data introduce inherent assumptions that require careful interpretation. Despite these constraints, the observed decline suggests a potential epidemiological shift in PCV2 prevalence, warranting further validation through standardized longitudinal surveillance programs.

**Table 2.** The frequency of PCV2 in different regions in Henan Province of China during 2020 and 2023.

| Region | No. of Tested Samples | No. of PCV2 Positive Samples | Frequency |
|---|---|---|---|
| Anyang | 38 | 30 | 78.95% |
| Puyang | 16 | 12 | 75.00% |
| Hebi | 8 | 5 | 62.50% |
| Xinxiang | 36 | 28 | 77.78% |
| Jiaozuo | 24 | 15 | 62.50% |
| Jiyuan | 9 | 6 | 66.67% |
| Zhenzhou | 11 | 7 | 63.64% |
| Kaifeng | 27 | 20 | 74.07% |
| Shangqiu | 22 | 16 | 72.73% |
| Zhoukou | 29 | 23 | 79.31% |
| Xuchang | 19 | 12 | 63.16% |
| Luohe | 15 | 10 | 66.67% |
| Zhumadian | 32 | 27 | 84.38% |
| Pingdingshan | 21 | 14 | 66.67% |
| Nanyang | 26 | 16 | 61.54% |
| Luoyang | 18 | 12 | 66.67% |
| Sanmenxia | 18 | 10 | 55.56% |
| Xinyang | 16 | 11 | 68.75% |
| Total | 385 | 274 | 71.17% |

### 3.2. Genome Sequence Analysis of PCV2

To characterize the genetic diversity of recently prevalent PCV2 strains in Henan Province, 34 PCV2-positive samples were randomly selected from multiple regions for full-genome amplification, sequencing, and analysis. All 34 PCV2 strains had a genome length of 1767 nt. The following sequences were deposited in GenBank: OQ749408-OQ749433, OR114683-OR114690 (Table 3). The ORF1 gene and ORF2 genes were 945 nt and 702 nt in length, respectively. Pairwise sequence comparisons revealed nucleotide homologies ranging from 94.10% to 100.00% for the complete genome, 96.60% to 100.00% for ORF1, and 89.00% to 100.00% for ORF2. At the amino acid level, homology was found to be 98.10% to 100.00% for Rep and 87.20% to 100.00% for Cap.

**Table 3.** Detailed information of PCV2 strains sequenced in this study.

| Strain | Collection Year | Isolation Origin | Genotype | Accession No. | Length (nt) |
|---|---|---|---|---|---|
| HNZK-2020 | 2020 | Zhoukou, Henan | PCV2b | OQ749408 | 1767 |
| HNXX-1-2020 | 2020 | Xinxiang, Henan | PCV2b | OQ749409 | 1767 |
| HNNY-2020 | 2020 | Nanyang, Henan | PCV2b | OQ749410 | 1767 |
| HNJZ-2020 | 2020 | Jiaozuo, Henan | PCV2b | OQ749411 | 1767 |
| HNXC-2020 | 2020 | Xuchang, Henan | PCV2b | OQ749412 | 1767 |
| HNAY-1-2020 | 2020 | Anyang, Henan | PCV2a | OQ749413 | 1767 |
| HNXX-2-2020 | 2020 | Xinxiang, Henan | PCV2b | OQ749414 | 1767 |
| HNHB-2020 | 2020 | Hebi, Henan | PCV2b | OQ749415 | 1767 |
| HNAY-2-2020 | 2020 | Anyang, Henan | PCV2d | OQ749416 | 1767 |
| HNJY-2020 | 2020 | Jiyuan, Henan | PCV2d | OQ749417 | 1767 |
| HNAY-2021 | 2021 | Anyang, Henan | PCV2b | OQ749418 | 1767 |
| HNZZ-2021 | 2021 | Zhengzhou, Henan | PCV2d | OQ749419 | 1767 |
| HNPDS-2021 | 2021 | Pingdingshan, Henan | PCV2d | OQ749420 | 1767 |
| HNSQ-2021 | 2021 | Shangqiu, Henan | PCV2d | OQ749421 | 1767 |
| HNLH-2021 | 2021 | Luohe, Henan | PCV2d | OQ749422 | 1767 |
| HNNY-2021 | 2021 | Nanyang, Henan | PCV2d | OQ749423 | 1767 |
| HNZMD-2021 | 2021 | Zhumadian, Henan | PCV2d | OQ749424 | 1767 |
| HNSMX-2021 | 2021 | Sanmenxia, Henan | PCV2d | OQ749425 | 1767 |
| HNXX-2022 | 2022 | Xinxiang, Henan | PCV2d | OQ749426 | 1767 |
| HNAY-2022 | 2022 | Anyang, Henan | PCV2d | OQ749427 | 1767 |
| HNXY-2022 | 2022 | Xinyang, Henan | PCV2b | OQ749428 | 1767 |
| HNXC-2022 | 2022 | Xuchang, Henan | PCV2b | OQ749429 | 1767 |
| HNLY-2022 | 2022 | Luoyang, Henan | PCV2d | OQ749430 | 1767 |
| HNNY-2022 | 2022 | Nanyang, Henan | PCV2d | OQ749431 | 1767 |
| HNFQ-2022 | 2022 | Xinxiang, Henan | PCV2b | OQ749432 | 1767 |
| HNLY-2022 | 2022 | Luoyang, Henan | PCV2d | OQ749433 | 1767 |
| HNJZ-2023 | 2023 | Jiaozuo, Henan | PCV2d | OR114683 | 1767 |
| HNZMD-2023 | 2023 | Zhumadian, Henan | PCV2d | OR114684 | 1767 |
| HNXC-2023 | 2023 | Xuchang, Henan | PCV2d | OR114685 | 1767 |
| HNSQ-2023 | 2023 | Shangqiu, Henan | PCV2d | OR114686 | 1767 |
| HNNY-2023 | 2023 | Nanyang, Henan | PCV2b | OR114687 | 1767 |
| HNAY-1-2023 | 2023 | Anyang, Henan | PCV2d | OR114688 | 1767 |
| HNSMX-2023 | 2023 | Sanmenxia, Henan | PCV2a | OR114689 | 1767 |
| HNAY-2-2023 | 2023 | Anyang, Henan | PCV2d | OR114690 | 1767 |

### 3.3. Phylogenetic Analysis of PCV2 Strains

Phylogenetic analysis of PCV2 strains based on complete PCV2 genome sequences and ORF2 amino acid sequences from the 34 strains showed that three sub-genotypes (PCV2a, PCV2b, and PCV2d) were identified in the current study (Figure 2), 58.82% (20/34) strains belonged to the sub-genotype PCV2d; and 35.30% (12/34) strains belonged to the sub-genotype PCV2b. Only two strains (5.88%, 2/34), HNAY-1-2020 and HNSMX-2023, were clustered with PCV2a strains. The percentages of PCV2d strains in 2021 (87.50%), 2022 (62.50%) and 2023 (75.00%) were higher than that in 2020 (20.00%) (Figure 3). Noticeably, 90.00% (18/20) of PCV2d strains were collected during 2021–2023, whereas 41.67% (5/12) PCV2b strains were collected during the same period.

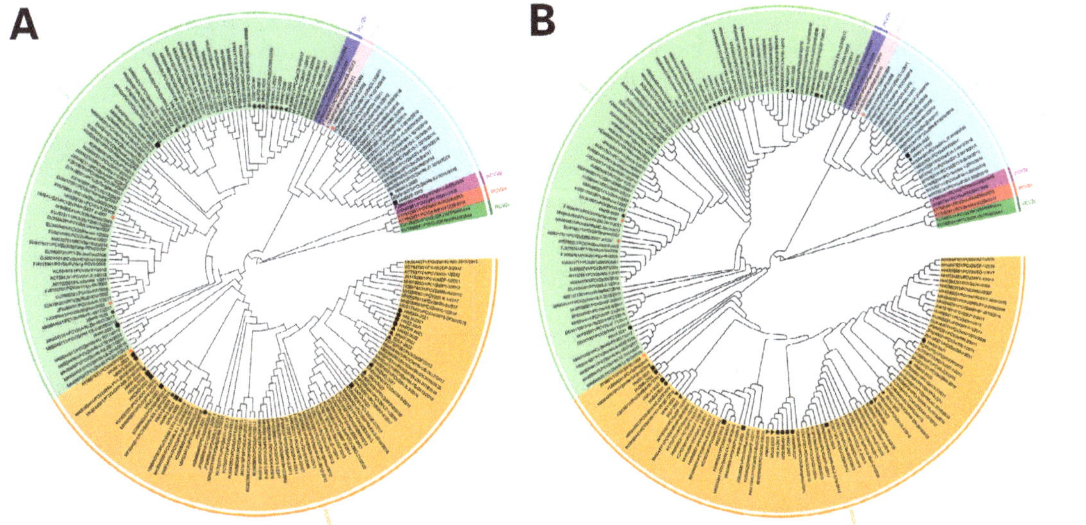


**Figure 2.** Phylogenetic trees based on the PCV2 strains obtained in this study and reference strains. (**A**) Phylogenetic tree of the complete genome sequences. (**B**) Phylogenetic tree of deduced amino acid sequences of ORF2. The phylogenetic reconstruction was performed in MEGA6.0 software using the neighbor-joining algorithm with 1000 bootstrap replicates. Red star represented PCV2 commercial vaccine strains. Black stars, black circles, black triangles and black squares represented PCV2 strains obtained in this study in 2020, 2021, 2022 and 2023, respectively.

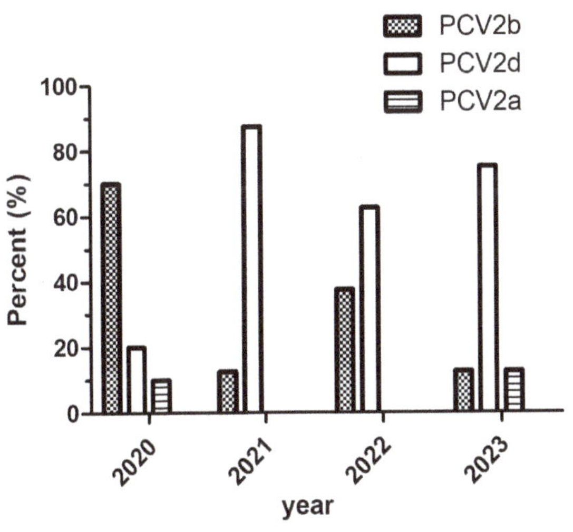


**Figure 3.** The proportion of PCV2 subtypes in different years in this study. A total of 385 clinical samples were collected from pig farms in 18 cities in Henan Province during February 2020 and May 2023, with 274 testing positive for PCV2. The percentages of PCV2d strains in 2021 (87.50%), 2022 (62.50%) and 2023 (75.00%) were higher than that in 2020 (20.00%).

### 3.4. Mutation Analysis of Cap Proteins

To explore the characteristics of the amino acid sequences of Cap of the PCV2 strains in this study, the amino acid sequence alignment was performed between the 34 PCV2 strains and the 28 reference PCV2 strains, including PCV2 commercial vaccine strains, LG (HM038034.1, PCV2a), WuHan (FJ5980441.1, PCV2b), and DBN-SX07 (HM641752.1, PCV2b) (Supplementary Figure S1). The results showed that the lengths of Cap amino acid sequences of 34 PCV2 strains were 234 amino acids. Genotype-specific substitutions were identified: PCV2b exhibited V57I, I89R/L, P134T, and D210E (absent in PCV2a/d), while PCV2d showed S68N, I89L, T/P134N, and S169G/R (absent in PCV2a/b). In addition, in this study, no amino acid deletions or insertions were observed among the 34 PCV2 strains. Thirty-one critical amino acid substitution sites were found in the Cap proteins, predominantly clustering in immunodominant regions (Table 4).

**Table 4.** Critical amino acid substitutions in the Cap proteins of PCV2 strains identified in this study.

| Position | PCV2a | PCV2b | PCV2d | Position | PCV2a | PCV2b | PCV2d |
|---|---|---|---|---|---|---|---|
| 8 | F | $F_{(1/12)}/Y_{(11/12)}$ | $F_{(16/20)}/Y_{(4/20)}$ | 121 | S | S | T |
| 47 | A | T | T | 131 | M | T | T |
| 53 | F | $F_{(11/12)}/I_{(1/12)}$ | I | 133 | V | A | A |
| 57 | V | I | V | 134 | P | T | N |
| 59 | A | $K_{(11/12)}/R_{(1/12)}$ | K | 136 | Q | L | L |
| 63 | S | $K_{(1/12)}/R_{(11/12)}$ | R | 151 | P | T | T |
| 68 | S | A | N | 169 | S | $S_{(11/12)}/C_{(1/12)}$ | $R_{(3/20)}/G_{(17/20)}$ |
| 73 | L | M | M | 185 | M | L | L |
| 77 | L | I | I | 190 | S | $A_{(1/12)}/T_{(11/12)}$ | T |
| 78 | D | N | N | 191 | R | G | G |
| 80 | V | L | L | 206 | T/K | I | I |
| 86 | T | S | S | 210 | D | E | D |
| 88 | K | P | P | 215 | V | V | I |
| 89 | I | $R_{(11/12)}/L_{(1/12)}$ | L | 232 | N | N | N |
| 90 | S | S | T | 234 | K | $K_{(1/12)}/_{(11/12)}$ | K |
| 91 | I | V | V | | | | |

### 3.5. Rates of Substitution

Bayesian MCMC analysis of the PCV2 ORF2 nucleotide sequences revealed evolutionary rates, which are summarized in Table 5. Convergence diagnostics confirmed stable posterior distributions across three independent runs (10% burn-in), with effective sample size (ESS) values exceeding 200 for all parameters. Genotype-specific analyses demonstrated distinct substitution rates: PCV2a ($4.658 \times 10^{-4}$ ssy), PCV2b ($6.476 \times 10^{-4}$ ssy), and PCV2d ($2.127 \times 10^{-3}$ ssy). When all three genotypes were analyzed collectively, the substitution rate for the PCV2 ORF2 nucleotide sequences was estimated to be $1.098 \times 10^{-3}$ ssy.

**Table 5.** Substitution rates of PCV2 ORF2 nucleotide sequences.

| PCV2 Genotype | Mean Rate of Substitution (Substitution/site/year) | 95% Highest Posterior Density | Effective Sample Size |
|---|---|---|---|
| PCV2a | $4.658 \times 10^{-4}$ | $8.643 \times 10^{-4}$ | 245,643 |
| PCV2b | $6.476 \times 10^{-4}$ | $1.063 \times 10^{-3}$ | 254,638 |
| PCV2d | $2.127 \times 10^{-3}$ | $2.853 \times 10^{-3}$ | 67,627 |
| Combined analysis for all genotypes | $1.098 \times 10^{-3}$ | $1.381 \times 10^{-3}$ | 81,829 |

## 4. Discussion

PCV2 has been endemic in Chinese swine populations for decades, with PCVADs posing significant economic and health challenges to the swine industry [1,34,40]. The rapid evolution of PCV2 is driven by international breeding pig transportation, natural viral evolution, and immune pressure from vaccination [36]. Two major genotype transitions have been documented in China: PCV2a dominated prior to 2003, PCV2b emerged as the primary genotype between 2003 and 2010, and PCV2d has become predominant since 2010 [11,36,41,42]. Notably, the co-circulation of multiple genotypes within individual farms highlights the genetic plasticity of PCV2 and complicates disease management [1,30, 43,44].

In this study, PCV2 was detected in 71.17% (274/385) of tissue samples collected from Henan Province (2020–2023), a rate comparable to historical regional data (62.4–72.9%, 2015–2019) [30,31], but higher than that reported in Shanghai (57.78%) [2], Yunnan (60.93%) [45], Shandong (36.98%) [11], and Hebei Province (39.96%) [32]. Taken together, these aforementioned studies indicated a high PCV2 detection rate in China. The consistently high detection rates suggest widespread PCV2 circulation in China. However, our findings indicate a gradual decline in detection rates from 2020 to 2023, likely due to structural transformations within the swine industry. Henan Province dominated China's swine production, with tens of millions of pigs raised in thousands of pig farms of various scale every year. Owing to the high pig price (from 2020 to 2021), millions of piglets were transferred to Henan Province, which may be the reason of high frequency of PCV2 from 2020 to 2021 in Henan Province. Nevertheless, post-ASF industry restructuring led to the closure of small-scale farms with poor biosecurity, reducing swine density and inter-regional transport—key drivers of PCV2 transmission—which may explain the declining detection rates from 2022 onward.

To date, phylogenetic analyses of PCV2 genomic sequences currently delineate nine genotypes (PCV2a–i), with China exhibiting sustained co-circulation of multiple genotypes [2,34,46]. Notably, evolutionary surveillance since 2020 has identified PCV2d as the dominant circulating genotype in Chinese swine populations, replacing previously prevalent strains [2]. In this study, phylogenetic trees based on whole genome sequences and ORF2 encoded amino acid sequences classified the 34 PCV2 strains PCV2a (5.88%, 2/34), PCV2b (35.30%, 12/34), and PCV2d (58.82%, 20/34), reinforcing PCV2d's predominance in Henan Province. This finding aligns with previous studies, including a 2018–2020 central China survey (PCV2d: 47.06%) [26] and a 2020–2021 Henan-specific analysis (PCV2d: 40.00%) [47], confirming the ongoing shift toward PCV2d dominance in China's swine populations.

The capsid (Cap) protein, the sole structural protein of PCV2, plays a crucial role in viral entry, replication, and host immune responses [46,48]. Specially, residues 1–41 are involved in the nuclear localization of the virus [49], while residues 47–63 are essential for the recognition of PCV2 epitopes [50]. Notably, PCV2 exhibits higher substitution rates ($10^{-3}$–$10^{-4}$ substitutions/site/year) than typical DNA viruses [51,52], as evidenced by genotype-specific rates in this study: PCV2a ($4.658 \times 10^{-4}$ ssy), PCV2b ($6.476 \times 10^{-4}$ ssy), and PCV2d ($2.127 \times 10^{-3}$ ssy). The accelerated evolution of PCV2 likely contributes to amino acid substitutions in Cap, potentially altering viral characteristics and immune escape potential [23,33].

Analysis of ORF2 sequences from the 34 PCV2 strains revealed substantial variability (nucleotide: 88.90–100.00%; amino acid: 88.60–100.00%). In addition, three variable regions (residues 53–91, 121–136, 190–206) (Table 4) overlapped immunodominant domains (residues 65–87, 113–147, and 193–207) in the Cap protein [53]. Notably, residue 59 (Alanine in PCV2a), a neutralizing epitope, exhibited mutations that may facilitate antigenic drift and variant emergence [54]. While previous studies have postulated a correlation between Cap protein variability and PCV2 pathogenicity [54], virulence determinants are likely multigenic. Therefore, comprehensive structural analyses of genotype-specific antigenic conformations are needed to elucidate pathogenesis mechanisms and guide the development of high-efficacy PCV2 vaccines.

This study has several limitations: first, its small sample size ($n = 34$) and geographic biases; second, temporal sampling discontinuities hindering evolutionary trend resolution. Despite these constraints, our findings provide insights into PCV2 evolution and epidemiology. Future studies should address these limitations by incorporating larger cohorts and longitudinal sampling strategies.

## 5. Conclusions

This study delineates critical epidemiological transitions in PCV2 within one of China's major swine-producing regions (Henan Province), where declining detection rates and genotype shifts (PCV2d dominance) mirror global trends, providing actionable insights for refining vaccine strategies and containment protocols in intensive pig production systems worldwide. Observed antigenic divergence in PCV2d Cap epitopes highlights potential gaps in PCV2a-based vaccine efficacy, urgently requiring PCV2d-targeted or multivalent vaccine development. Despite post-African swine fever (ASF) biosecurity improvements, persistent co-circulation of multiple genotypes highlights the necessity to integrate real-time genomic surveillance with enhanced containment measures, including optimized farm biosecurity protocols and regionally tailored vaccination programs. Collectively, these measures are critical to mitigate economic losses and preempt PCVAD resurgence in China's swine industry.

**Supplementary Materials:** The following supporting information can be downloaded at https://www.mdpi.com/article/10.3390/vetsci12040343/s1: Supplementary Table S1: Background of clinical cases of porcine circovirus-associated diseases from pig farms distributed in Henan Province, China; Supplementary Table S2: Accession numbers of PCV2 strains isolated in Henan Province and other reference PCV2 strains used in this study; Supplementary Figure S1: Amino acid sequence alignment of the PCV2 Cap protein.

**Author Contributions:** Conceptualization, C.B. and Y.S.; methodology, H.Z., H.L. (Huifang Lv) and L.Z.; software, Y.H. and H.L. (Huawei Li); validation, Z.P., K.Z. and H.L. (Huifang Lv); formal analysis, Z.P.; investigation, Z.P. and H.L. (Huifang Lv); resources, Z.P.; data curation, L.Z. and H.L. (Huifang Lv); writing—original draft preparation, Z.P., H.L. (Huifang Lv) and H.Q.; writing—review and editing, Z.P., H.L. (Huifang Lv), L.Z., H.L. (Huawei Li), Y.S. and C.B.; visualization, H.L. (Huifang Lv); supervision, H.Q.; project administration, Z.P.; funding acquisition, C.B. All authors have read and agreed to the published version of the manuscript.

**Funding:** This work was supported by the grants from Key Discipline of Veterinary Medicine of Henan Province (serial number: 312), Key Discipline of Preventive Veterinary Medicine of Henan University of Animal Husbandry and Economy (XJK202202), Scientific and Technological Project of Henan Province (242102110002; 232102111048) and Henan Province key research and development project (241111113400).

**Institutional Review Board Statement:** This study was approved by Animal Ethical Committee of Henan University of Animal Husbandry and Economy, code: HNUAHEER2425D228 and date: 2020-02-06.

**Informed Consent Statement:** Informed consent was obtained from all subjects involved in the study.

**Data Availability Statement:** All data generated or analyzed during this study are included in this published article [and its Supplementary Information Files].

**Acknowledgments:** We are very grateful to Yongsheng Wang (Henan Dahenong Animal Husbandry Co. Ltd.) for kindly providing tissue samples and epidemic information.

**Conflicts of Interest:** The authors declare no conflicts of interest.

## References


1. Yang, Y.; Xu, T.; Wen, J.; Yang, L.; Lai, S.; Sun, X.; Xu, Z.; Zhu, L. Prevalence and phylogenetic analysis of porcine circovirus type 2 (PCV2) and type 3 (PCV3) in the Southwest of China during 2020–2022. *Front. Vet. Sci.* **2022**, *9*, 1042792. [CrossRef]
2. Kang, L.; Wahaab, A.; Shi, K.; Mustafa, B.E.; Zhang, Y.; Zhang, J.J.; Li, Z.J.; QIU, Y.F. Molecular Epidemic Characteristics and Genetic Evolution of Porcine Circovirus Type 2 (PCV2) in Swine Herds of Shanghai, China. *Viruses* **2022**, *14*, 289. [CrossRef]
3. Li, N.; Liu, J.; Qi, J.; Hao, F.; Xu, L.; Guo, K. Genetic Diversity and Prevalence of Porcine Circovirus Type 2 in China During 2000–2019. *Front. Vet. Sci.* **2021**, *8*, 788172. [CrossRef]
4. Tegegne, D.; Tsegaye, G.; Aman, S.; Faustini, G.; Franzo, G. Molecular Epidemiology and Genetic Characterization of PCV2 Circulating in Wild Boars in Southwestern Ethiopia. *J. Trop. Med.* **2022**, *2022*, 5185247. [CrossRef]
5. Liu, Y.; Gong, Q.L.; Nie, L.B.; Wang, Q.; Ge, G.Y.; Li, D.L.; Ma, B.Y.; Sheng, C.Y.; Su, N.; Zong, Y.; et al. Prevalence of porcine circovirus 2 throughout China in 2015-2019: A systematic review and meta-analysis. *Microb. Pathog.* **2020**, *149*, 104490. [CrossRef] [PubMed]
6. Wang, Y.; Noll, L.; Lu, N.; Porter, E.; Stoy, C.; Zheng, W.; Liu, X.; Peddireddi, L.; Niederwerder, M.; Bai, J. Genetic diversity and prevalence of porcine circovirus type 3 (PCV3) and type 2 (PCV2) in the Midwest of the USA during 2016-2018. *Transbound. Emerg. Dis.* **2020**, *67*, 1284–1294. [CrossRef]
7. de Sousa Moreira, A.; Santos-Silva, S.; Mega, J.; Palmeira, J.D.; Torres, R.T.; Mesquita, J.R. Epidemiology of Porcine Circovirus Type 2 Circulating in Wild Boars of Portugal during the 2018-2020 Hunting Seasons Suggests the Emergence of Genotype 2d. *Animals* **2022**, *12*, 451. [CrossRef]
8. Ouyang, T.; Zhang, X.W.; Liu, X.H.; Ren, L.Z. Co-Infection of Swine with Porcine Circovirus Type 2 and Other Swine Viruses. *Viruses* **2019**, *11*, 185. [CrossRef] [PubMed]
9. Harding, J.C. The clinical expression and emergence of porcine circovirus 2. *Vet. Microbiol.* **2004**, *98*, 131–135. [CrossRef]
10. Rose, N.; Opriessnig, T.; Grasland, B.; Jestin, A. Epidemiology and transmission of porcine circovirus type 2 (PCV2). *Virus Res.* **2012**, *164*, 78–89. [CrossRef]
11. Ma, Z.; Liu, M.; Liu, Z.; Meng, F.; Wang, H.; Cao, L.; Li, Y.; Jiao, Q.; Han, Z.; Liu, S. Epidemiological investigation of porcine circovirus type 2 and its coinfection rate in Shandong province in China from 2015 to 2018. *BMC Vet. Res.* **2021**, *17*, 17. [CrossRef]
12. Chae, C. Porcine respiratory disease complex: Interaction of vaccination and porcine circovirus type 2, porcine reproductive and respiratory syndrome virus, and Mycoplasma hyopneumoniae. *Vet. J.* **2016**, *212*, 1–6. [CrossRef]
13. Tischer, I.; Mields, W.; Wolff, D.; Vagt, M.; Griem, W. Studies on epidemiology and pathogenicity of porcine circovirus. *Arch. Virol.* **1986**, *91*, 271–276. [CrossRef]
14. Franzo, G.; Cortey, M.; de Castro, A.M.; Piovezan, U.; Szabo, M.P.; Drigo, M.; Segalés, J.; Richtzenhain, L.J. Genetic characterisation of Porcine circovirus type 2 (PCV2) strains from feral pigs in the Brazilian Pantanal: An opportunity to reconstruct the history of PCV2 evolution. *Vet. Microbiol.* **2015**, *178*, 158–162. [CrossRef] [PubMed]
15. Xiao, C.T.; Halbur, P.G.; Opriessnig, T. Global molecular genetic analysis of porcine circovirus type 2 (PCV2) sequences confirms the presence of four main PCV2 genotypes and reveals a rapid increase of PCV2d. *J. Gen. Virol.* **2015**, *96*, 1830–1841. [CrossRef] [PubMed]
16. Zhai, S.L.; He, D.S.; Qi, W.B.; Chen, S.N.; Deng, S.F.; Hu, J.; Li, X.P.; Li, L.; Chen, R.A.; Luo, M.L.; et al. Complete genome characterization and phylogenetic analysis of three distinct buffalo-origin PCV2 isolates from China. *Infect. Genet. Evol.* **2014**, *28*, 278–282. [CrossRef]
17. Guo, L.J.; Lu, Y.H.; Wei, Y.W.; Huang, L.P.; Liu, C.M. Porcine circovirus type 2 (PCV2): Genetic variation and newly emerging genotypes in China. *Virol. J.* **2010**, *7*, 273. [CrossRef]
18. Jantafong, T.; Boonsoongnern, A.; Poolperm, P.; Urairong, K.; Lekcharoensuk, C.; Lekcharoensuk, P. Genetic characterization of porcine circovirus type 2 in piglets from PMWS-affected and -negative farms in Thailand. *Virol. J.* **2011**, *8*, 88. [CrossRef]
19. Bao, F.; Mi, S.; Luo, Q.; Guo, H.; Tu, C.; Zhu, G.; Gong, W. Retrospective study of porcine circovirus type 2 infection reveals a novel genotype PCV2f. *Transbound. Emerg. Dis.* **2018**, *65*, 432–440. [CrossRef]
20. Yang, X.; Chen, F.; Cao, Y.; Pang, D.; Ouyang, H.; Ren, L. Complete genome sequence of porcine circovirus 2b strain CC1. *J. Virol.* **2012**, *86*, 9536. [CrossRef]
21. Mankertz, A.; Mankertz, J.; Wolf, K.; Buhk, H.J. Identification of a protein essential for replication of porcine circovirus. *J. Gen. Virol.* **1998**, *79*, 381–384. [CrossRef]
22. Finsterbusch, T.; Mankertz, A. Porcine circoviruses--small but powerful. *Virus Res.* **2009**, *143*, 177–183. [CrossRef] [PubMed]
23. Nawagitgul, P.; Morozov, I.; Bolin, S.R.; Harms, P.A.; Sorden, S.D.; Paul, P.S. Open reading frame 2 of porcine circovirus type 2 encodes a major capsid protein. *J. Gen. Virol.* **2000**, *81*, 2281–2287. [CrossRef]
24. Shi, R.; Hou, L.; Liu, J. Host immune response to infection with porcine circoviruses. *Anim. Dis.* **2021**, *1*, 23. [CrossRef]
25. Olvera, A.; Cortey, M.; Segalés, J. Molecular evolution of porcine circovirus type 2 genomes: Phylogeny and clonality. *Virology* **2007**, *357*, 175–185. [CrossRef]

26. Xu, T.; Zhang, Y.H.; Tian, R.B.; Hou, C.Y.; Li, X.S.; Zheng, L.L.; Wang, L.Q.; Chen, H.Y. Prevalence and genetic analysis of porcine circovirus type 2 (PCV2) and type 3 (PCV3) between 2018 and 2020 in central China. *Infect. Genet. Evol.* **2021**, *94*, 105016. [CrossRef] [PubMed]
27. Alarcon, P.; Rushton, J.; Wieland, B. Cost of post-weaning multi-systemic wasting syndrome and porcine circovirus type-2 subclinical infection in England—An economic disease model. *Prev. Vet. Med.* **2013**, *110*, 88–102. [CrossRef]
28. Lang, H.W.; Zhang, G.C.; Wu, F.Q.; Zhang, C.G. Detection of serum antibody against postweaning multisystemic wasting syndrome in pigs. *Chin. J. Vet. Sci. Technol.* **2000**, *30*, 3–5.
29. Wang, S.Y.; Sun, Y.F.; Wang, Q.; Yu, L.X.; Zhu, S.Q.; Liu, X.M.; Yao, Y.; Wang, J.; Shan, T.L.; Zheng, H.; et al. An epidemiological investigation of porcine circovirus type 2 and porcine circovirus type 3 infections in Tianjin, North China. *PeerJ* **2020**, *8*, e9735. [CrossRef]
30. Zheng, G.; Lu, Q.; Wang, F.; Xing, G.; Feng, H.; Jin, Q.; Guo, Z.; Teng, M.; Hao, H.; Li, D.; et al. Phylogenetic analysis of porcine circovirus type 2 (PCV2) between 2015 and 2018 in Henan Province, China. *BMC Vet. Res.* **2020**, *16*, 6. [CrossRef]
31. Jia, Y.; Zhu, Q.; Xu, T.; Chen, X.; Li, H.; Ma, M.; Zhang, Y.; He, Z.; Chen, H. Detection and genetic characteristics of porcine circovirus type 2 and 3 in Henan province of China. *Mol. Cell Probes* **2022**, *61*, 101790. [CrossRef]
32. Han, L.; Yuan, G.F.; Chen, S.J.; Dai, F.; Hou, L.S.; Fan, J.H.; Zuo, Y.Z. Porcine circovirus type 2 (PCV2) infection in Hebei Province from 2016 to 2019: A retrospective study. *Arch. Virol.* **2021**, *166*, 2159–2171. [CrossRef] [PubMed]
33. Franzo, G.; Cortey, M.; Segalés, J.; Hughes, J.; Drigo, M. Phylodynamic analysis of porcine circovirus type 2 reveals global waves of emerging genotypes and the circulation of recombinant forms. *Mol. Phylogenetics Evol.* **2016**, *100*, 269–280. [CrossRef]
34. Xiao, Q.; Wen, L.; Yin, L.; Zhang, F.; Li, T.; Banma, Z.; He, K.; Suolang, S.; Nan, W.; Wu, J.; et al. Prevalence and genetic diversity of porcine circovirus type 2 in northern Guangdong Province during 2016–2021. *Front. Vet. Sci.* **2022**, *9*, 932612. [CrossRef]
35. Xu, G.; Sarkar, A.; Qian, L.; Shuxia, Z.; Rahman, M.A.; Yongfeng, T. The impact of the epidemic experience on the recovery of production of pig farmers after the outbreak-Evidence from the impact of African swine fever (ASF) in Chinese pig farming. *Prev. Vet. Med.* **2022**, *199*, 105568. [CrossRef] [PubMed]
36. Xu, P.L.; Zhao, Y.; Zheng, H.H.; Tian, R.B.; Han, H.Y.; Chen, H.Y.; Zheng, L.L. Analysis of genetic variation of porcine circovirus type 2 within pig populations in central China. *Arch. Virol.* **2019**, *164*, 1445–1451. [CrossRef]
37. Hu, Y.; Zhan, Y.; Wang, D.; Xie, X.; Liu, T.; Liu, W.; Wang, N.; Deng, Z.; Lei, H.; Yang, Y.; et al. Evidence of natural co-infection with PCV2b subtypes in vivo. *Arch. Virol.* **2017**, *162*, 2015–2020. [CrossRef]
38. Drummond, A.J.; Suchard, M.A.; Xie, D.; Rambaut, A. Bayesian phylogenetics with BEAUti and the BEAST 1.7. *Mol. Biol. Evol.* **2012**, *29*, 1969–1973. [CrossRef]
39. Hill, V.; Baele, G. Bayesian Estimation of Past Population Dynamics in BEAST 1.10 Using the Skygrid Coalescent Model. *Mol. Biol. Evol.* **2019**, *36*, 2620–2628. [CrossRef]
40. Suh, J.; Oh, T.; Park, K.; Yang, S.; Cho, H.; Chae, C. A Comparison of Virulence of Three Porcine Circovirus Type 2 (PCV2) Genotypes (a, b, and d) in Pigs Singularly Inoculated with PCV2 and Dually Inoculated with PCV2 and Porcine Reproductive and Respiratory Syndrome Virus. *Pathogens* **2021**, *10*, 891. [CrossRef]
41. Constans, M.; Ssemadaali, M.; Kolyvushko, O.; Ramamoorthy, S. Antigenic Determinants of Possible Vaccine Escape by Porcine Circovirus Subtype 2b Viruses. *Bioinform. Biol. Insights* **2015**, *9*, 1–12. [CrossRef] [PubMed]
42. Hou, Z.; Wang, H.; Feng, Y.; Song, M.; Li, Q.; Li, J. Genetic variation and phylogenetic analysis of Porcine circovirus type 2 in China from 2016 to 2018. *Acta Virol.* **2019**, *63*, 459–468. [CrossRef]
43. Yue, W.; Li, Y.; Zhang, X.; He, J.; Ma, H. Prevalence of Porcine Circoviruses in Slaughterhouses in Central Shanxi Province, China. *Front. Vet. Sci.* **2022**, *9*, 820914. [CrossRef]
44. Jang, G.; Yoo, H.; Kim, Y.; Yang, K.; Lee, C. Genetic and phylogenetic analysis of porcine circovirus type 2 on Jeju Island, South Korea, 2019-2020: Evidence of a novel intergenotypic recombinant. *Arch. Virol.* **2021**, *166*, 1093–1102. [CrossRef] [PubMed]
45. Lv, N.; Zhu, L.; Li, W.; Li, Z.; Qian, Q.; Zhang, T.; Liu, L.; Hong, J.; Zheng, X.; Wang, Y.; et al. Molecular epidemiology and genetic variation analyses of porcine circovirus type 2 isolated from Yunnan Province in China from 2016–2019. *TBMC Vet. Res.* **2020**, *16*, 96. [CrossRef]
46. Huang, Y.; Chen, X.; Long, Y.; Yang, L.; Song, W.; Liu, J.; Li, Q.; Liang, G.; Yu, D.; Huang, C.; et al. Epidemiological Analysis From 2018 to 2020 in China and Prevention Strategy of Porcine Circovirus Type 2. *Front. Vet. Sci.* **2021**, *8*, 753297. [CrossRef]
47. Xu, T.; Hou, C.Y.; Zhang, Y.H.; Li, H.X.; Chen, X.M.; Pan, J.J.; Chen, H.Y. Simultaneous detection and genetic characterization of porcine circovirus 2 and 4 in Henan province of China. *Gene* **2022**, *808*, 145991. [CrossRef]
48. Zhan, Y.; Yu, W.; Cai, X.; Lei, X.; Lei, H.; Wang, A.; Sun, Y.; Wang, N.; Deng, Z.; Yang, Y. The Carboxyl Terminus of the Porcine Circovirus Type 2 Capsid Protein Is Critical to Virus-Like Particle Assembly, Cell Entry, and Propagation. *J. Virol.* **2020**, *94*, e00042-00020. [CrossRef]
49. Hou, Q.; Hou, S.; Chen, Q.; Jia, H.; Xin, T.; Jiang, Y.; Guo, X.; Zhu, H. Nuclear localization signal regulates porcine circovirus type 2 capsid protein nuclear export through phosphorylation. *Virus Res.* **2018**, *246*, 12–22. [CrossRef]

50. Lekcharoensuk, P.; Morozov, I.; Paul, P.S.; Thangthumniyom, N.; Wajjawalku, W.; Meng, X.J. Epitope mapping of the major capsid protein of type 2 porcine circovirus (PCV2) by using chimeric PCV1 and PCV2. *J. Virol.* **2004**, *78*, 8135–8145. [CrossRef]
51. Tan, C.Y.; Thanawongnuwech, R.; Arshad, S.S.; Hassan, L.; Fong, M.W.C.; Ooi, P.T. Genotype Shift of Malaysian Porcine Circovirus 2 (PCV2) from PCV2b to PCV2d within a Decade. *Animals* **2022**, *12*, 1849. [CrossRef] [PubMed]
52. Duffy, S.; Shackelton, L.A.; Holmes, E.C. Rates of evolutionary change in viruses: Patterns and determinants. *Nat. Rev. Genet.* **2008**, *9*, 267–276. [CrossRef] [PubMed]
53. Mahé, D.; Blanchard, P.; Truong, C.; Arnauld, C.; Le Cann, P.; Cariolet, R.; Madec, F.; Albina, E.; Jestin, A. Differential recognition of ORF2 protein from type 1 and type 2 porcine circoviruses and identification of immunorelevant epitopes. *J. Gen. Virol.* **2000**, *81*, 1815–1824. [CrossRef]
54. Larochelle, R.; Magar, R.; D'Allaire, S. Genetic characterization and phylogenetic analysis of porcine circovirus type 2 (PCV2) strains from cases presenting various clinical conditions. *Virus Res.* **2002**, *90*, 101–112. [CrossRef] [PubMed]

veterinary sciences

MDPI

Article

# STEAP3 Inhibits Porcine Reproductive and Respiratory Syndrome Virus Replication by Regulating Fatty Acid and Lipid Droplet Synthesis

**Chenyang Yuan [1,2], Kaifeng Guan [3,*] and Gaiping Zhang [1,2,3,*]**

1 College of Veterinary Medicine, Northwest A&F University, Yangling 712100, China; yuanchenyang@nwafu.edu.cn
2 International Joint Research Center of National Animal Immunology, College of Veterinary Medicine, Henan Agricultural University, Zhengzhou 450046, China
3 School of Advanced Agricultural Sciences, Peking University, Beijing 100871, China
* Correspondence: guankaifeng1121@pku.edu.cn (K.G.); zhanggaiping@pku.edu.cn (G.Z.)

**Simple Summary:** Porcine Reproductive and Respiratory Syndrome (PRRS) caused by PRRSV has caused significant economic losses to the pig industry. Focusing on host antiviral genes and exploring the interaction between PRRSV and the host are crucial to understand the pathogenesis of PRRSV. This study takes STEAP3, which is significantly downregulated after PRRSV infection, as the entry point and finds that lipid metabolism plays a key role in the replication of PRRSV. We found that STEAP3 inhibits PRRSV replication by regulating fatty acid and lipid droplet synthesis. This provides a new strategy for PRRS prevention and control.

**Abstract:** Porcine Reproductive and Respiratory Syndrome (PRRS) is a contagious disease that impacts swine health worldwide. Lipid metabolism plays a vital role in energy production and is regulated by various genes involved in lipogenesis and lipolysis. In this study, we found that PRRSV infection significantly reduced the protein expression of STEAP3. The overexpression of STEAP3 can notably inhibit PRRSV replication. Additionally, we utilized transcriptomics and metabolomics to examine the effects of STEAP3 on PRRSV replication, identifying important pathways associated with energy metabolism and lipogenesis. We subsequently found that STEAP3 can suppress PRRSV replication by regulating fatty acid synthesis and enhancing lipid droplet formation. Overall, these findings indicate that STEAP3 could be a potential target for developing strategies to manage PRRSV infection by modulating lipid metabolism.

**Keywords:** STEAP3; PRRSV; transcriptomics; metabolomics; lipogenesis

## 1. Introduction

Porcine Reproductive and Respiratory Syndrome (PRRS) represents a highly contagious animal disease that significantly affects swine health on a global scale [1]. The causative agent, Porcine Reproductive and Respiratory Syndrome Virus (PRRSV), is a single-stranded positive-sense RNA virus characterized by a genome approximately 15 KB in length that encodes various structural and non-structural proteins [2]. Infection with PRRSV can result in respiratory complications in piglets and reproductive challenges in sows, leading to substantial economic repercussions for the swine industry annually [1]. Given its inability to independently metabolize, the virus is dependent on host cellular mechanisms to complete its lifecycle [3]. Following viral infection, the metabolic processes of host cells can be modulated to either facilitate viral persistence or promote its elimination. Research has indicated that PRRSV infection can induce modifications in the lipid metabolism of host cells, and conversely, changes in host lipid metabolism can influence the replication of PRRSV [4]. Consequently, investigating the interplay between host lipid metabolism and PRRSV replication is crucial for comprehending virus replication strategies and for developing effective prevention and control measures.

Lipid metabolism constitutes a critical cellular metabolic process that is modulated by a variety of essential genes associated with lipid metabolism. These include genes involved in lipogenesis, such as Acetyl-CoA Carboxylase 1 (ACC1), Fatty Acid Synthase (FASN), and Stearoyl-CoA Desaturase-1 (SCD-1), as well as those implicated in lipolysis, including Adipose Triglyceride Lipase (ATGL), Hormone-Sensitive Lipase (HSL), and Monoglyceride Lipase (MGL) [5,6]. As a primary source of energy within cells, lipid metabolism significantly influences the viral replication cycle [7,8]. Empirical studies have demonstrated that Dengue Virus (DENV) infection leads to the release of free fatty acids (FFA) through the degradation of lipid droplets (LDs), thereby generating ATP that facilitates viral replication [9]. Furthermore, Sterol Regulatory Element Binding Protein (SREBP), a pivotal regulatory factor for lipogenesis, is upregulated following Hepatitis C Virus (HCV) infection, which in turn enhances viral replication [10]. Additionally, the PRRSV is also dependent on the host's lipid metabolic pathways for its replication [11]. Research findings suggest that the deletion of Sphingomyelin Phosphodiesterase Acid-Like 3B (SMPDL3B) can effectively impede PRRSV infection by modulating lipid metabolism [12]. Moreover, Yin Yang 1 (YY1) has been shown to inhibit PRRSV replication through the regulation of lipid metabolism and the reprogramming of intracellular lipid droplet synthesis [3].

The STEAP (Six-Transmembrane Epithelial Antigen of the Prostate) family comprises a distinct class of transmembrane proteins that function as reductive enzymes exclusive to mammals. These proteins are integral to redox reactions within the organism and are implicated in various metabolic disorders, including obesity, insulin resistance, and cardiac hypertrophy [13,14]. Among this family, STEAP3 serves as the principal intracellular reductase and is involved in numerous biological processes. STEAP3 facilitates the reduction in intracellular ferric ions to ferrous ions, subsequently promoting lipid peroxidation and cellular ferroptosis via the Fenton reaction [15,16]. The absence of STEAP3 has been shown to impede the progression of non-alcoholic fatty liver disease (NAFLD), whereas its overexpression can lead to lipid accumulation within hepatic cells [17]. Furthermore, STEAP3 has been demonstrated to inhibit the replication of human cytomegalovirus [18]. Considering the significant association between STEAP3 and lipid metabolism, as well as the critical role of lipid metabolism in viral replication, we posited that STEAP3 might regulate the replication of PRRSV through the modulation of lipid metabolism. In our investigation, we observed a marked decrease in the protein expression levels of STEAP3 over time following PRRSV infection. Notably, the overexpression of STEAP3 had a substantial impact on lipid metabolism, ultimately leading to the inhibition of PRRSV replication, which was mediated by the regulation of fatty acid and LD synthesis.

## 2. Materials and Methods

### Cell, Virus, Plasmids, and Transfections

MARC-145 cells (monkey embryonic kidney epithelial cells) were obtained from the Key Laboratory of Animal Immunology at the Henan Academy of Agricultural Sciences and maintained in DMEM (Dulbecco's modified Eagle medium) (Solarbio, Beijing, China), supplemented with 10% FBS (fetal bovine serum) (TransGen Biotech, Beijing, China). The HP-PRRSV heN-07 strain, and the pcDNA3.1 plasmid, are all derived from the above laboratory. The pcDNA3.1-STEAP3 and the empty pcDNA3.1 vector were individually introduced into MARC-145 cells via electroporation utilizing GenePulser Xcell (Bio-Rad, Richmond, CA, USA). The electroporation was conducted at a voltage of 180V with a capacitance of 950 μF. The cells were transferred to culture plates and then infected with PRRSV (MOI = 1) after being cultured for 12 h. The cells were collected at different times post-infection.

**Western blot (Original Images in the Supplementary Materials)**

Cell samples were lysed using RIPA (Beyotime, Shanghai, China) to extract total protein. Following denaturation, the proteins were subjected to separation via SDS-PAGE and subsequently transferred to PVDF membrane (Millipore, MA, USA). The membrane was then blocked with a 5% non-fat milk solution for a duration of 1–2 h. The primary antibody (Anti-STEAP3 antibody, Abcam, Cambridge, UK, ab151566, 1:1000; PRRS virus Nucleocapsid protein antibody, GeneTex, Irvine, CA, USA, GTX637650, 1:1000; GAPDH Polyclonal antibody, Proteintech, 10494-1-AP, 1:10,000), prepared in PBST, was incubated with the membrane overnight at 4 °C. After the removal of the primary antibody, the membrane was washed three times for 10 min each. The secondary antibody (HRP-conjugated Goat Anti-Rabbit IgG(H+L), Proteintech, Wuhan, China, SA00001-2, 1:10,000), also diluted in PBST, was applied and incubated at room temperature for 1 h. Following three additional washes, the membrane was developed using ECL chemiluminescent reagent (Beyotime, Shanghai, China).

**Quantitative Real-Time PCR**

Total RNA from the cells was extracted using RNAiso Plus (TAKARA, Tokyo, Japan), followed by reverse transcription to obtain cDNA samples. Quantitative PCR was performed using cDNA as the template to assess the mRNA levels of the relevant genes. The reaction mixture consisted of the following components: 5 μL of SYBR Green, 3.2 μL of dd $H_2O$, 0.4 μL of upstream primer, 0.4 μL of downstream primer, and 1 μL of cDNA. The specific quantitative primers used are detailed in Table 1.

**Table 1.** List of primers used in the quantitative real-time PCR analysis.

| Primers | Primer Sequences (5′→3′) |
|---|---|
| ORF7 (N)-F | AAAACCAGTCCAGAGGCAAG |
| ORF7 (N)-R | CGGATCAGACGCACAGTATG |
| STEAP3-F | CCAATGCTGAGTACCTGGC |
| STEAP3-R | ATCTCCGAGACAGCACGC |
| FASN-qF | CACATCGTTCGAGCAGCATG |
| FASN-qR | AATTTCCAGGAAGCGACCGT |
| ACC1-qF | TAGTCTGCCACGGATCCAGA |
| ACC1-qR | GGGAGGGATCTCTGAGGGTT |
| PPARγ-qF | TGAGTTCGCTGTGAAGTT |
| PPARγ-qR | CAATCTGTCTGAGGTCTGT |
| TNF-α-qF | CTGTAGGTTGCTCCCACCTG |
| TNF-α-qR | CCAGTAGGGCGGTTACAGAC |
| LPIN1-qF | CAAGCAAGTAGGAGTGTCT |
| LPIN1-qR | GCGGAGGCAGAATGAATA |
| GAPDH-F | TGACAACAGCCTCAAGATCG |
| GAPDH-R | GTCTTCTGGGTGGCAGTGAT |

**Transmission Electron Microscope (TEM)**

MARC-145 cells were infected with PRRSV that overexpresses STEAP3, alongside the establishment of a negative control group. Following a 24-h incubation period, the cells were fixed using glutaraldehyde (Solarbio, Beijing, China). Transmission electron microscopy (HITACHI H-7000FA, Tokyo, Japan) was then conducted to assess the morphology, size, and quantity of the mitochondria, which were subsequently analyzed statistically.

**RNA-seq and Analysis**

MARC-145 cells with or without STEAP3 overexpression were infected with PRRSV, and total RNA was extracted from the cells 24 h post-infection. The data were quality-controlled using fastp (version 0.23.0) software to obtain high-quality clean data. Subsequently, the clean data were aligned to the *Chlorocebus sabaeus* reference genome (ChlSab1.1_release108) to obtain comprehensive transcript information. Finally, transcriptome analysis was performed, including differential expression gene screening and functional pathway analysis of differentially expressed genes.

**Metabolomics and Analysis**

MARC-145 cells with or without STEAP3 overexpression were infected with PRRSV, and cell samples were collected 24 h post-infection. All small-molecule metabolites in the samples were detected using a high-resolution LC-MS/MS platform. After constructing the metabolome library and quality control, the identified metabolites were annotated and analyzed using the HMDB and KEGG databases. Fisher's exact test and GSEA method were employed to calculate significantly enriched pathways of differentially abundant metabolites.

**Flow Cytometry Analysis**

MARC-145 cells were seeded in a 6-well plate, and then STEAP3 overexpression was induced in these cells before infecting them with PRRSV. After 24 h, the cells were trypsinized and resuspended in PBS. Then, the cells were treated with the BODIPY 493/503 Staining Kit (Beyotime, Shanghai, China) and the fluorescence intensity of the lipid droplets was measured using a flow cytometer.

**Microscopy**

PRRSV-infected MARC-145 cells with or without STEAP3 overexpression were stained using the BODIPY 493/503 Staining Kit (Beyotime, Shanghai, China). Subsequently, green puncta of LDs were observed on confocal microscopy (Zeiss LSM 800, Oberkochen, Germany).

**Statistical Analysis**

Data analysis and plotting were performed using GraphPad Prism 10.1.2 software, and significance difference analysis was conducted. All data are presented as mean ± standard deviation (SD). *: $p < 0.05$; **: $p < 0.01$.

## 3. Results

### *3.1. STEAP3 Suppresses PRRSV Replication*

To verify the relationship between STEAP3 and PRRSV, we conducted experiments using MARC-145 cells infected with PRRSV at various time intervals. The findings revealed that the protein expression of STEAP3 notably declined following PRRSV infection, with this reduction being dependent on the time elapsed (Figure 1A,B). Additionally, when STEAP3 was overexpressed, there was a significant decrease in the expression levels of PRRSV *ORF7* mRNA and N protein (Figure 1C–E). This suggests that STEAP3 plays a crucial role in inhibiting PRRSV replication.

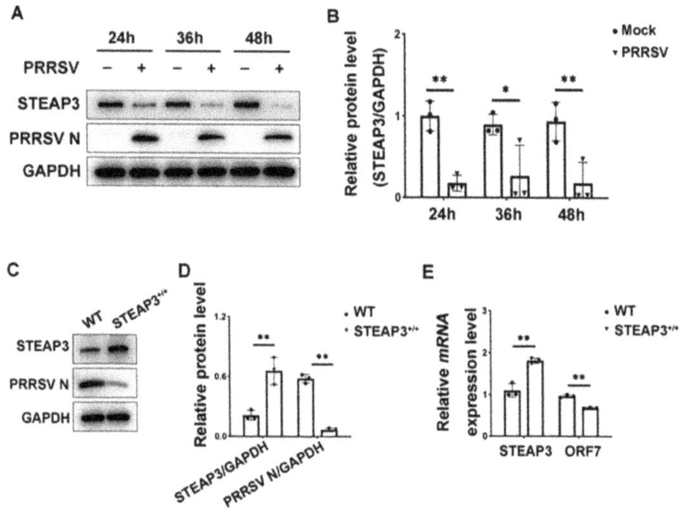


**Figure 1.** STEAP3 inhibits PRRSV replication. (**A**) and (**B**) show the expression of STEAP3 and PRRSV N protein in MARC-145 cells. Cells were infected at different time points of PRRSV (MOI = 1). (**C**) and (**D**) show the expression of STEAP3 and PRRSV N protein in MARC-145 cells. Cells were infected with PRRSV (MOI = 1) after being electroporated with pcDNA3.1-STEAP3 or the empty pcDNA3.1 vector. (**E**) Expression of *STEAP3* and PRRSV *ORF7* mRNA in MARC-145 cells. * $p < 0.05$; ** $p < 0.01$; two-way ANOVA test; n = 3.

### 3.2. Transcriptomics Sequencing and Analysis

To explore the mechanism by which STEAP3 inhibits PRRSV replication, MARC-145 cells that overexpress STEAP3 were infected with PRRSV and subsequently underwent transcriptome sequencing. The differential expression analysis revealed a total of 400 differentially expressed genes (DEGs) (fold change > 1.5 and $p < 0.05$), with 176 DEGs showing increased expression and 224 DEGs showing decreased expression (Figure 2A,B). The functional enrichment analysis of differentially expressed genes indicated significant enrichment in terms of the response to oxygen levels, response to decreased oxygen levels, cellular response to oxygen levels, and cellular response to decreased oxygen levels in biological processes (BP) (Figure 2C). In KEGG pathways, significant enrichment was observed in fatty acid biosynthesis, fatty acid metabolism, glycolysis/gluconeogenesis, amino sugar and nucleotide sugar metabolism, carbohydrate digestion and absorption, and carbon metabolism (Figure 2D). These findings suggest that STEAP3 may inhibit PRRSV replication by influencing fatty acid synthesis and energy metabolism.

### 3.3. Metabolomics and Analysis

Lipid synthesis and energy metabolism are accompanied by changes in small-molecule metabolites. Therefore, non-targeted metabolomics sequencing was conducted using the aforementioned materials. The results of the OPLS-DA analysis indicated that the experimental group and the treatment group were significantly separated into two regions (Figure 3A). The analysis of differentially expressed metabolites (DEMs) revealed a total of 86 DEMs ($p < 0.05$, VIP > 1.0), of which 52 were upregulated and 34 were downregulated (Figure 3B). The KEGG enrichment results showed that the DEMs were mainly enriched in pathways such as glycolysis/gluconeogenesis, ether lipid metabolism, and amino sugar and nucleotide sugar metabolism (Figure 3C). The chemical classification results of the DEMs indicated that 5% of the metabolites belonged to the fatty acyls category (main class); 2% and 3% of the metabolites were in the fatty acids and conjugates and fatty acid esters categories (sub class), respectively; and 15% of the metabolites were in the lipids and lipid-like molecules category (super class) (Figure 3D).

*3.4. Combined Analysis of Transcriptomics and Metabolomics*

To further investigate the effect of STEAP3 on lipogenesis, a combined analysis was performed using the previously mentioned transcriptomic and metabolomic results. The analysis revealed that 19 DEMs and 49 DEGs were linked to 13 KEGG pathways. Notably, amino sugar and nucleotide sugar metabolism, glycolysis/gluconeogenesis, and glycerophospholipid metabolism are significant contributors to lipogenesis and energy metabolism (Figure 4). Consequently, STEAP3 may hinder PRRSV replication by influencing lipogenesis and energy metabolism.

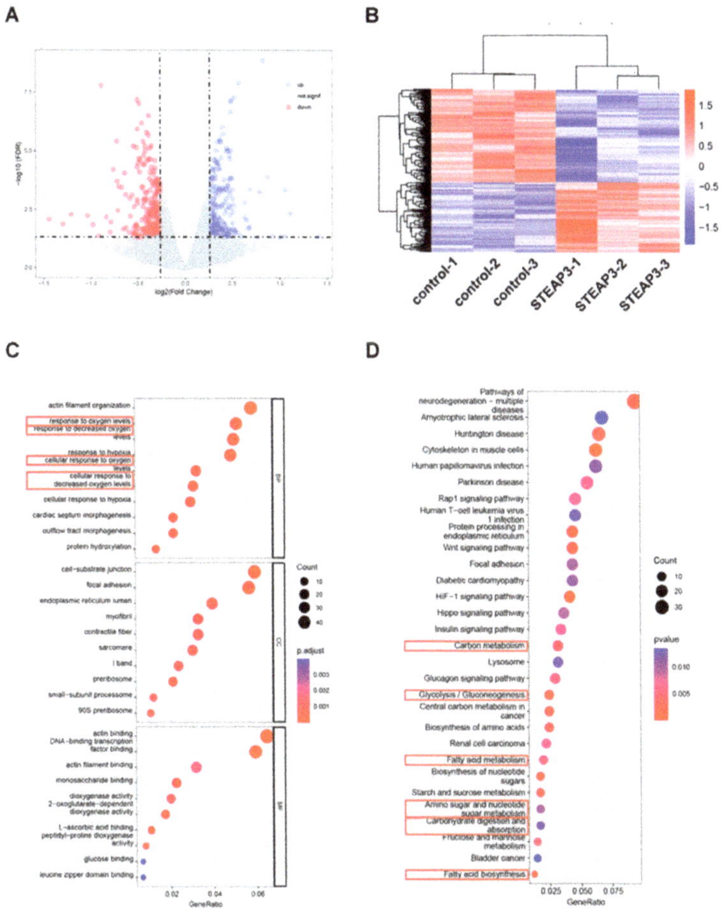


**Figure 2.** The impact of STEAP3 on the transcriptomic profile in MARC-145 cells with PRRSV infection. (**A**) The volcano plot of DEGs. (**B**) The heat map of DEGs. (**C**) Gene Ontology (GO) and (**D**) Kyoto Encyclopedia of Genes and Genomes (KEGG) pathways analysis of DEGs. Red box: GO and KEGG pathways related to lipid metabolism.

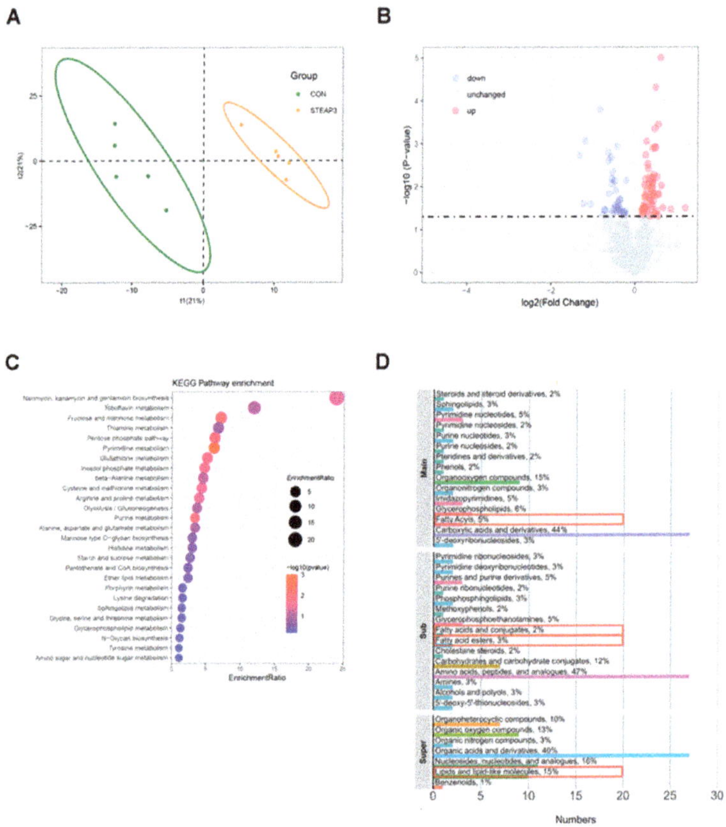


**Figure 3.** The effect of STEAP3 on metabolomics in MARC-145 cells with PRRSV infection. (**A**) The OPLS-DA plot of the STEAP3 overexpression group and the control group. (**B**) The volcano plot of DEMs. (**C**) KEGG pathways analysis of DEMs. (**D**) Chemical classification analysis of DEMs. Red box: proportion of small-molecule metabolites related to fatty acids and lipid metabolism.

**Figure 4.** The co-effect of STEAP3 on transcriptomics and metabolomics in MARC-145 cells with PRRSV infection. Left: DEMs; middle: KEGG shared by DEMs and DEGs; right: DEGs.

### 3.5. STEAP3 Inhibits FASN Expression and Promotes LD Synthesis

This study, based on the combined analysis results, examines lipogenesis and assesses the effects of STEAP3 overexpression on important genes related to this process. The quantitative real-time PCR findings revealed that STEAP3 overexpression notably reduced the mRNA levels of *FASN*, *ACC1*, and *PLIN1* while increasing the mRNA levels of *PPAR*$\gamma$ and *TNF-*$\alpha$ (Figure 5A). Furthermore, the transmission electron microscopy results showed that STEAP3 overexpression facilitated the development of intracellular LDs (Figure 5B). In addition, we also observed that the overexpression of STEAP3 in PRRSV-infected MARC-145 cells enhanced the staining of LDs (Figure 5C). The flow analysis results indicated that STEAP3 overexpression significantly increased the fluorescence intensity of intracellular LDs (Figure 5D,E), suggesting that STEAP3 effectively enhances LD synthesis. These findings imply that STEAP3 could inhibit FASN expression and alter LD synthesis, thereby reducing PRRSV replication.

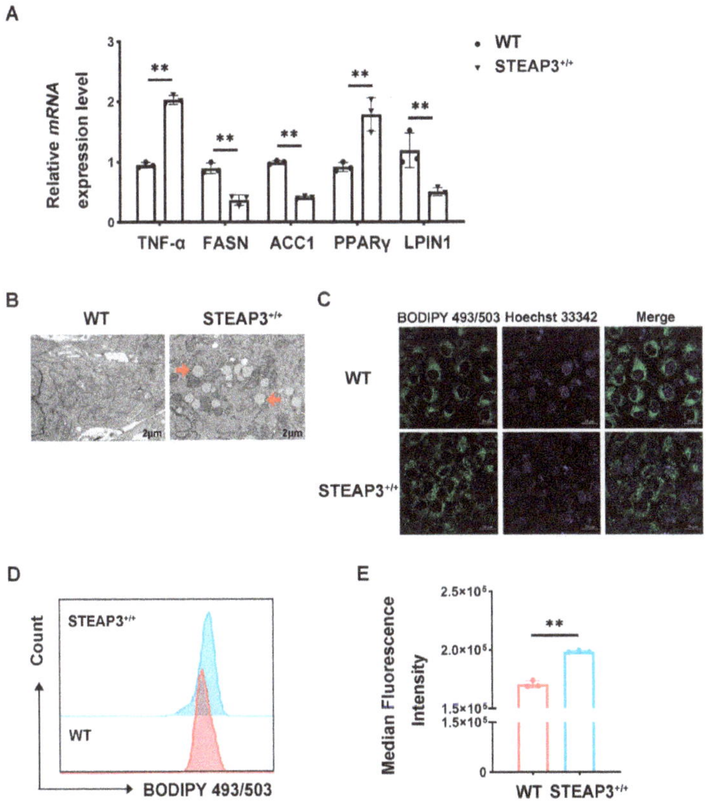


**Figure 5.** STEAP3 inhibits FASN expression and promotes LD synthesis. (**A**) Expression of *FASN*, *ACC1*, *PLIN1*, *PPAR*$\gamma$, and *TNF-*$\alpha$ mRNA in MARC-145 cells. Cells were infected with PRRSV (MOI = 1) after being electroporated with pcDNA3.1-STEAP3 and the empty pcDNA3.1 vector. (**B**) Representative TEM images of MARC-145 cells. The cells' treatment was the same as above. Scale bars: 2 μm. (**C**) Confocal images of PRRSV-infected MARC-145 cells with or without STEAP3 overexpression for 24 h. BODIPY$^{+}$ puncta represented LDs. Scale bars: 20 μm. (**D**) and (**E**) show the cytometry analysis of LDs' fluorescence intensity in MARC-145 cells. The cells' treatment was the same as above. ** $p < 0.01$; two-way ANOVA test; n $\geq$ 3.

## 4. Discussion

Viral infections can significantly impact various essential cellular functions and metabolic processes, including autophagy, apoptosis, mitochondrial activity, glucose metabolism, and lipid metabolism [19,20]. Lipids, which are vital components of cell membranes and serve as energy reserves, are crucial for the invasion and replication of RNA viruses [21]. Lipid droplets (LDs), which are organelles in the cytoplasm that store energy and are rich in neutral lipids, can interact with the endoplasmic reticulum, mitochondria, and peroxisomes [22,23]. These LDs release fatty acids by breaking down neutral lipids and transporting them to mitochondria or peroxisomes for β-oxidation, playing a role in the infection and replication of various RNA viruses [24]. Research has indicated that the structural proteins of DENV bind to LDs in a potassium ion-dependent manner, inhibiting viral replication [25]. Additionally, DENV can reduce the accumulation of intracellular LDs through autophagy (lipophagy), which aids in its replication [9]. In our study, we observed that PRRSV infection leads to a time-dependent decrease in STEAP3 expression, and that overexpressing STEAP3 can inhibit viral replication. Transmission electron microscopy revealed that the overexpression of STEAP3 significantly enhances LDs' formation in MARC-145 cells infected with PRRSV, highlighting the importance of LDs in PRRSV replication. Previous research has shown that following PRRSV infection, host cells modulate the synthesis of intracellular LDs via YY1, which contributes to antiviral activity [3]. Additionally, EGCG has been found to inhibit PRRSV replication by interfering with lipid metabolism, while PRRSV activates lipid synthesis through the ROS-dependent AKT/PCK1/INSIG/SREBPs pathway [4,26]. This indicates that lipid or lipid droplet synthesis is crucial for PRRSV replication, offering new avenues for anti-PRRS research.

STEAP3 is recognized as a marker of ferroptosis, with lipid peroxidation being its most significant characteristic. Previous research using iron-bound red blood cells has shown that while STEAP3 appears to enhance reductase activity, it also seems to paradoxically increase oxidative damage [27]. However, fatty acids need oxidation for energy production. Studies related to STEAP, including those on non-alcoholic fatty liver disease, cardiac hypertrophy, and insulin resistance, suggest its involvement in lipid metabolism [13,14,17]. This study utilized transcriptomics and metabolomics to reveal that STEAP3 can significantly influence fatty acid synthesis, lipogenesis, and various oxidative processes during PRRSV infection. Furthermore, the QPCR results indicated that STEAP3 notably downregulated the expression of lipogenesis-related genes such as ACC1, FASN, and PLIN1 while significantly upregulating peroxisome proliferator-activated receptor gamma (PPARγ). These findings suggest that STEAP3 plays a crucial role in regulating lipid metabolism. Additionally, the observation that STEAP3 can significantly increase TNF-α levels and inhibit PRRSV replication implies that during PRRSV infection, STEAP3 may negatively regulate fatty acid synthesis, positively regulate PPARγ, promote LDs' formation, and suppress PRRSV replication. Previous studies have demonstrated that the host gene YY1 can suppress FASN expression and reduce fatty acid synthesis while simultaneously promoting PPARγ expression to increase LD synthesis, which in turn hinders PRRSV replication [3]. Additionally, the host gene STEAP3 is capable of modulating intracellular inflammatory responses [28]. These results align with our findings. This presents a new target for PRRS prevention and control. The results indicate that managing LDs' generation or lipid metabolism could serve as an effective antiviral treatment strategy. Previous studies have demonstrated that Pep14-23, a peptide that targets the DENV capsid, disrupts the interaction between the capsid and LDs, thereby reducing DENV replication [29]. Additionally, triacsin C, an ACSL inhibitor, has proven itself to be highly effective in limiting rotavirus (RV) replication [30].

## 5. Conclusions

This research demonstrates that the upregulation of STEAP3 plays a significant role in modulating lipid metabolism and the formation of LDs upon PRRSV infection. The underlying mechanism appears to involve the effect of STEAP3 on the expression of several regulatory genes and metabolites associated with lipid metabolism, which in turn affects LDs' formation and ultimately inhibits the replication of PRRSV. Our findings contribute novel insights and potential targets regarding the function of STEAP3 in the regulatory mechanisms of PRRSV replication, thereby providing a theoretical foundation for the development of innovative strategies aimed at preventing and controlling PRRS through the modulation of lipid metabolism.

**Supplementary Materials:** The following supporting information can be downloaded at: https://www.mdpi.com/article/10.3390/vetsci12020147/s1.

**Author Contributions:** K.G., G.Z. and C.Y. designed this study. C.Y. and K.G. wrote the manuscript. All authors have read and agreed to the published version of the manuscript.

**Funding:** This study was supported by the China Postdoctoral Science Foundation (2024M750114) and the National Natural Science Foundation of China (32302844).

**Institutional Review Board Statement:** Not applicable.

**Informed Consent Statement:** Not applicable.

**Data Availability Statement:** The datasets presented in this study can be found in online repositories. The names of the repository/repositories and accession number(s) can be found below: https://www.ncbi.nlm.nih.gov/, NO. PRJNA1208659.

**Conflicts of Interest:** All authors declare no conflicts of interest.

## References

1. Guo, Z.; Chen, X.X.; Li, R.; Qiao, S.; Zhang, G. The prevalent status and genetic diversity of porcine reproductive and respiratory syndrome virus in China: A molecular epidemiological perspective. *Virol. J.* **2018**, *15*, 2. [CrossRef]
2. Dea, S.; Gagnon, C.A.; Mardassi, H.; Pirzadeh, B.; Rogan, D. Current knowledge on the structural proteins of porcine reproductive and respiratory syndrome (PRRS) virus: Comparison of the North American and European isolates. *Arch. Virol.* **2000**, *145*, 659–688. [CrossRef]
3. Zheng, Z.; Ling, X.; Li, Y.; Qiao, S.; Zhang, S.; Wu, J.; Ma, Z.; Li, M.; Guo, X.; Li, Z.; et al. Host cells reprogram lipid droplet synthesis through YY1 to resist PRRSV infection. *mBio* **2024**, *15*, e0154924. [CrossRef] [PubMed]
4. Ma, Y.X.; Han, Y.Q.; Wang, P.Z.; Wang, M.Y.; Yang, G.Y.; Li, J.L.; Wang, J.; Chu, B.B. Porcine reproductive and respiratory syndrome virus activates lipid synthesis through a ROS-dependent AKT/PCK1/INSIG/SREBPs axis. *Int. J. Biol. Macromol.* **2024**, *282*, 136720. [CrossRef]
5. Currie, E.; Schulze, A.; Zechner, R.; Walther, T.C.; Farese, R.V., Jr. Cellular fatty acid metabolism and cancer. *Cell Metab.* **2013**, *18*, 153–161. [CrossRef] [PubMed]
6. Ducharme, N.A.; Bickel, P.E. Lipid droplets in lipogenesis and lipolysis. *Endocrinology* **2008**, *149*, 942–949. [CrossRef]
7. Stapleford, K.A.; Miller, D.J. Role of cellular lipids in positive-sense RNA virus replication complex assembly and function. *Viruses* **2010**, *2*, 1055–1068. [CrossRef] [PubMed]
8. Lorizate, M.; Kräusslich, H.G. Role of lipids in virus replication. *Cold Spring Harb. Perspect. Biol.* **2011**, *3*, a004820. [CrossRef]
9. Heaton, N.S.; Randall, G. Dengue virus-induced autophagy regulates lipid metabolism. *Cell Host Microbe* **2010**, *8*, 422–432. [CrossRef]
10. Li, Q.; Pène, V.; Krishnamurthy, S.; Cha, H.; Liang, T.J. Hepatitis C virus infection activates an innate pathway involving IKK-α in lipogenesis and viral assembly. *Nat. Med.* **2013**, *19*, 722–729. [CrossRef] [PubMed]
11. Jiang, D.; Yang, L.; Meng, X.; Xu, Q.; Zhou, X.; Liu, B. Let-7f-5p Modulates Lipid Metabolism by Targeting Sterol Regulatory Element-Binding Protein 2 in Response to PRRSV Infection. *Vet. Sci.* **2024**, *11*, 392. [CrossRef] [PubMed]
12. Shen, H.H.; Zhao, Q.; Wen, Y.P.; Wu, R.; Du, S.Y.; Huang, X.B.; Wen, X.T.; Cao, S.J.; Zeng, L.; Yan, Q.G. Porcine reproductive and respiratory syndrome virus upregulates SMPDL3B to promote viral replication by modulating lipid metabolism. *iScience* **2023**, *26*, 107450. [CrossRef] [PubMed]

13. Arner, P.; Stenson, B.M.; Dungner, E.; Näslund, E.; Hoffstedt, J.; Ryden, M.; Dahlman, I. Expression of six transmembrane protein of prostate 2 in human adipose tissue associates with adiposity and insulin resistance. *J. Clin. Endocrinol. Metab.* **2008**, *93*, 2249–2254. [CrossRef] [PubMed]
14. Chen, X.; Zhu, C.; Ji, C.; Zhao, Y.; Zhang, C.; Chen, F.; Gao, C.; Zhu, J.; Qian, L.; Guo, X. STEAP4, a gene associated with insulin sensitivity, is regulated by several adipokines in human adipocytes. *Int. J. Mol. Med.* **2010**, *25*, 361–367. [CrossRef]
15. Gao, J.; Wang, Q.; Tang, Y.D.; Zhai, J.; Hu, W.; Zheng, C. When ferroptosis meets pathogenic infections. *Trends Microbiol.* **2023**, *31*, 468–479. [CrossRef]
16. Li, J.; Cao, F.; Yin, H.L.; Huang, Z.J.; Lin, Z.T.; Mao, N.; Sun, B.; Wang, G. Ferroptosis: Past, present and future. *Cell Death Dis.* **2020**, *11*, 88. [CrossRef]
17. Ding, T.; Chen, S.; Xiao, W.; Liu, Z.; Tu, J.; Yu, Y.; Dong, B.; Chen, W.; Zeng, Y. Six-Transmembrane Epithelial Antigen of Prostate 3 Promotes Hepatic Insulin Resistance and Steatosis. *J. Lipid Res.* **2023**, *64*, 100318. [CrossRef]
18. Li, Y.; Shang, W.; Xiao, G.; Zhang, L.K.; Zheng, C. A Comparative Quantitative Proteomic Analysis of HCMV-Infected Cells Highlights pUL138 as a Multifunctional Protein. *Molecules* **2020**, *25*, 2520. [CrossRef] [PubMed]
19. Reyes, A.; Duarte, L.F.; Farías, M.A.; Tognarelli, E.; Kalergis, A.M.; Bueno, S.M.; González, P.A. Impact of Hypoxia over Human Viral Infections and Key Cellular Processes. *Int. J. Mol. Sci.* **2021**, *22*, 7954. [CrossRef]
20. Quarleri, J.; Cevallos, C.; Delpino, M.V. Apoptosis in infectious diseases as a mechanism of immune evasion and survival. *Adv. Protein Chem. Struct. Biol.* **2021**, *125*, 1–24. [CrossRef] [PubMed]
21. Farías, M.A.; Diethelm-Varela, B.; Kalergis, A.M.; González, P.A. Interplay between lipid metabolism, lipid droplets and RNA virus replication. *Crit. Rev. Microbiol.* **2024**, *50*, 515–539. [CrossRef]
22. Cui, L.; Liu, P. Two Types of Contact Between Lipid Droplets and Mitochondria. *Front. Cell Dev. Biol.* **2020**, *8*, 618322. [CrossRef]
23. Olzmann, J.A.; Carvalho, P. Dynamics and functions of lipid droplets. *Nat. Rev. Mol. Cell Biol.* **2019**, *20*, 137–155. [CrossRef] [PubMed]
24. Qu, Y.; Wang, W.; Xiao, M.Z.X.; Zheng, Y.; Liang, Q. The interplay between lipid droplets and virus infection. *J. Med. Virol.* **2023**, *95*, e28967. [CrossRef]
25. Carvalho, F.A.; Carneiro, F.A.; Martins, I.C.; Assunção-Miranda, I.; Faustino, A.F.; Pereira, R.M.; Bozza, P.T.; Castanho, M.A.; Mohana-Borges, R.; Da Poian, A.T.; et al. Dengue virus capsid protein binding to hepatic lipid droplets (LD) is potassium ion dependent and is mediated by LD surface proteins. *J. Virol.* **2012**, *86*, 2096–2108. [CrossRef] [PubMed]
26. Yu, P.W.; Fu, P.F.; Zeng, L.; Qi, Y.L.; Li, X.Q.; Wang, Q.; Yang, G.Y.; Li, H.W.; Wang, J.; Chu, B.B.; et al. EGCG Restricts PRRSV Proliferation by Disturbing Lipid Metabolism. *Microbiol. Spectr.* **2022**, *10*, e0227621. [CrossRef]
27. Howie, H.L.; Hay, A.M.; de Wolski, K.; Waterman, H.; Lebedev, J.; Fu, X.; Culp-Hill, R.; D'Alessandro, A.; Gorham, J.D.; Ranson, M.S.; et al. Differences in Steap3 expression are a mechanism of genetic variation of RBC storage and oxidative damage in mice. *Blood Adv.* **2019**, *3*, 2272–2285. [CrossRef]
28. Zhang, F.; Tao, Y.; Zhang, Z.; Guo, X.; An, P.; Shen, Y.; Wu, Q.; Yu, Y.; Wang, F. Metalloreductase Steap3 coordinates the regulation of iron homeostasis and inflammatory responses. *Haematologica* **2012**, *97*, 1826–1835. [CrossRef]
29. Faustino, A.F.; Guerra, G.M.; Huber, R.G.; Hollmann, A.; Domingues, M.M.; Barbosa, G.M.; Enguita, F.J.; Bond, P.J.; Castanho, M.A.; Da Poian, A.T.; et al. Understanding dengue virus capsid protein disordered N-Terminus and pep14-23-based inhibition. *ACS Chem. Biol.* **2015**, *10*, 517–526. [CrossRef]
30. Kim, Y.; George, D.; Prior, A.M.; Prasain, K.; Hao, S.; Le, D.D.; Hua, D.H.; Chang, K.O. Novel triacsin C analogs as potential antivirals against rotavirus infections. *Eur. J. Med. Chem.* **2012**, *50*, 311–318. [CrossRef] [PubMed]

*veterinary sciences*

MDPI

*Article*

# Evaluation of the Immunogenicity of a Pool of Recombinant *Lactococcus lactis* Expressing Eight Antigens of African Swine Fever Virus in a Mouse Model

**Jingshan Huang [1,2], Tianqi Gao [1], Zhanhao Lu [1], Dailang Zhong [1], Mingzhi Li [1], Hua-Ji Qiu [1], Yongfeng Li [1], Hongxia Wu [1,*] and Yuan Sun [1,2,*]**

[1] State Key Laboratory for Animal Disease Control and Prevention, National African Swine Fever Para-Reference Laboratory, Harbin Veterinary Research Institute, Chinese Academy of Agricultural Sciences, Harbin 150069, China; hjs45333@163.com (J.H.); 19990736668@163.com (T.G.); luzhanhao106@163.com (Z.L.); hsrzlxcrb2021zdl@163.com (D.Z.); lmz824925@163.com (M.L.); qiuhuaji@caas.cn (H.-J.Q.); liyongfeng@caas.cn (Y.L.)

[2] College of Animal Science and Technology, Jilin Provincial Engineering Research Center of Animal Probiotics, Key Laboratory of Animal Production and Product Quality Safety of Ministry of Education, Jilin Agricultural University, Changchun 130118, China

[*] Correspondence: wuhongxia@caas.cn (H.W.); sunyuan@caas.cn (Y.S.); Tel.: +86-451-51051708 (Y.S.)

**Simple Summary:** In light of the urgent need to combat the global threat posed by African swine fever (ASF) to the pig industry, this study generated a pool of recombinant *Lactococcus lactis* expressing eight distinct proteins of African swine fever virus (ASFV), the causative agent of ASF, and evaluated its immunogenicity in mice via both oral gavage and intramuscular injection. The results demonstrate that intramuscular immunization induced high levels of serum IgG antibodies in the mice, while oral immunization resulted in a transient increase in secretory IgA (sIgA) antibodies in mouse feces.

**Abstract:** African swine fever (ASF), caused by African swine fever virus (ASFV), poses a great threat to the global pig industry. There is an urgent demand for effective and safe vaccines to address this threat. This study reports the generation and evaluation of a recombinant *Lactococcus lactis* pool, each strain expressing one of eight ASFV antigens (F317L, H171R, D117L, E120R, B602L, CD2v, p54, and p72). We evaluated the immune responses in mice through oral gavage and intramuscular immunization to the recombinant *L. lactis* pool. The results show that the mice immunized via intramuscular injection induced high-level serum IgG antibodies within 7 d post-primary immunization, which was maintained over an extended period. Additionally, there was a marked increase in the interferon gamma (IFN-$\gamma$) and interleukin 10 (IL-10) levels in the sera. In contrast, the mice immunized via oral gavage did not induce obvious serum IgG antibodies. However, they experienced a transient peak of secretory IgA (sIgA) antibodies in fecal samples within 7 d post-primary immunization, which subsequently decreased to levels that were statistically similar to those in the control group. In addition, this study also found that the multi-antigen cocktail vaccination was safe for mice. This study provides a reference for the development and immunization of ASF vaccines with *L. lactis* as live carriers.

**Keywords:** African swine fever virus; *Lactococcus lactis*; immunogenicity; mouse model

## 1. Introduction

African swine fever (ASF), caused by African swine fever virus (ASFV), is a highly contagious and hemorrhagic disease. It exhibits a diverse spectrum of clinical manifestations, ranging from chronic to peracute forms, and poses a significant threat to the stability and development of the global swine industry stability [1,2]. To mitigate ASF-associated losses, the development of safe and efficacious vaccines is imperative. The current landscape of ASF vaccine development encompasses various strategies, such as inactivated vaccines, live-attenuated vaccines, subunit vaccines, and live vector vaccines [3–7]. However, these vaccine strategies still face some limitations that necessitate addressing. For instance, while live-attenuated ASF vaccines are perceived to exhibit better efficacy, they entail higher safety risks due to potential residual virulence [8,9]. Conversely, subunit ASF vaccines have higher safety, but their protective efficacy still requires improvement [10,11]. Therefore, there is a need for continued research and development to address these challenges and ultimately produce a safe and effective ASF vaccine.

ASFV is a double-stranded DNA virus and the sole known tick-borne DNA virus [12]. The genome of ASFV spans a size range of 170 to 194 kilobase pairs (kb) and contains between 151 and 167 open reading frames (ORFs) [13]. ASFV encodes over 60 structural proteins and 100 non-structural proteins, which provide significant assistance in its assembly, replication, repair, and immune evasion [14]. Although the functions of some of these proteins remain unclear, numerous immunogenic candidate protective antigens have been identified. Among them, the ASFV p72 (encoded by the *B646L* gene), p54 (encoded by the *E183L* gene), p30 (encoded by the *CP204L* gene), and CD2v (encoded by the *EP402R* gene) proteins are the most commonly used protective antigen targets for ASF [15–17]. Furthermore, researchers systematically evaluated the immune responses of pigs to individual ASFV antigens by priming with a DNA vaccine via delivery of individual genes, followed by booster immunization using a recombinant vaccinia virus vector. They conducted an exhaustive screening of approximately 30% of ASFV antigens. The results indicated that the D117L and E120R proteins have potential as significant inducers of protective antibody responses or as serological markers of infection. Additionally, the F317L, B602L, p72, p54, and CD2v proteins were found to effectively stimulate specific lymphocytes in immunized pigs, eliciting a robust cellular immune response [18].

*Lactococcus lactis*, owing to its safety and ability to induce mucosal immune responses, has been widely employed in the development of mucosal vaccines. *L. lactis*, a well-studied lactic acid bacteria (LAB), holds the generally recognized as safe (GRAS) status due to its extensive history of use in human food fermentations and products [19]. This makes *L. lactis* widely used, including as a carrier for mucosal vaccines to present bacterial and viral antigens [20]. The nisin-controlled gene expression (NICE) system, a tightly regulated inducible expression platform, enables precise heterologous protein production in *L. lactis*, facilitating its genetic engineering for vaccinology [21]. Numerous studies have demonstrated the potential of *L. lactis* as a vaccine vector. For instance, recombinant *L. lactis* (r*L. lactis*) expressing multi-epitope antigens of foot-and-mouth disease virus (FMDV) serotype A was found to effectively induce the production of secretory IgA (sIgA) and IgG in mucosal tissues while also eliciting robust cellular immune responses [22]. In addition, *L. lactis* activates innate immune pathways, as evidenced by strain *L. lactis*-pNZ8148-BLS-Usp45, which expresses the *Brucella* LS protein, induced IgG antibodies and high levels of cytokines, such as interferon gamma (IFN-$\gamma$), tumor necrosis factor alpha (TNF-$\alpha$), interleukin 4 (IL-4), and IL-10 [23]. Another study showed that using *L. lactis* as an oral vaccine delivery vector to express the VP6 protein of grass carp reovirus type II resulted in significant increases in the survival rate (46.7%) and protection rate (42.9%) compared with the control groups [24]. Researchers have also constructed r*L. lactis* strains capable of expressing *Eimeria tenella* immune mapped protein-1 (EtIMP1), a protective antigen against coccidiosis in chickens. Animal trials showed that compared to *L. lactis* carrying the parental vector, all three strains expressing EtIMP1 significantly reduced oocyst shedding, mitigated pathological changes in the cecum, and improved weight gain [25].

In this study, we selected eight immunogenic proteins of ASFV (F317L, H171R, D117L, E120R, B602L, CD2v, p54, and p72) to construct ASF vaccine candidates using *L. lactis* as a vector, and then evaluated the immune responses of these vaccine candidates in mice through both oral gavage and intramuscular injection routes. The objective was to establish a foundation for developing an oral vaccine using *L. lactis* as a carrier. By exploring the immune response capabilities induced by these r*L. lactis* strains, we aim to present new ideas and directions for advancing the development of an ASF vaccine.

## 2. Materials and Methods

### 2.1. Bacterial Strains, Plasmids, and Target DNAs

The nisin-inducible expression plasmid pNZ8149 and the *L. lactis* NZ3900 strain used in this study were generously provided by Dr. Hongyu Cui at the Harbin Veterinary Research Institute (HVRI) of the Chinese Academy of Agricultural Sciences (CAAS). The pNZ8149 plasmid is non-resistant, and the recombinant bacterial strains harboring pNZ8149 plasmid were screened by using a lactose-containing medium. The gene sequence of the ASFV HLJ/18 strain (accession number: MK333180.1) was retrieved from the GenBank database and optimized based on the whole genome sequence of the *L. lactis* MG1363 strain (the parental strain of *L. lactis* NZ3900). The optimized ASFV gene sequences were then synthesized by Nanjing GenScript.

### 2.2. Construction of the Recombinant Plasmids

The primers listed in Table 1 were used to amplify the linearized vector fragments from pNZ8149 and the target genes for ASFV. The resulting DNA fragments were separated by agarose gel electrophoresis and purified using a gel extraction kit (Omega, Norcross, GA, USA). The purified fragments were then subjected to homologous recombination using the ClonExpress II One Step cloning kit (Vazyme Biotech Co., Nanjing, China) to construct the recombinant plasmid [26].

**Table 1.** The primers used for constructing the recombinant plasmids.

| Name | Sequences (5′-3′) |
|---|---|
| F317L-F | GCACTCACCATGGTTGAAACACAAATGGAT |
| H171R-F | GCACTCACCATGGTTGTTTATGATCTTCTT |
| D117L-F | GCACTCACCATGGATACAGAAACTTCACCT |
| E120R-F | GCACTCACCATGGCTGATTTAATTCACCA |
| B602L-F | GCACTCACCATGGCTGAATTTAATATT |
| CD2v-F | GCACTCACCATGATTATTCTTATTTTTCTT |
| p54-F | GCACTCACCATGGATTCAGAATTT |
| p72-F | GCACTCACCATGGCTTCAGGTGGA |
| Common-R | AAGCTTGAGCTCTCACTTCTCGAACTGGGG |
| pNZ8149-F | GGTACCACTAGTTCTAGAGAGCTC |
| pNZ8149-R | GCATGCCTGCAGTACCCATGGTGA |

Common-R represents the same reverse primer for all the target genes.

### 2.3. Construction of the Recombinant L. lactis

The electrocompetent cells of the *L. lactis* NZ3900 strain were prepared following the method described previously [27]. Subsequently, *L. lactis* electrocompetent cells were chilling on ice until they were fully thawed. A total of 10 µL of the recombinant plasmid DNA was gently mixed with 50 µL of electrocompetent cells and incubated on ice for 2 min. Thereafter, the mixture was transferred into a pre-cooled 0.2-cm electroporation cuvette.

The electroporation settings were adjusted to deliver a single pulse of 2000 V, 25 μF, and 200 Ω, with a single electrical pulse duration of 4 to 5 ms. Post-electroporation, the cells were promptly resuspended in 700 μL of pre-cooled GM17 broth (supplemented with 20 mM of $MgCl_2$ and 2 mM of $CaCl_2$), and placed on ice for 5 min. Activation ensued in a 30°C incubator over a period of 1 h. Following this, the culture was centrifuged at 6000× *g* for 2 min and resuspended in 50 μL of fresh GM17 medium. The resuspended cells were plated onto lactose-supplemented EM agar medium and statically incubated in a 30°C incubator for 24 to 48 h. Subsequently, single colonies were streaked onto fresh selective plates for purity verification prior to overnight culture in GM17 at 30°C. The plasmids were extracted for gene sequencing, and the r*L. lactis* strains with correctly matched sequences were preserved at −80°C for future use.

### 2.4. Western Blotting

To determine the subcellular localization of the recombinant proteins expression, the r*L. lactis* strains were induced with nisin to express the ASFV proteins, lysed by ultrasonication, and then fractionated into soluble (supernatant) and insoluble (precipitate) fractions by centrifugation. Equal volumes of 40 μL of the supernatant and precipitate were mixed with 10 μL of 5× sodium dodecyl sulfate-polyacrylamide gel electrophoresis (SDS-PAGE) loading buffer and boiled for 10 min. The samples were then subjected to SDS-PAGE separation and subsequently transferred onto a nitrocellulose membrane. The membrane was incubated with mouse anti-Strep antibodies (1:2000) as the primary antibody, followed by a goat anti-mouse IgG antibody (1:10,000) (Abcam, Cambridge, UK) as the secondary antibody. The blots were subsequently scanned using the Odyssey scanning imaging system.

### 2.5. Growth Curve of r*L. lactis* Strains

In order to evaluate the growth characteristics of the eight r*L. lactis* strains, we selected r*L. lactis*-F317L as a model organism. Overnight cultures of r*L. lactis*-F317L, *L. lactis*-pNZ8149 (strain containing an empty vector plasmid), and *L. lactis* NZ3900 (plasmid-free strain) were inoculated into EM broth at a 1:20 dilution ratio. Both induced and non-induced groups were established. The optical density at 600 nm ($OD_{600nm}$) values of the cultures were measured initially, and subsequently recorded at hourly intervals until the bacterial stationary phase was reached. Utilizing these data points, growth curves were drawn for the r*L. lactis*-F317L, *L. lactis*-pNZ8149, and *L. lactis* NZ3900 under both the induced and non-induced conditions.

### 2.6. Immunization of Mice

A total of 20 specific-pathogen-free (SPF) six-week-old BALB/c mice were randomly divided into 4 groups, with 5 mice per group, as follows: (A) in the r*L. lactis* oral gavage group, each mouse was oral gavage-immunized with $8 \times 10^9$ colony forming units (CFU) of a mixture of r*L. lactis*-F317L, r*L. lactis*-H171R, r*L. lactis*-D117L, r*L. lactis*-E120R, r*L. lactis*-B602L, r*L. lactis*-CD2v, r*L. lactis*-p54, and r*L. lactis*-p72 at a dose of $10^9$ CFU per strain; this mixture was termed the r*L. lactis*-mix; (B) in the r*L. lactis* intramuscular group, the immunogen and immunization doses were the same as in group A, but immunization was carried out through intramuscular injection; (C) in the *L. lactis*-pNZ8149 group, each mouse was oral gavage-immunized with $8 \times 10^9$ CFU of *L. lactis*-pNZ8149; (D) in the PBS group, each mouse oral gavage-immunized with 200 μL of PBS. Among them, the r*L. lactis* intramuscular group was intramuscularly injected after the strains were inactivated, and the other groups of strains were immunized with live bacteria via oral gavage immunization. The immunization schedules are shown in Table 2.

**Table 2.** The immunization doses and routes of r*L. lactis* in mice.

| Group | Amount | Immunogens | Immunizing dose | Immunization route |
|---|---|---|---|---|
| A | 5 | r*L. lactis*-mix | $8 \times 10^9$ CFU | oral gavage |
| B | 5 | *L. lactis*-pNZ8149 | $8 \times 10^9$ CFU | oral gavage |
| C | 5 | PBS | 200 μL | oral gavage |
| D | 5 | r*L. lactis*-mix | $8 \times 10^9$ CFU | intramuscular injection |

The immunization schedule for the mice comprised the following steps: primary immunization, followed by oral gavage immunization on the subsequent day for a continuous period of 14 d. Two weeks later, a booster immunization was administered via oral gavage for three consecutive days. After another two weeks, a second booster immunization was given for three days. Each immunization was accompanied by a single intramuscular injection. Blood samples were collected from the orbital venous plexus on Day 0, Day 21 (7 d post-primary immunization), Day 37 (7 d post-first booster immunization), Day 44 (14 d post-first booster immunization), and Day 54 (7 days post-second booster immunization). On Day 54, the mice were anesthetized with $CO_2$ and euthanized by cervical dislocation. Their spleens were harvested for subsequent lymphocyte proliferation assays.

### 2.7. Enzyme-Linked Immunosorbent Assay (ELISA)

Blood and feces samples of the mice were collected at various time points post-immunization. The collected blood samples were placed in a 37°C incubator for 2 h and centrifuged at 1000× *g* for 15 min. The supernatant obtained, which is the serum, was collected for the next detection. Regarding the feces samples, they were first weighed, and 0.4 mL of feces extract (prepared with 0.05 M EDTA-Na2 solution in PBS) was added to each 0.1 g feces. After shaking for 10 min, it was placed overnight at 4°C, centrifuged at 12,000× *g* for 5 min, and the supernatant was collected for the next detection. We used the prokaryotic expression system to express and purify the eight ASFV proteins in this study, with *Escherichia coli* Rosetta as the expression host and the pCold-TF plasmid as the expression plasmid.

The 96-well microplates were coated with the purified ASFV protein and incubated at 4°C overnight. Then, each well was washed three times with 200 μL of PBS containing 0.05% tween 20 (PBST). After washing, 150 μL of blocking solution (5% skimmed milk) was added to each well and incubated in a 37°C incubator for 2 h. The blocking solution was then discarded, and the microplates were dried. Next, 100 μL of the treated serum samples or fecal supernatants were added to the microplates (using reference anti-ASFV sera as the positive control and SPF pig sera as the negative control), and they were incubated in a 37°C incubator for 1 h. After incubation, the microplates were washed five times with PBST, 3 min each time, and then dried. Subsequently, 100 μL of HRP-conjugated rabbit anti-mouse IgG or goat anti-mouse IgA (Abcam, Cambridge, UK) was added into each well (the control sera were used with goat anti-pig IgG), and the microplates were incubated in a 37°C incubator for 45 min. After incubation, the microplates were washed five times with PBST for 3 min each time. The tetramethylbenzidine (TMB) chromogenic solution was added, 100 μL per well, and the microplates were kept at room temperature in the dark for 15 min. The reaction was stopped by adding 50 μL of 2 M $H_2SO_4$ to each well [28]. The concentrations of cytokines IL-2, IL-4, IL-10, and IFN-$\gamma$ in the sera were quantified using a commercially available ELISA kit (Cloud-Clone, Wuhan, China).

### 2.8. Spleen Lymphocytes Proliferation Test

Spleen lymphocytes separation medium (TBD, Tianjin, China) was used to isolate the spleen lymphocytes of the mice after euthanasia. In brief, the splenic lymphocytes were suspended in RPMI-1640 medium supplemented with 10% fetal bovine serum (FBS) and 5% penicillin/streptomycin (Gibco, Waltham, MA, USA). The cells were then cultured in 96-well microplates at a density of $5 \times 10^6$ cells per mL with 50 μL/well. The splenic lymphocytes were stimulated with $10^5$ $TCID_{50}$ ASFV as a specific antigen stimulation, concanavalin A (ConA) (10 μg/mL) as a positive control, and RPMI-1640 medium as a negative control, respectively. The plates were then incubated for 72 h in a 5% $CO_2$ incubator at 37°C. To assess the proliferation responses of the splenic lymphocytes, the cell counting kit-8 (CCK-8) assay (Ape×Bio, Houston, TX, USA) was used, after incubation at 37°C in an incubator in the dark for 3 h, the absorbance at 450 nm was read on a microplate reader. The data were presented as stimulation indices (SI), calculated as the ratio of the (stimulated value − blank value)/(unstimulated value − blank value).

### 2.9. Statistical Analysis

The results are presented as the means ± standard deviations (SDs). Statistical evaluations were conducted using GraphPad Prism version 9. To discern statistically significant variations among the various groups, a one-way analysis of variance (ANOVA) was employed, and in the text, ns denotes $p \geq 0.05$, indicating no significance, * denotes $p < 0.05$, ** denotes $p < 0.01$, *** denotes $p < 0.001$, and **** denotes $p < 0.0001$.

## 3. Results

### 3.1. Eight rL. lactis Strains Enabled Soluble Expression of Corresponding ASFV Antigens

The plasmid used to express the antigens of ASFV is pNZ8149, which can induce the expression of exogenous proteins in the presence of nisin. Given that the fusion of eight genes encoding ASFV antigens into a single plasmid might complicate the plasmid construction process and potentially impair the expression of the proteins, we constructed eight recombinant plasmids, each expressing only one ASFV antigen. However, the construction methods for these plasmids were largely similar. In summary, we obtained a linearized plasmid, pNZ8149, through inverse PCR and subsequently generated recombinant plasmids by performing homologous recombination between the amplified eight ASFV genes and the linearized pNZ8149, respectively (Figure 1a). To assess the expression levels and localization of the eight ASFV antigens, the r*L. lactis* strains expressing the respective ASFV antigens were induced with nisin. Subsequently, we analyzed the expression of the ASFV antigens in both the supernatant and precipitate of the ultrasonically lysed samples by Western blotting. The results demonstrate that all eight proteins, ASFV F317L (40 kDa), H171R (24 kDa), D117L (16 kDa), E120R (20 kDa), B602L (71 kDa), CD2v (43 kDa), p54 (22 kDa), and p72 (75 kDa), could be expressed in a soluble form (Figure 1b).

### 3.2. Stable Inheritance of Recombinant Plasmids Without Affecting Bacterial Strain Growth

In order to evaluate the growth characteristics of the eight constructed r*L. lactis* strains, the growth curve of r*L. lactis*-F317L was assessed as the model strain. The $OD_{600nm}$ values of r*L. lactis*-F317L, *L. lactis*-pNZ8149, and *L. lactis* NZ3900 were measured and recorded at 1 h intervals. This monitoring was continued until the bacteria reached the stationary phase, following which growth curves were plotted based on the collected data. The results show that there was no significant difference in the growth characteristics between the recombinant strain and the parental strain (Figure 2a). To verify the genetic stability of the r*L. lactis* strains, serial passages were conducted. At the 22nd generation, the r*L. lactis* strains were induced to express the corresponding ASFV antigens, and the expressions of these proteins were confirmed by Western blotting. The results indicate that the eight r*L. lactis* strains exhibited robust genetic stability. Specifically, at the 22nd generation, none of the recombinant plasmids had been lost, and stable expression of the eight ASFV antigens (F317L, H171R, D117L, E120R, B602L, CD2v, p54, and p72) was maintained (Figure 2b).

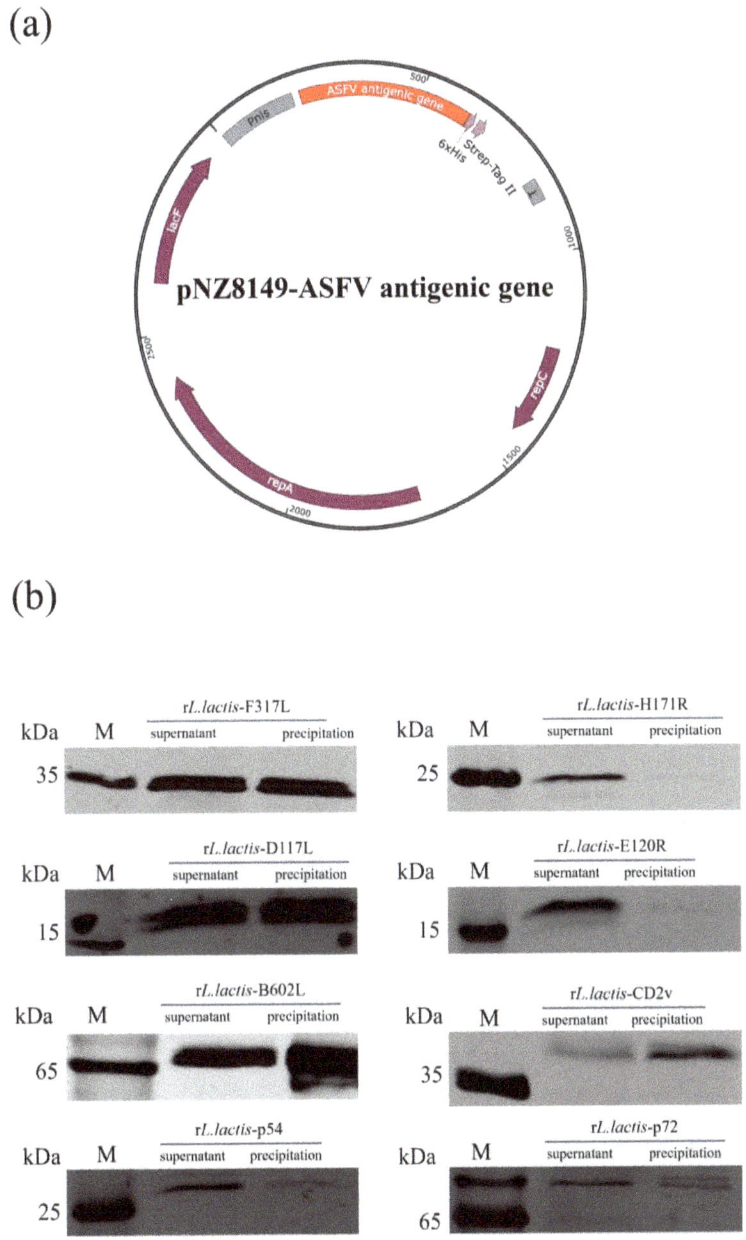


**Figure 1.** The construction of the recombinant plasmids and identification of soluble expressions of the eight ASFV antigens. (**a**) A schematic diagram of the recombinant plasmid expressing ASFV antigen. (**b**) Identification of the ASFV protein expressions of the r*L. lactis* strains by Western blotting. The r*L. lactis* strains were induced to express the proteins. Following ultrasonic disruption, the supernatant and precipitate were collected and subjected to SDS-PAGE electrophoresis to assess the expression status of the target proteins in these fractions.

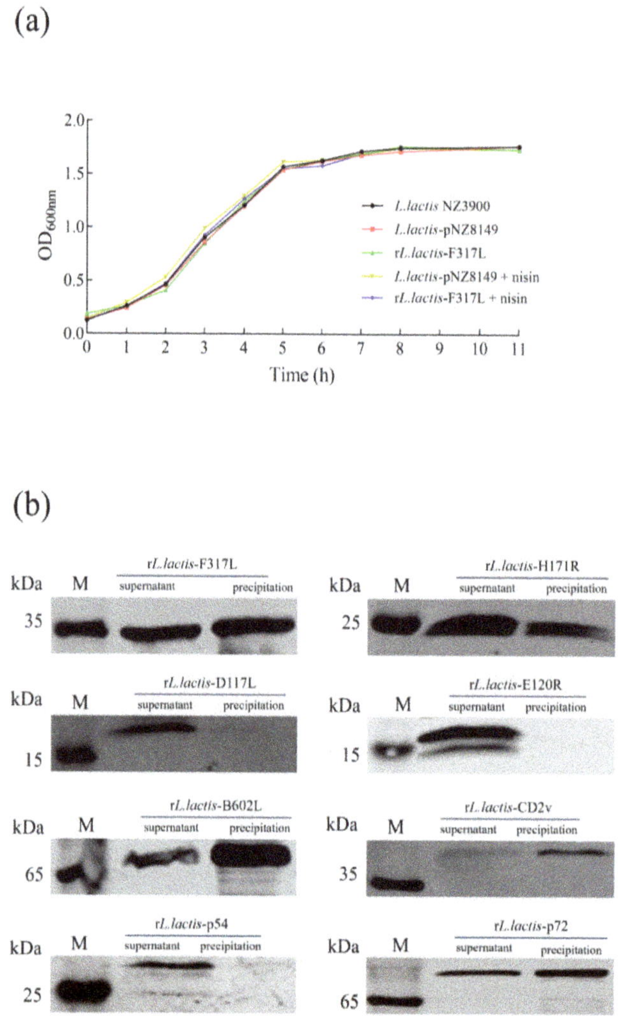


**Figure 2.** The growth characteristics and genetic stability of the r*L. lactis* strains. (**a**) The growth curve of the r*L. lactis* strains. Initially, the $OD_{600nm}$ values of the cultures were measured. Subsequently, the $OD_{600nm}$ values were recorded at hourly intervals until the bacterial stationary phase was reached. Utilizing these data points, the growth curves were constructed. (**b**) The detection of ASFV protein expressions in the 22nd-generation r*L. lactis* strains by Western blotting. The eight r*L. lactis* strains were subcultured daily, and upon reaching the 22nd generation, protein expressions were induced, and the supernatant and precipitate of the strain were collected using ultrasound to detect the presence of the target protein in these components.

### *3.3. Intramuscular Injection of* r*L. lactis-Mix Induced High-Level Serum IgG Antibodies in Immunized Mice*

To evaluate the immunogenic efficacy of the r*L. lactis*-mix, experimental mice were immunized with the $8 \times 10^9$ CFU of r*L. lactis*-mix through intramuscular injection and oral gavage. Additionally, the *L. lactis*-pNZ8149 group mice were immunized with $8 \times 10^9$ CFU of *L. lactis*-pNZ8149 via oral gavage, while the PBS group mice were immunized with 200 μL of PBS administered oral gavage. Sera and fecal samples were collected from the immunized mice at various time points (Figure 3a). In this study, the eight ASFV antigens were expressed and purified by prokaryotic expression system, and an indirect ELISA detection method was established to detect the specific IgG antibodies in the sera of immunized mice. The results show that the r*L. lactis* intramuscular immunization induced high-level serum IgG antibodies 7 d post-primary immunization, which were maintained until 7 d post-second booster immunization. Unfortunately, the mice in the oral gavage group did not exhibit a significant increase in sera antibodies compared with the control group (Figure 3b).

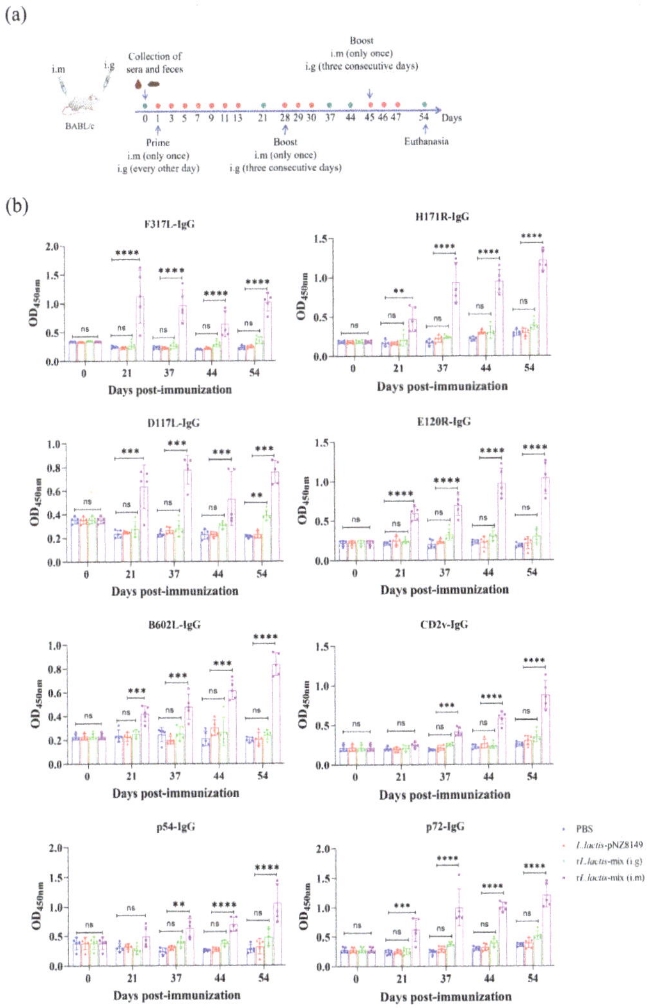

**Figure 3.** The mouse immunization schedule and sera IgG antibodies levels. (**a**) The mouse immunization schedule. The primary immunization, followed by oral gavage immunization on the subsequent day for a continuous period of 14 d. Two weeks later, a booster immunization was administered via oral gavage for three consecutive days. After another two weeks, a second booster immunization was given for three days. Each immunization was accompanied by a single intramuscular injection. (**b**) The level of specific IgG antibodies in the sera of the immunized mice. Each ASFV antigen protein was coated in 96-well microplates, and the serum specific IgG antibodies against different antigens in the immunized mice were detected by ELISA. The data were analyzed by a one-way analysis of variance. ns, $p \geq 0.05$; **, $p < 0.01$; ***, $p < 0.001$; ****, $p < 0.0001$.

*3.4. Oral Gavage of rL. lactis-Mix Induced High-Level sIgA Antibodies in Immunized Mice at the Early Stage of Immunization*

In order to evaluate the mucosal immune responses induced by the r*L. lactis*-mix, the feces of the immunized mice were collected at various time points, and the level of sIgA antibodies in the feces were detected by ELISA. The results show that sIgA antibodies were present in the feces of mice oral gavage with the r*L. lactis*-mix at 7 d after the primary immunization. However, in the fecal samples collected at the subsequent time points, the level of sIgA in the fecal samples of the r*L. lactis*-mix oral gavage group decreased to having no significant difference from that of the control group. Furthermore, there was no significant difference in the level of sIgA antibodies in feces between the r*L. lactis*-mix intramuscular injection group and the control group (Figure 4). Given that the intestinal mucosa provides a more precise indication of mouse intestinal sIgA levels, intestinal mucosal samples were collected from euthanized mice, and their sIgA levels were evaluated. The results show no significant difference in the level of intestinal mucosal sIgA between the mice in the r*L. lactis*-mix oral gavage group and the control group (Figure S1).

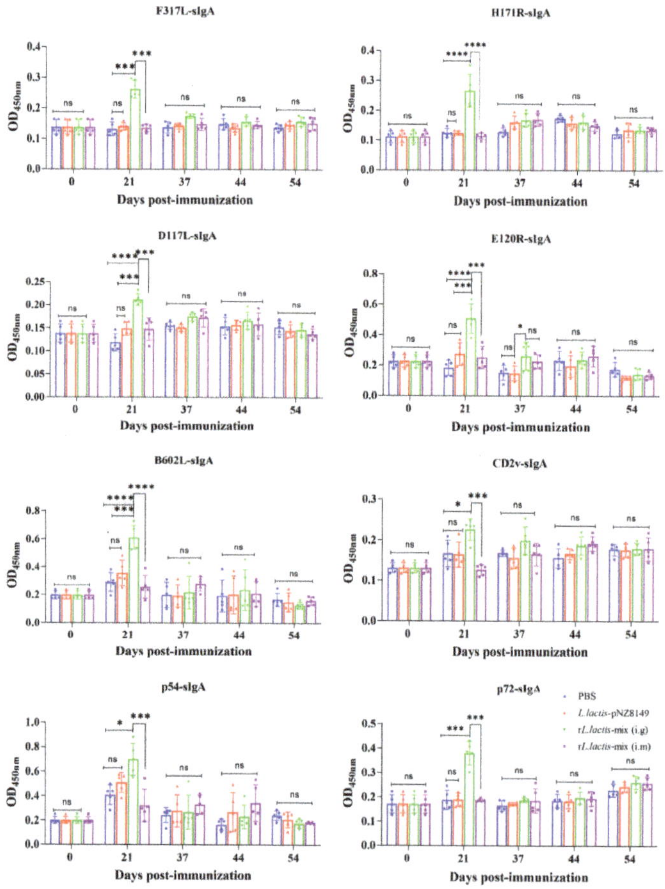


**Figure 4.** The sIgA antibodies in the feces of the immunized mice. Each ASFV antigen protein was coated in a 96-well microtiter plate, the levels of sIgA antibodies in the mice feces were detected by ELISA, and the statistical differences in the results were analyzed by a one-way analysis of variance. ns, $p \geq 0.05$; *, $p < 0.05$; ***, $p < 0.001$; ****, $p < 0.0001$.

### 3.5. Intramuscular Injection of rL. lactis-Mix Elevated Sera IFN-γ and IL-10 in Immunized Mice

In order to evaluate the effect of the r*L. lactis*-mix immunization on the innate immunity of mice, we evaluated the levels of cytokines, including IL-2, IL-4, IL-10, and IFN-γ in the sera of mice. IFN-γ and IL-2 are cytokines produced by Th1 cells, which can promote the activation and proliferation of cytotoxic T lymphocytes (CTLs), macrophages, and NK cells and enhance the number of $CD8^+$ T cells and innate immune responses. IL-4 and IL-10 belong to Th2 cytokine subset, which promote B cell proliferation and differentiation into plasma cells, thereby initiating antibody secretion, enhancing humoral immune response, and inhibiting inflammation. The results show that the levels of both IFN-γ and IL-10 in the sera of the mice in the r*L. lactis*-mix intramuscular injection group were significantly higher compared with those in the other groups (Figure 5).

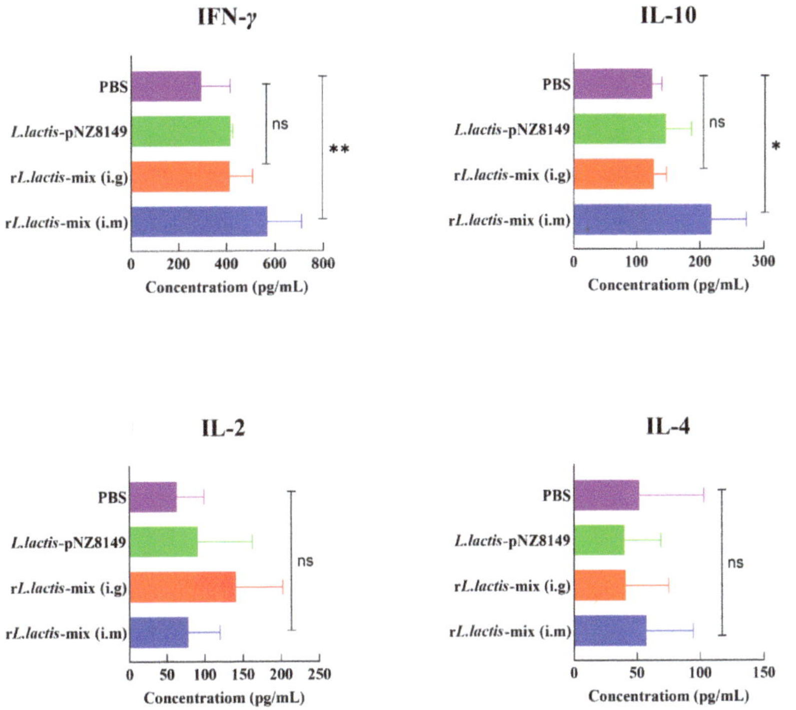


**Figure 5.** The levels of the Th1 cytokine subset (IFN-γ, IL-2) and the Th2 cytokine subset (IL-4, IL-10) in the sera of immunized mice. ns, $p \geq 0.05$; *, $p < 0.05$; **, $p < 0.01$.

To further evaluate the level of cellular immune response mediated by the mice immunized with the r*L. lactis*-mix, we determined the proliferation of spleen lymphocytes in the mice. Lymphocyte proliferation serves as an indicator of lymphocyte activity and functional status in response to relevant stimuli. The results show that there was no significant difference in the stimulation index of spleen lymphocytes among the four groups of mice spleen lymphocyte stimulated with $10^5$ $TCID_{50}$ ASFV (Figure S2).

## 4. Discussion

Given the persistent threat posed by ASF to the global pig industry, the development of effective and safe vaccines has become a matter of urgency. Various strategies have been explored for developing ASF vaccines, yet each approach harbors distinct limitations [29]. ASFV naturally infects pigs primarily through contact or aerosols, breaching the mucosal barrier and leading to disease onset [30]. Currently, ASF vaccine candidates are predominantly designed for intramuscular administration, which typically fails to elicit a mucosal immune response. Consequently, recombinant ASF vaccines employing *L. lactis* may offer a promising approach to prevent ASFV invasion during the early stages of infection. In this study, we constructed eight r*L. lactis* strains expressing ASFV antigens by utilizing the NICE expression system. These eight r*L. lactis* strains included r*L. lactis*-F317L, r*L. lactis*-H171R, r*L. lactis*-D117L, r*L. lactis*-E120R, r*L. lactis*-B602L, r*L. lactis*-CD2v, r*L. lactis*-p54, and r*L. lactis*-p72. Subsequently, a mixture of these eight r*L. lactis* strains was administered to mice via both intramuscular injection and oral gavage for immunization.

The results indicate that intramuscular injection of the r*L. lactis*-mix successfully elicited high-level IgG antibodies in the sera of mice, whereas oral gavage administration of the same r*L. lactis*-mix failed to induce specific IgG antibody production in the sera. This discrepancy could be attributed to the intracellular localization of the recombinant proteins, which might hinder efficient recognition and capture of the vaccine antigens by the intestinal mucosal epithelium, thereby impeding the induction of an effective immune response. In contrast, the r*L. lactis*-mix in the intramuscular injection group was disrupted by ultrasound, making the proteins more accessible to immune cells in the bloodstream and subsequently stimulating an immune response. Consistent with our findings, previous researchers constructed the r*L. lactis*-OptiVP2-RCK, which co-expressed the primary antigen VP2 of infectious bursal disease virus and the RCK protein of an intestinal *Salmonella* strain. Despite detecting no specific IgG antibodies in the sera following oral gavage immunization, neutralizing antibodies were present and conferred protection to chickens [31]. The authors speculated that differences in protein structure, as well as distinctions between mucosal and systemic immunity, could account for these observations.

The utilization of *L. lactis* as a vaccine vector is notably advantageous due to its capacity to induce mucosal immune responses via mucosal immunization. sIgA, the predominant antibody in mucosal secretions, offers protection against pathogen adhesion and intestinal barrier penetration. In this study, we observed a significant elevation in sIgA antibody levels in the feces of the mice immunized via oral gavage with the r*L. lactis*-mix on the 7 d post-primary immunization. However, following booster immunization, the sIgA levels in the feces of these mice decreased, showing no significant difference compared with the control groups. This decline may be attributed to the early-secreted nature of sIgA, as studies suggest its primary role during the initial secretory phase of immunization [32]. In contrast to systemic immunity, the mucosal immune system not only defends against pathogenic microorganisms but also tolerates a plethora of antigens and normal flora present in food [33]. The quantity, nature of the antigen, and the frequency of immunizations are crucial factors influencing the development of immune tolerance [34]. Oral immunization with small, soluble antigen doses has been demonstrated to potentially induce a reduction in host cellular immunity and the production of immune tolerance [35,36]. Therefore, the low-level expression of the ASFV antigen protein by r*L. lactis*, combined with the prolonged immune stimulation from the primary immunization, may have led to the induction of immune tolerance in the mouse intestinal tract. This could explain why the sIgA antibody levels in the feces of the immunized mice decreased to levels indistinguishable from those of the control group during the later stages of immunization and failed to stimulate an increase in lymphocyte proliferation.

Cytokines are small molecular proteins exhibiting diverse biological activities, produced by both immune and certain non-immune cells [37]. Serving as cellular signaling molecules, they regulate immune responses, participate in the differentiation and development of immune cells, mediate inflammatory reactions, stimulate hematopoiesis, and facilitate tissue repair. Specifically, cytokines such as IL-2 and IFN-$\gamma$, secreted by Th1 cells, augment the activation and proliferation of CTLs, macrophages, and NK cells, thereby augmenting cytotoxicity against infected cells, enhancing phagocytic activity, and strengthening $CD8^+$ T cell and innate immune responses. Conversely, IL-10 and IL-4 produced by Th2 cells promote the proliferation and differentiation of B cells into plasma cells, initiating antibody secretion and fostering humoral immune responses. Numerous studies have demonstrated the potential of utilizing *L. lactis* as a carrier to enhance both Th1 and Th2 immune responses. For instance, oral administration of pNZ8149-S1/NZ3900 expressing the S1 protein of porcine epidemic diarrhea virus (PEDV) can induce high levels of IL-4 and IFN-$\gamma$ in the sera of immunized mice [38], and using *L. lactis* as a carrier to express the gD protein of Herpes simplex virus type 1 (HSV-1) can also significantly elevate the levels of IFN-$\gamma$ in the serum of immunized mice [39]. Similar to previous studies, oral gavage of the r*L. lactis*-mix induced higher levels of IFN-$\gamma$ and IL-10 production in the sera of mice.

Concurrently, it is recognized as a limitation of this study that the immunogenicity and protective efficacy of these eight r*L. lactis* strains were not assessed in pigs. However, our findings provide valuable insights into the development of ASF mucosal vaccines based on *L. lactis*. The results indicate that oral gavage immunization can elicit the production of sIgA, potentially providing early protection to pigs, whereas intramuscular immunization more effectively induces the generation of serum-specific IgG antibodies. Furthermore, more experimental animals are needed to obtain more accurate and statistically significant data. Increasing the number of immunized mice may enhance the immune data mediated by r*L. lactis*. We hypothesize that surface-displayed antigens may enhance the immunogenicity by presenting antigens on the surface of the *L. lactis*, thereby increasing the chances of antigen capture by antigen-presenting cells (APCs) and potentially amplifying the initiation of immune responses. Additionally, increasing the expression levels of ASFV antigens could further boost the immunogenic properties of the vaccines.

## 5. Conclusions

In this study, we constructed eight pools of r*L. lactis* strains, each expressing different antigens of ASFV, and evaluated the immune responses induced by these strains through oral gavage and intramuscular injection in a mouse model. The results demonstrate that the intramuscular injection group induced high levels of serum-specific IgG antibodies and significantly elevated the levels of IFN-$\gamma$ and IL-10 in the serum. In the oral gavage immunization group, a transient increase in the sIgA antibody level was observed in fecal samples 7 d after the primary immunization. This indicates that these antigens possess good immunogenicity and that the multi-antigen cocktail vaccination is safe for mice. This study provides a reference for the development and immunization of ASF vaccines using *Lactobacillus* as a live vector.

**Supplementary Materials:** The following supporting information can be downloaded at: https://www.mdpi.com/article/10.3390/vetsci12020140/s1, Figure S1: The sIgA antibodies in the intestinal mucosa of the immunized mice; Figure S2: T lymphocyte proliferation of the immunized mice.

**Author Contributions:** Conceptualization, Y.S., H.W. and J.H.; validation, J.H. and H.W.; data curation, Y.S.; formal analysis, H.-J.Q., Y.S., J.H. and H.W.; funding acquisition, H.W. and Y.S.; investigation, J.H., H.W., T.G., D.Z., M.L., Z.L. and Y.L.; methodology, Y.S., H.W. and J.H.; project administration, H.W. and Y.S.; supervision, Y.S. and H.W.; writing—original draft, J.H.; writing—review and editing, H.-J.Q., J.H., Y.S., H.W., T.G., D.Z., M.L., Z.L. and Y.L. All authors have read and agreed to the published version of the manuscript.

**Funding:** This research was supported by the National Key Research and Development Program of China in the 14th Five-Year Period (grant no. 2021YFD1801403).

**Institutional Review Board Statement:** The animal study protocol was approved by the Committee on the Ethics of Animal Experiments of the Harbin Veterinary Research Institute (HVRI) of the Chinese Academy of Agricultural Sciences (CAAS) of China (No. 220802-02-GR).

**Informed Consent Statement:** Not applicable.

**Data Availability Statement:** The dataset is available from the corresponding author upon reasonable request.

**Acknowledgments:** Thank you to the staff of the Animal Experiments of the Harbin Veterinary Research Institute of the Chinese Academy of Agricultural Sciences for their help in our research, and especially to Hongyu Cui for gifting the host strain.

**Conflicts of Interest:** The authors declare no conflicts of interest.

## References

1. Zhao, D.; Sun, E.; Huang, L.; Ding, L.; Zhu, Y.; Zhang, J.; Shen, D.; Zhang, X.; Zhang, Z.; Ren, T.; et al. Highly lethal genotype I and II recombinant African swine fever viruses detected in pigs. *Nat. Commun.* **2023**, *14*, 3096. [CrossRef] [PubMed]
2. Penrith, M.L.; Kivaria, F.M. One hundred years of African swine fever in Africa: Where have we been, where are we now, where are we going? *Transbound. Emerg. Dis.* **2022**, *69*, e1179–e1200. [CrossRef] [PubMed]
3. Blome, S.; Gabriel, C.; Beer, M. Modern adjuvants do not enhance the efficacy of an inactivated African swine fever virus vaccine preparation. *Vaccine* **2014**, *32*, 3879–3882. [CrossRef] [PubMed]
4. Chen, W.; Zhao, D.; He, X.; Liu, R.; Wang, Z.; Zhang, X.; Li, F.; Shan, D.; Chen, H.; Zhang, J.; et al. A seven-gene-deleted African swine fever virus is safe and effective as a live attenuated vaccine in pigs. *Sci. China Life Sci.* **2020**, *63*, 623–634. [CrossRef] [PubMed]
5. Borca, M.V.; Ramirez-Medina, E.; Silva, E.; Vuono, E.; Rai, A.; Pruitt, S.; Holinka, L.G.; Velazquez-Salinas, L.; Zhu, J.; Gladue, D.P. Development of a highly effective African swine fever virus vaccine by deletion of the *I177l* gene results in sterile immunity against the current epidemic Eurasia strain. *J. Virol.* **2020**, *94*, e02017-19. [CrossRef] [PubMed]
6. Lopera-Madrid, J.; Medina-Magues, L.G.; Gladue, D.P.; Borca, M.V.; Osorio, J.E. Optimization in the expression of ASFV proteins for the development of subunit vaccines using poxviruses as delivery vectors. *Sci. Rep.* **2021**, *11*, 23476. [CrossRef]
7. Lokhandwala, S.; Waghela, S.D.; Bray, J.; Martin, C.L.; Sangewar, N.; Charendoff, C.; Shetti, R.; Ashley, C.; Chen, C.H.; Berghman, L.R.; et al. Induction of robust immune responses in swine by using a cocktail of adenovirus-vectored African swine fever virus antigens. *Clin. Vaccine Immunol.* **2016**, *23*, 888–900. [CrossRef] [PubMed]
8. Chathuranga, K.; Lee, J.S. African swine fever virus (ASFV): Immunity and vaccine development. *Vaccines* **2023**, *11*, 199. [CrossRef] [PubMed]
9. Kosowska, A.; Cadenas-Fernandez, E.; Barroso, S.; Sanchez-Vizcaino, J.M.; Barasona, J.A. Distinct African swine fever virus shedding in wild boar infected with virulent and attenuated isolates. *Vaccines* **2020**, *8*, 767. [CrossRef] [PubMed]
10. Zajac, M.D.; Trujillo, J.D.; Yao, J.; Kumar, R.; Sangewar, N.; Lokhandwala, S.; Sang, H.; Mallen, K.; McCall, J.; Burton, L.; et al. Immunization of pigs with replication-incompetent adenovirus-vectored African swine fever virus multi-antigens induced humoral immune responses but no protection following contact challenge. *Front. Vet. Sci.* **2023**, *10*, 1208275. [CrossRef]
11. Wang, Z.; Ai, Q.; Huang, S.; Ou, Y.; Gao, Y.; Tong, T.; Fan, H. Immune escape mechanism and vaccine research progress of African swine fever virus. *Vaccines* **2022**, *10*, 344. [CrossRef]
12. Bosch-Camos, L.; Lopez, E.; Rodriguez, F. African swine fever vaccines: A promising work still in progress. *Porcine Health Manag.* **2020**, *6*, 17. [CrossRef]
13. Iyer, L.M.; Balaji, S.; Koonin, E.V.; Aravind, L. Evolutionary genomics of nucleo-cytoplasmic large DNA viruses. *Virus Res.* **2006**, *117*, 156–184. [CrossRef]
14. Yang, S.; Miao, C.; Liu, W.; Zhang, G.; Shao, J.; Chang, H. Structure and function of African swine fever virus proteins: Current understanding. *Front. Microbiol.* **2023**, *14*, 1043129. [CrossRef] [PubMed]
15. Lu, W.; Bai, Y.; Zhang, S.; Zhao, X.; Jin, J.; Zhu, X.; Wang, R.; Wu, Y.; Zhang, A.; Zhang, G.; et al. An intracellular epitope of ASFV CD2v protein elicits humoral and cellular immune responses. *Animals* **2023**, *13*, 1967. [CrossRef]
16. Liu, W.; Li, H.; Liu, B.; Lv, T.; Yang, C.; Chen, S.; Feng, L.; Lai, L.; Duan, Z.; Chen, X.; et al. A new vaccination regimen using adenovirus-vectored vaccine confers effective protection against African swine fever virus in swine. *Emerg. Microbes Infect.* **2023**, *12*, 2233643. [CrossRef] [PubMed]

17. Song, J.; Wang, M.; Zhou, L.; Tian, P.; Sun, Z.; Sun, J.; Wang, X.; Zhuang, G.; Jiang, D.; Wu, Y.; et al. A candidate nanoparticle vaccine comprised of multiple epitopes of the African swine fever virus elicits a robust immune response. *J. Nanobiotechnol.* **2023**, *21*, 424. [CrossRef]
18. Jancovich, J.K.; Chapman, D.; Hansen, D.T.; Robida, M.D.; Loskutov, A.; Craciunescu, F.; Borovkov, A.; Kibler, K.; Goatley, L.; King, K.; et al. Immunization of pigs by DNA prime and recombinant vaccinia virus boost to identify and rank African swine fever virus immunogenic and protective proteins. *J. Virol.* **2018**, *92*, e02219-17. [CrossRef]
19. Dieye, Y.; Usai, S.; Clier, F.; Gruss, A.; Piard, J.C. Design of a protein-targeting system for lactic acid bacteria. *J. Bacteriol.* **2001**, *183*, 4157–4166. [CrossRef] [PubMed]
20. Cano-Garrido, O.; Rueda, F.L.; Sanchez-Garcia, L.; Ruiz-Avila, L.; Bosser, R.; Villaverde, A.; Garcia-Fruitos, E. Expanding the recombinant protein quality in *Lactococcus lactis*. *Microb. Cell Fact.* **2014**, *13*, 167. [CrossRef]
21. Zhou, X.X.; Li, W.F.; Ma, G.X.; Pan, Y.J. The nisin-controlled gene expression system: Construction, application and improvements. *Biotechnol. Adv.* **2006**, *24*, 285–295. [CrossRef]
22. Zhang, F.; Zhang, Z.; Li, X.; Li, J.; Lv, J.; Ma, Z.; Pan, L. Immune responses to orally administered recombinant *Lactococcus lactis* expressing multi-epitope proteins targeting M cells of foot-and-mouth disease virus. *Viruses* **2021**, *13*, 2036. [CrossRef] [PubMed]
23. Fatehi, Z.; Doosti, A.; Jami, M.S. Oral vaccination with novel *Lactococcus lactis* mucosal live vector-secreting *Brucella* lumazine synthase (BLS) protein induces humoral and cellular immune protection against *Brucella abortus*. *Arch. Microbiol.* **2023**, *205*, 122. [CrossRef] [PubMed]
24. Wang, N.; Li, J.; Wang, Y.; Zhang, D.; Shi, C.; Li, Y.; Bergmann, S.M.; Mo, X.; Yin, J.; Wang, Q. Recombinant *Lactococcus lactis* expressing grass carp reovirus VP6 induces mucosal immunity against grass carp reovirus infection. *Front. Immunol.* **2022**, *13*, 914010. [CrossRef] [PubMed]
25. Ma, C.; Li, G.; Chen, W.; Jia, Z.; Yang, X.; Pan, X.; Ma, D. *Eimeria tenella*: IMP1 protein delivered by *Lactococcus lactis* induces immune responses against homologous challenge in chickens. *Vet. Parasitol.* **2021**, *289*, 109320. [CrossRef] [PubMed]
26. Tian, Y.; Feng, T.; Hou, T.; Zhu, W.G. Protocol to purify the histone deacetylase sirt6 and assess its activity *in vitro*. *Star. Protoc.* **2023**, *4*, 102206. [CrossRef]
27. Liu, X.; Qi, L.; Lv, J.; Zhang, Z.; Zhou, P.; Ma, Z.; Wang, Y.; Zhang, Y.; Pan, L. The immune response to a recombinant *Lactococcus lactis* oral vaccine against foot-and-mouth disease virus in mice. *Biotechnol. Lett.* **2020**, *42*, 1907–1917. [CrossRef]
28. Huang, J.; Wu, H.; Gao, T.; Zhai, H.; Moon, A.; Song, X.; Li, S.; Lu, Z.; Lan, J.; Zhong, D.; et al. A pool of bacterium-like particles displaying African swine fever virus antigens induces both humoral and cellular immune responses in pigs. *Vaccines* **2025**, *13*, 5. [CrossRef] [PubMed]
29. Munoz-Perez, C.; Jurado, C.; Sanchez-Vizcaino, J.M. African swine fever vaccine: Turning a dream into reality. *Transbound. Emerg. Dis.* **2021**, *68*, 2657–2668. [CrossRef]
30. Podgorski, T.; Pepin, K.M.; Radko, A.; Podbielska, A.; Lyjak, M.; Wozniakowski, G.; Borowik, T. How do genetic relatedness and spatial proximity shape African swine fever infections in wild boar? *Transbound. Emerg. Dis.* **2022**, *69*, 2656–2666. [CrossRef] [PubMed]
31. Wang, W.; Song, Y.; Liu, L.; Zhang, Y.; Wang, T.; Zhang, W.; Li, K.; Qi, X.; Gao, Y.; Gao, L.; et al. Neutralizing-antibody-mediated protection of chickens against infectious bursal disease via one-time vaccination with inactivated recombinant *Lactococcus lactis* expressing a fusion protein constructed from the RCK protein of *Salmonella enterica* and VP2 of infectious bursal disease virus. *Microb. Cell Fact.* **2019**, *18*, 21. [PubMed]
32. Jiang, H.Q.; Bos, N.A.; Cebra, J.J. Timing, localization, and persistence of colonization by segmented filamentous bacteria in the neonatal mouse gut depend on immune status of mothers and pups. *Infect. Immun.* **2001**, *69*, 3611–3617. [CrossRef] [PubMed]
33. Pietrzak, B.; Tomela, K.; Olejnik-Schmidt, A.; Mackiewicz, A.; Schmidt, M. Secretory IgA in intestinal mucosal secretions as an adaptive barrier against microbial cells. *Int. J. Mol. Sci.* **2020**, *21*, 9254. [CrossRef] [PubMed]
34. Friedman, A.; Weiner, H.L. Induction of anergy or active suppression following oral tolerance is determined by antigen dosage. *Proc. Natl. Acad. Sci. USA* **1994**, *91*, 6688–6692. [CrossRef]
35. Chen, Y.; Kuchroo, V.K.; Inobe, J.; Hafler, D.A.; Weiner, H.L. Regulatory T cell clones induced by oral tolerance: Suppression of autoimmune encephalomyelitis. *Science* **1994**, *265*, 1237–1240. [CrossRef]
36. Nagler-Anderson, C.; Bhan, A.K.; Podolsky, D.K.; Terhorst, C. Control freaks: Immune regulatory cells. *Nat. Immunol.* **2004**, *5*, 119–122. [CrossRef] [PubMed]
37. Churiso, G.; Husen, G.; Bulbula, D.; Abebe, L. Immunity cell responses to RSV and the role of antiviral inhibitors: A systematic review. *Infect. Drug Resist.* **2022**, *15*, 7413–7430. [CrossRef] [PubMed]

38. Guo, M.; Yi, S.; Guo, Y.; Zhang, S.; Niu, J.; Wang, K.; Hu, G. Construction of a recombinant *Lactococcus lactis* strain expressing a variant porcine epidemic diarrhea virus S1 gene and its immunogenicity analysis in mice. *Viral. Immunol.* **2019**, *32*, 144–150. [CrossRef] [PubMed]
39. Qian, S.; Li, R.; He, Y.; Wang, H.; Zhang, D.; Sun, A.; Yu, L.; Song, X.; Zhao, T.; Chen, Z.; et al. Immunogenicity and protective efficacy of a recombinant *Lactococcus lactis* vaccine against HSV-1 infection. *Microb. Cell Fact.* **2024**, *23*, 244. [CrossRef]

*veterinary sciences*

MDPI

*Article*

# Paeoniflorin Inhibits Porcine Circovirus Type 2 Replication by Inhibiting Autophagy and Targeting AKT/mTOR Signaling

**Zhengchang Wu [1,2,†], Luchen Yu [1,†], Yueqing Hu [1], Wenbin Bao [1] and Shenglong Wu [1,*]**

1 Key Laboratory for Animal Genetics, Breeding, Reproduction and Molecular Design of Jiangsu Province, College of Animal Science and Technology, Yangzhou University, Yangzhou 225009, China; zcwu@yzu.edu.cn (Z.W.); luchenyu202209@163.com (L.Y.); 18595384877@163.com (Y.H.); wbbao@yzu.edu.cn (W.B.)

2 International Research Laboratory of Prevention and Control of Important Animal Infectious Diseases and Zoonotic Diseases of Jiangsu Higher Education Institutions, Yangzhou University, Yangzhou 225009, China

* Correspondence: slwu@yzu.edu.cn

† These authors contributed equally to this work.

**Simple Summary:** Porcine circovirus type 2 (PCV2) is the main pathogen causing post-weaning multisystemic wasting syndrome, which leads to enormous losses for the porcine industry. Due to the instability of commercial vaccines, it is necessary to identify effective antiviral drugs during PCV2 infection. In this study, we focused on the effect of paeoniflorin on PCV2 infection in porcine kidney cells (PK15). We found that paeoniflorin inhibits porcine circovirus type 2 replication by inhibiting autophagy by targeting AKT/mTOR signaling. Our study provides new information that could help the implementation of the antiviral drug paeoniflorin in PCV2 infection in the future.

**Abstract:** Porcine circovirus type 2 (PCV2) is an important pathogen that leads to great economic losses to the swine industry. Paeoniflorin (PF), a novel plant extract, has been reported to have antiviral properties. However, the role of paeoniflorin in regulating PCV2 replication remains unclear. Here, we used the CCK8 assay to demonstrate that PF within safe concentrations (0–275 mM) significantly inhibits PCV2 replication in a dose-dependent manner in porcine kidney cells. Subsequently, comparative transcriptome and functional verification revealed that PF probably inherits PCV2 replication via targeting AKT/mTOR signaling. Further experimental data show that the AKT/mTOR signaling pathway is highly relevant to autophagy. Thus, experimental data from Western blot, qPCR, and the indirect immunofluorescence test indicate that PF inhibits PCV2 replication by inhibiting autophagy by targeting the AKT/mTOR signaling pathway. Together, our results provide insight into the mechanism of paeoniflorin in regulating PCV2 replication and offer new ideas for the treatment of PCV2 infection in pigs.

**Keywords:** porcine circovirus type 2; paeoniflorin; autophagy; AKT/mTOR signaling

## 1. Introduction

Porcine circovirus (PCV) belongs to the genus circovirus within the family Circoviridae. It has no envelope and contains covalently closed single-stranded circular negative-sense DNA. Currently, four serotypes of porcine circovirus have been identified, namely PCV1, PCV2, PCV3, and PCV4 [1]. In 1991, a Canadian scholar, Clark, isolated a circovirus from diseased pigs suffering from postweaning multisystemic wasting syndrome (PMWS). Subsequently, Meehan et al. named this pathogenic circovirus porcine circovirus type 2 (PCV2). PCV2 is the smallest single-stranded circular DNA virus known to infect mammals, and the replication of PCV2 is heavily dependent on host cellular mechanisms [2]. PCV2 is also a major pathogen causing dermatitis and nephrotic syndrome, porcine interstitial pneumonia, sow reproductive disorders, infectious congenital tremors in piglets, etc. [3]. Since its discovery in the late 1990s, PCV2 has become a major pathogen causing porcine circovirus disease (PCVD) in pig production worldwide, thus leading to huge economic losses to the global pig industry [4–6]. However, the pathogenic mechanism of PCV2 is still unclear, and the main natural transmission route remains oronasal contact [7,8]. Lymphocyte exhaustion and histiocyte replacement in lymphoid tissues are the hallmark lesions of PCV2 infection [9]. PCV2 infection and replication in lymphoid tissues can disrupt the structure of lymphoid follicles, leading to lymphocyte exhaustion and subsequent replacement by histiocytes [10]. Thus, lymphoid follicle destruction and leukopenia associated with PCV2 infection can lead to immunosuppression in pigs. In addition, studies have demonstrated that PCV2 can replicate in bronchial and inguinal lymph nodes, the tonsils, lungs, liver, kidneys, spleen, and thymus tissue [11–14]. Numerous studies have shown that when PCV2 infects cells, the expression of cellular genes is generally altered, including the regulation of cellular signaling pathways, the induction of oxidative stress, apoptosis, and autophagy, the enhancement of the expression of inflammation-related genes, etc. [2]. Although the widespread use of commercial vaccines has reduced the subclinical symptoms of viral infections, the rate of antigen positivity remains high in the face of mixed infections with different PCV2 genotypes, constant genotypic variation, and different clinical manifestations of infection [15]. However, it is clear that vaccination cannot provide complete protection for the host against PCV2 infection.

Nowadays, a number of drugs with antiviral effects have been explored and developed. Natural products are a rich source of bioactive molecules and bear potential pharmacotherapeutic functions. It has been shown that natural extracts of scopoletin, artemisinin B, and artemisinic acid (a single fraction isolated from Artemisia annua) can interfere with the viral insertion and replication process by interacting with the 3CLpro and Spike proteins of SARS-CoV-2, which can help to combat the SARS-CoV-2 variant and other possible coronaviruses [16]. It has also been shown that polyphenol-rich sugarcane extract (PRSE), an extract from sugarcane treated with PRSE in influenza a virus (IAV)-infected cells, can inhibit the early stages of infection and reduce viral genome replication, mRNA transcription, and viral protein expression [17]. In addition, PRSE treatment also attenuates the replication of multiple H3N2 and H1N1 subtypes of IAV strains [18]. As a result, various plant extracts and natural components are continuously being exploited for their potential value and become novel antiviral agents. However, there are still relatively few studies on PCV2 at present; this lack of in-depth research has hindered the development of more targeted and efficient strategies to deal with PCV2-related issues.

Paeoniflorin (PF) is a monoterpene glycoside compound, which is the main bioactive component of *Paeonia lactiflora* [19]. During the cultivation of medicinal paeonies, the above-ground parts are often discarded as waste, resulting in a huge waste of resources. It has been recorded that PF, as an important active ingredient in paeonia lactiflora, has the ability to regulate immunity, antioxidant, anti-inflammatory, neuroprotective, inhibit tumor cells, and carry out other pharmacological activities [20–22]. Studies have shown that PF also has some therapeutic effects on viral infections. For example, some researchers have investigated the effects of PF on myocardial injury in coxsackie B3 virus (CVB3)-induced viral myocarditis (VMC) mice and its possible mechanism of action. The results showed that PF attenuated the inflammatory response of cardiomyocytes in VMC mice, reduced cardiomyocyte apoptosis, and ameliorated the myocardial tissue injury caused by CVB3 by inhibiting the activation of NLRP3 inflammatory vesicles [23–25]. Food rich in PF would be potential health-oriented foods, while natural plant extracts such as PF also have the prospect of applications as functional feed additives for livestock and poultry. As the main therapeutic component of paeonia lactiflora, PF has a wide range of applications and is a drug with great market prospects. However, studies on the effect of paeoniflorin on PCV2 are scarce, and its mechanism of action remains unclear, thus urgently awaiting our exploration.

To elucidate the role and underlying mechanism of PF in PCV2 replication, we use porcine kidney cells (PK15) as a model to investigate the effect of PF on PCV2 infection. This study first confirmed that PF significantly inhibited PCV2 replication in PK-15 cells. We then verified that the AKT/mTOR signaling pathway is closely associated with PCV2 replication. Finally, we investigated the regulation of the AKT/mTOR pathway by PF and elucidated the role of PF in regulating PCV2 replication through the AKT/mTOR pathway. Herein, this study facilitates the use of PF in the treatment of PCV2 infection in the future and provides a basis for further research on the mechanism of PCV2 replication.

## 2. Materials and Methods

### 2.1. Cell Culture, RAPA Treatment, and PCV2 Infection

PK15 cells (ATCC, CCL-33) were cultured in Dulbecco's modified Eagle's medium (DMEM) (Thermo, Waltham, MA, USA) supplemented with 10% fetal bovine serum (FBS) (Bio-Channel, Nanjing, China) and penicillin/streptomycin (100 U/mL) (Solarbio, Beijing, China) in a humidified atmosphere of 5% $CO_2$ and 95% air at 37 °C. PF ($\geq$98%, HPLC) (DASF, Nanjing, China) was dissolved in DMEM for 48 h. Rapamycin (RAPA) (MedChem-Express, Monmouth Junction, NJ, USA) was dissolved in DMEM to a concentration of 0.1 nM for 48 h. Additionally, PCV2d was preserved in our laboratory. PK15 cells were infected with PCV2d for a multiplicity of infection (MOI) of 1 for all experiments conducted.

### 2.2. CCK-8 Assay

PF was dissolved in the dimethyl sulfoxide (DMSO) (Solarbio, Beijing, China). A total of 20,000 cells (100 μL) per well were uniformly inoculated into 96-well plates containing 10% FBS medium. After incubation in a 37 °C, 5% $CO_2$ incubator for 24 h, the medium was replaced with the medium containing different concentrations of PF (15 mM, 30 mM, 60 mM, 125 mM, 250 mM, 500 mM, 1000 mM, 100 mM, 125 mM, 150 mM, 175 mM, 200 mM, 225 mM, 250 mM, 275 mM, 300 mM, 400 mM, 425 mM, 450 mM) in DMEM for 48 h. The Cell Counting Kit-8 (CCK-8) (Vazyme, Nanjing, China) was used to detect cell viability. The CCK-8 solution (Vazyme, Nanjing, China) was added to 96-well plates at 10 μL per well. Optical density values were detected at 450 nm using a Tecan Infinite 200 microplate reader (Sunrise, Tecan, Zurich, Switzerland).

### 2.3. Indirect Immunofluorescence Assay

Cells were infected with PCV2 and collected at different time points for the experiments. Cells were washed with phosphate-buffered saline (PBS), fixed by adding 4% paraformaldehyde, and incubated at 37 °C for 30 min. The cells were then treated with 0.5% Triton X-100 (Vazyme, Nanjing, China) for 15 min to rupture the cell membrane, and then 5% bovine serum albumin (BSA) (Solarbio, Beijing, China) was added and the cells were closed for 2 h at 37 °C. This was followed by incubation with PCV2 capsid (PCV2 *CAP* Antibody, VMRD, Pullman, WA, USA) and secondary Alexa Fluor 555 Conjugated Goat anti-mouse IgG (Huabio, Beijing, China). Finally, 4′,6-diamidino-2-phenylindole (DAPI) was added and incubated for 5 min, and then cells were treated by adding an antifade mounting medium for fluorescence (Biosharp, Beijing, China). The cells that underwent different treatments were then observed under a fluorescence microscope (Leica Microsystems, Wetzlar, Germany).

### 2.4. qPCR Analysis

The RNA extraction was isolated using the TRIzol reagent (Thermo, Waltham, MA, USA), and cDNA was synthesized using the HiScript II Q Select RT SuperMix with the gDNA Remover or miRNA 1st Strand cDNA Synthesis Kit (Vazyme, Nanjing, China) according to the manufacturer's instructions. The DNA extraction was isolated using the FastPure Cell/Tissue DNA Isolation Mini Kit (Vazyme, Nanjing, China) according to the manufacturer's instructions. AceQ qPCR SYBR Green Master Mix (Vazyme, Nanjing, China) was used to perform qPCR. Three replicates were set for each sample, and the CT value was average. Based on the gene sequences published in the GenBank database, qPCR primers were designed using Premier 6.0 software with *β-actin* as the internal reference gene (Table S1). All primers were prepared using TsingKe (TsingKe, Beijing, China).

### 2.5. Western Blot Analysis

After PCV2 infection, cells were collected and washed with PBS, pre-cooled at 4 °C, and a mixture containing the protein lysis solution (Applygen, Beijing, China), 100× phosphatase inhibitor (TargetMOI, Boston, MA, USA), and 100× protease inhibitor (TargetMOI, Boston, MA, USA) was added. After lysing the cells for 20 min at low temperature, the supernatant was obtained by centrifugation at 4 °C, 12000 rpm for 15 min. A bicinchoninic acid kit (Yeason, Shanghai, China) was used to normalize protein levels. A 5 × loading buffer (Vazyme, Nanjing, China) was added and denatured at 98 °C for 10 min. Finally, the denatured samples were stored at −20 °C.

The treated samples were separated using 10% sodium dodecyl sulfate–polyacrylamide gels and electro-transferred onto polyvinylidene fluoride (PVDF) membranes (Millipore, Billerica, MA, USA). After blocking for 2 h at room temperature, primary antibodies (PCV2 *CAP* antibody, GeneTex, San Antonio, TX, USA; HSP90 antibody, Proteintech, Wuhan, China; P62 antibody, Proteintech, Wuhan, China; LC3 antibody, Proteintech, Wuhan, China; mTOR antibody, Abcepta, Hangzhou, China; p-mTOR antibody, Huabio, Hangzhou, China; AKT antibody, Santa Cruz, Shanghai, China; p-AKT antibody, Santa Cruz, Shanghai, China) were incubated overnight at 4 °C. The PVDF membranes were washed by TBST and incubated with secondary antibodies (Proteintech, Wuhan, China) at room temperature for 2 h. The concentration of the primary antibody was 1 μg/mL, and that of the secondary antibody was 0.02 μg/mL. After being washed with TBST, samples were analyzed by ECL development (Tanon, Shanghai, China).

### 2.6. Confocal Immunofluorescence Microscopy

PK-15 cells stably expressing lentiviral pLV-GFP-mCherry-LC3 were inoculated on the slides of 12-well plates. After treatment with PF and RAPA, the cells were inoculated with PCV2 for 2 h and then incubated with 2% FBS DMEM, respectively, for 48 h. After fixation and staining of the nuclei with DAPI, the cells were observed under a confocal laser scanning microscope.

### 2.7. Flow Cytometry Analysis

The Annexin V-FITC/PI Apoptosis Detection Kit (Solarbio, Beijing, China) was adopted to detect changes in the apoptosis levels of PK15 cells and PF-treated cells. The treated cells were digested with trypsin into centrifuge tubes without EDTA and centrifuged to remove the supernatant. The single-stained tube group was used as a positive control, and 5 μL of the Annexin V/FITC mix was added and incubated at room temperature in the dark for 5 min. Solution without Annexin V/FITC and propidium iodide was used as the negative control group. Annexin V/FITC and PI were added to the treatment group.

Apoptosis was detected using a FACScan flow cytometer (Becton Dickinson, Franklin Lake, CA, USA). Analysis was performed using CytExpert 2.3 and FlowJo 7.6 software.

### 2.8. RNA Sequencing

After 48 h of PF treatment, PK15 cells were collected from the blank control group (NC, n = 4) and PF treatment group (PF, n = 4). Total RNA was extracted using the Trizol reagent, and the quality of RNA was assessed by 1% formaldehyde denaturing agarose gel electrophoresis. The RNA concentration was determined using an ND-1000 nucleic acid/protein concentration analyzer. Subsequently, all RNA samples were converted into double-stranded cDNA and sequenced on the Illumina Hiseq2500 platform from Oebiotech (Shanghai, China). Differential expression genes (DEGs) were identified based on the principles of corrected $p$-values < 0.05 and $|\log_2(\text{foldchange})| < 1$. The Gene Ontology (GO) function of DEGs was labeled on the returned data using GOseq. The enrichment of DEGs in the Kyoto Encyclopedia of Genes and Genomes (KEGG) pathway was detected using KOBAS software (version 3.0). GO terms and KEGG pathway analysis were performed using a corrected cut-off value of 0.05. The raw high-throughput RNA-seq data from this analysis were uploaded to the NCBI SRA (Sequence Read Archive) database (accession number PRJNA1173311).

### 2.9. Statistical Analysis

SPSS 18.0 software (Chicago, IL, USA) was employed to conduct statistical analyses. Relative quantitative results were examined via the $2^{-\Delta\Delta Ct}$ method. Results were represented as mean ± standard deviation (SD), followed by a comparison via Student's $t$-test. $p < 0.05$ or $p < 0.01$ indicated statistical significance. Semi-quantification was performed using ImageJ software (National Institutes of Health, Bethesda, MD, USA).

## 3. Results

### 3.1. Safety Assessment of PF in PK15 Cells

In order to investigate the safe working concentration of PF, we treated PK15 cells with different concentrations of PF (100 mM, 125 mM, 150 mM, 175 mM, 200 mM, 225 mM, 250 mM, 275 mM, 300 mM, 325 mM, 350 mM, 375 mM, 400 mM, and 425 mM, 450 mM) before incubation for 48 h. After that, the effects of different concentrations of PF on the activity of PK15 cells were detected according to the CCK8 method. The results showed that there was no clear toxic response to PK15 cells when the PF concentration was 0–250 mM (Figure 1A); therefore, we further subdivided the concentration gradient, and the results showed that there was no obvious toxic response to PK15 cells when the PF concentration was 0–275 mM; moreover, the addition of PF from 300 mM onwards significantly decreased the cell activity (Figure 1B). Therefore, the three acting concentrations (100 and 200 mM), which did not cause a toxic response to PK15 cells, were taken for subsequent experiments. A blank control and three concentrations of PF-treated groups were set up, and an apoptosis assay was performed after 48 h of drug treatment. The results of flow cytometry showed that there was no significant difference in the apoptosis level of PK15 cells between the drug-treated group and the blank control group (Figure 1C), indicating that PF does not affect the apoptosis level of PK15 cells (Figure 1D). To conclude, when we used PF at concentrations of 0–275 mM, there was no significant inhibition of PK15 cell viability or apoptosis.

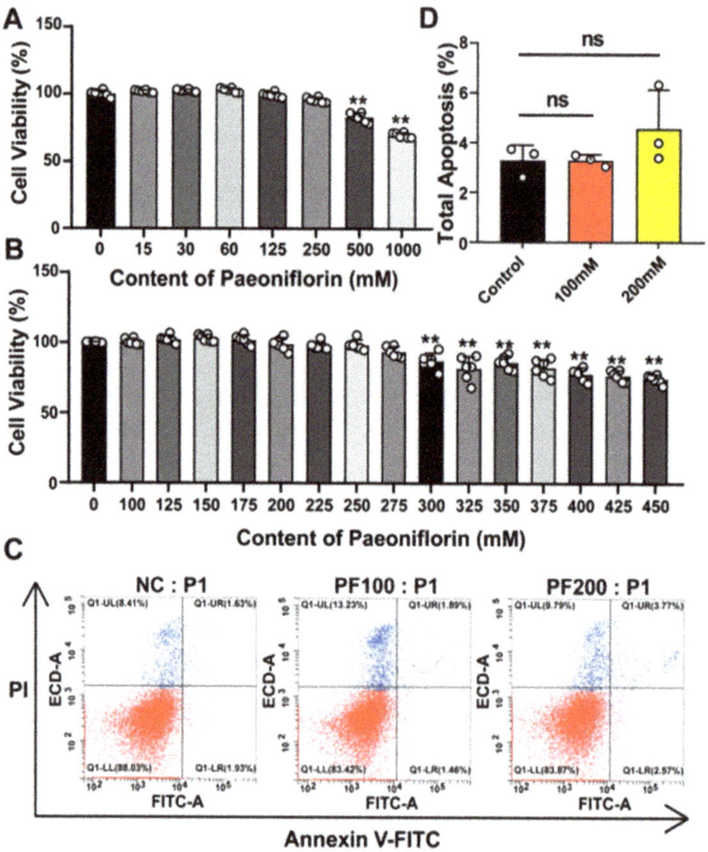


**Figure 1.** The safety assessment of PF. (**A**) The gradient dilution method was used to determine the approximate working concentration of PF. (**B**) The effect of 0–450 mM PF on the activity of PK15 cells was detected through the CCK8 method. (**C**) The apoptosis levels in cells treated with 100 mM and 200 mM PF were detected by flow cytometry. (**D**) Statistical analysis of the flow cytometry was performed to detect the apoptosis levels in PK15 cells after PF treatment. All data are presented as the mean ± SD, $^{ns}$ $p > 0.05$, ** $p < 0.01$.

### *3.2. Construction of PCV2-Infected PK15 Cell Model*

In order to detect the proliferation pattern of PCV2 (MOI = 1) on PK15 cells and to determine a time point for the subsequent collection of samples, we used qPCR, Western blot, and the indirect immunofluorescence assay (IFA) to detect the expression of the PCV2 *CAP* gene at different time points. As is shown in Figure 2, after the PCV2 infection of PK15 cells, the replication level of the PCV2 *CAP* gene in PK15 cells from 0 to 48 h showed an increasing trend with time. However, the expression of the PCV2 *CAP* gene in PK15 cells slightly decreased over a 48–72 h period (Figure 2A). Both Western blot (Figure 2B) and IFA (Figure 2C) results were consistent with the quantitative results. Thus, we successfully established a model of PCV2-infecting PK15 cells and determined that 48 h is the peak time point for PCV2 replication.

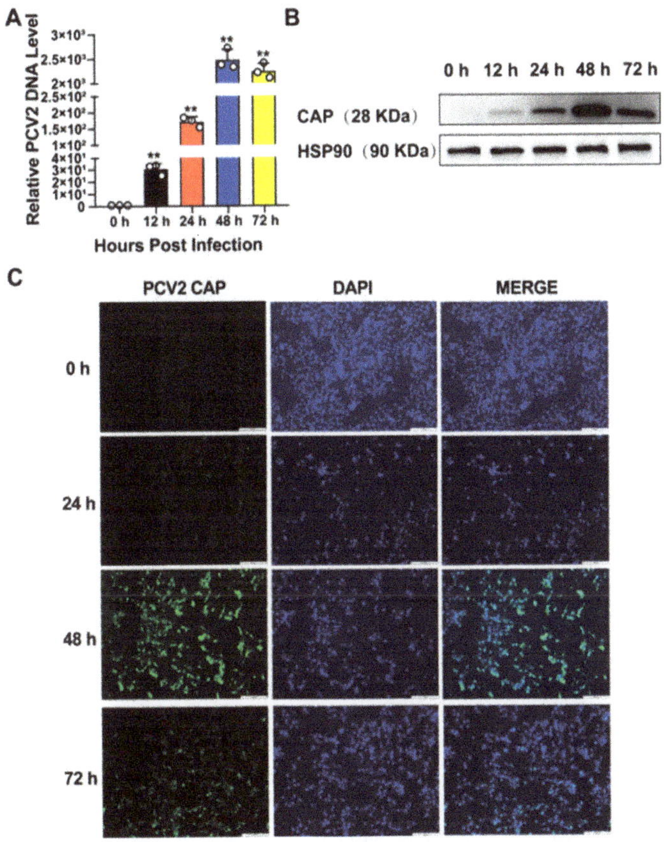


**Figure 2.** The establishment of the PCV2 infection model. (**A**) PK15 cells were infected with PCV2 for designated lengths of time (0, 12, 24, 48, 72 h), and the relative expression of PCV2 CAP was measured using qPCR. (**B**) The protein expression of the PCV2 capsid protein in PK15 cells within the specified time periods was analyzed. (**C**) The results of the fluorescent expression of the PCV2 capsid protein in PK15 cells within the specified time periods are shown. The image scale is 100 μm. All data are presented as the mean ± SD, ** $p < 0.01$.

### 3.3. *Effect Analysis of PF on PCV2 Replication*

To investigate whether PF inhibits the replication of PCV2 in PK15 cells, qPCR, Western blot, and IFA were employed to detect the expression of the PCV2 *CAP* gene. The experiment was designed with different concentrations (0 mM, 5 mM, 15 mM, 25 mM, 50 mM, 100 mM, 200 mM) and treatment methods (pre-treatment (PF was added 24 h before PCV2 addition), simultaneous treatment (PF was added simultaneously with PCV2), continuous treatment (pre-treatment + simultaneous treatment), and post-treatment (PF was added 24 h after PCV2 addition)), and the conditions at the same time points (24 h, 48 h, 72 h) were compared. In the concentration experiment, the addition of PF reduced the expression of the PCV2 *CAP* gene, suggesting its inhibitory effect on PCV2. Moreover, it was found that the inhibitory effect of paeoniflorin on PCV2 replication was enhanced with the increase in treatment concentration, showing a dose-dependent manner (Figure 3A). The Western blot results were consistent with the quantitative results (Figures 3B and S1). Subsequently, cells infected with PCV2 were treated with the same concentration of PF but with different treatment methods. The results indicate that when the treatment method was not used post-treatment, the expression of the PCV2 CAP gene in the PF-treated group was significantly reduced compared with that in the negative control group (Figure 3C), and the Western blot results were consistent with the quantitative results (Figures 3D nd S1). In particular, the PF pre-treatment had the most obvious inhibitory effect on PCV2, suggesting the great potential of paeoniflorin as a drug for preventing PCV2. When the concentration and treatment method of PF were the same, the inhibition rate of the PCV2 *CAP* gene expression was the highest at 48 h (Figure 3E), and the Western blot results were consistent with the quantitative results (Figures 3F and S1). Samples were collected after the specified treatment time and subjected to IFA experiments. The results showed that the fluorescence expression in the PF-treated group was significantly lower than that in the negative control group (Figure 3G). Therefore, PF has a significant anti-PCV2 replication effect in PK15 cells.

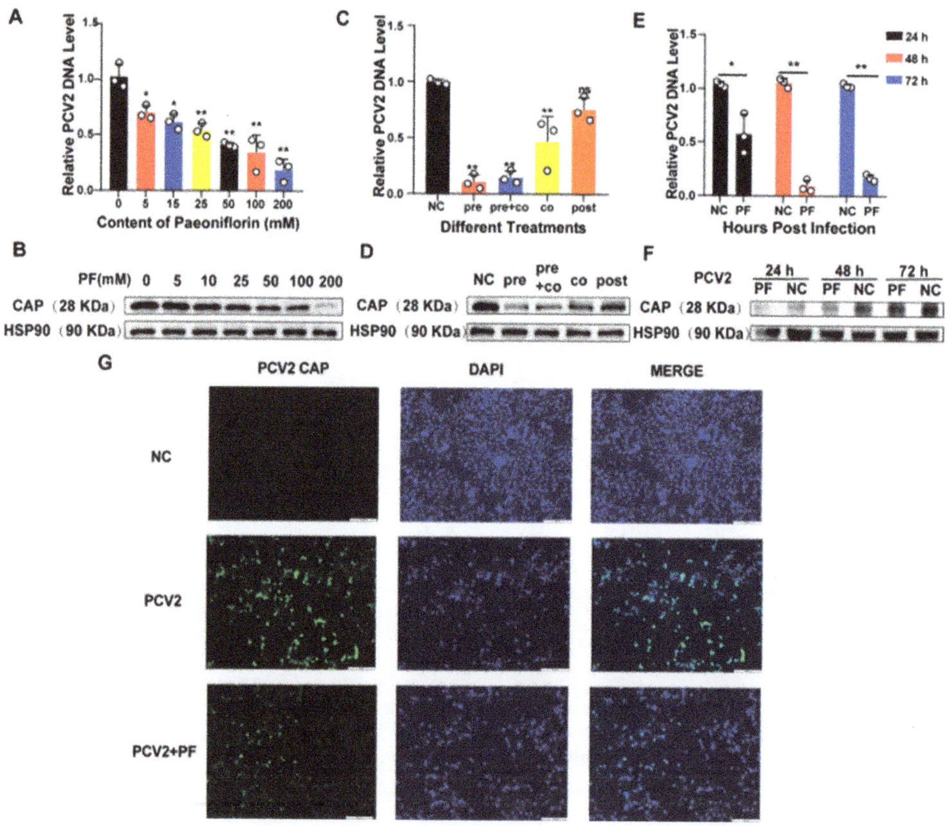


**Figure 3.** The effect of PF on PCV2 replication. (**A**,**B**) Different concentrations of PF (0 mM, 5 mM, 15 mM, 25 mM, 50 mM, 100 mM, 200 mM) were used to treat PCV2-infected cells. Subsequently, the DNA expression level of PCV2 CAP was detected by qPCR, and the expression levels of PCV2 CAP were determined by Western blot. (**C**,**D**) PCV2-infected cells were treated with different PF addition methods (pre: adding PF 24 h before adding PCV2; co: adding PF simultaneously with adding PCV2; post: adding PF 24 h after adding PCV2). Then, the DNA expression level of PCV2 CAP was detected by qPCR, and the expression levels of PCV2 CAP were detected by Western blot. (**E**,**F**) PCV2-infected cells were treated with PF for the indicated times. After that, the DNA expression level of PCV2 CAP was detected by qPCR, and the expression levels of PCV2 CAP were detected by Western blot. (**G**) The expression of PCV2 CAP on PK15 cells after 48 h of pre-treatment of PF and the protein fluorescence expression on PK15 cells are presented. The scale of the image is 100 μm. All data are presented as the mean ± SD, $^{ns}$ $p > 0.05$, * $p < 0.05$, ** $p < 0.01$.

### 3.4. Transcriptome Analysis of PK15 Cells with PF Treatment and PCV2 Infection

To explore potential regulators associated with the cellular responses to PF (200 mM) treatment, cell samples treated by PF for 48 h and corresponding controls were collected for transcriptome analysis with the RNA-seq technique. After quality control, an average of 48.9 million clean reads per sample was obtained (Table S2). Differential expression analysis was then performed to identify transcriptomic differences between the two groups. The principal component analysis plot showed a high degree of similarity between the samples and a high degree of reliability in the sequencing results (Figure 4A). Hierarchical clustering analysis was then performed based on expression levels to show the differences in the expression patterns of differentially expressed genes between the two groups. Differential gene grouping cluster plots showed significant differences between the blank-treated group and the PF-treated group (Figure 4B). Differential gene expression analysis showed that a total of 1787 differentially expressed genes were detected, with 742 differentially upregulated and 1045 differentially downregulated genes (Figure 4C and Table S3). KEGG enrichment analysis of all the differentially expressed genes in the combined analysis showed that DEGs were mainly enriched in cytokine–cytokine receptor interactions and the PI3K/AKT signaling pathway (Figure 4D).

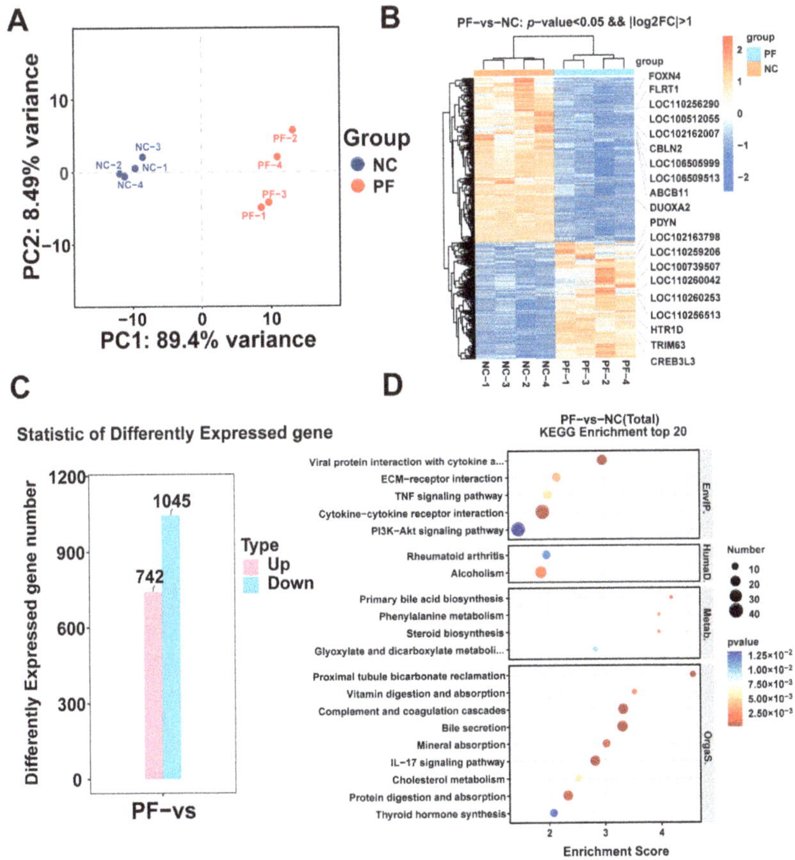


**Figure 4.** Sequencing results of PF-treated (200 mM) and negative control groups. (**A**) Principal component analysis plot showing the similarity between sequencing groups. (**B**) Clustering plot of differential gene grouping, where red indicates the relatively high expression of protein-coding genes and blue indicates the relatively low expression of protein-coding genes. (**C**) Statistical histogram of

differentially expressed genes showing the number of differentially expressed genes that were up-regulated and down-regulated. (**D**) KEGG bubble map: the bubbles represent the top 20 KEGG pathways enriched in DEGs. The size of the dots indicates the degree of DEG enrichment. The color of the dots indicates the importance of DEG enrichment.

We then used the previous sequencing results of the subject group [26] and jointly analyzed the negative control group, the PF-treated group, and the PCV2 infection group. Differential expression analysis was then performed to identify transcriptomic differences between the two groups. We found that a total of 455 genes co-increased, 489 genes co-decreased, and 31 genes showed different trends in the comparison between the two groups. In addition, the results showed that gene expression underwent similar changes in these two groups. (Figure 5A and Table S4). The KEGG enrichment analysis of all the differentially expressed genes in the combined analysis showed that DEGs were mainly enriched in cytokine–cytokine receptor interactions, the PI3K/AKT signaling pathway, and the MAPK signaling pathway (Figure 5B and Table S5). After reading the relevant literature and conducting preliminary experiments, we selected the PI3K/AKT signaling pathway, which ranks second in KEGG, to complete the subsequent verification and further experiments.

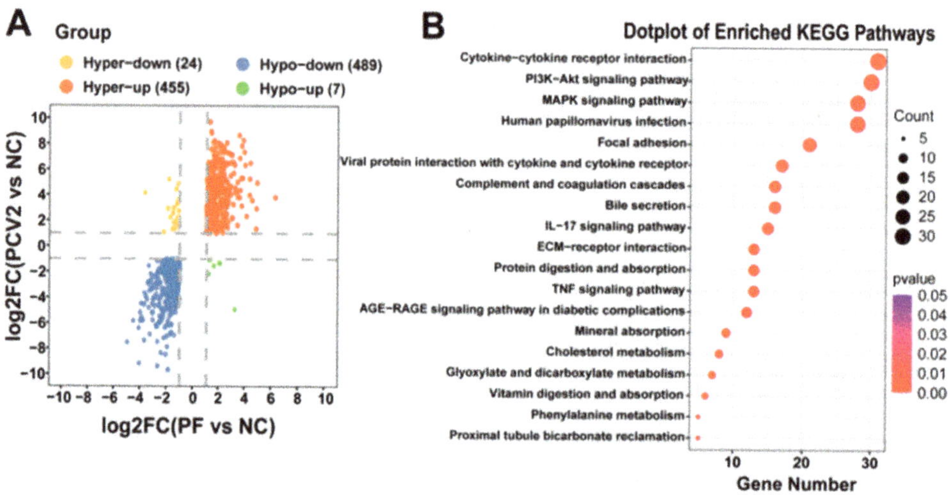


**Figure 5.** Sequencing results of PF-treated, PCV2-infected, and negative control groups. (**A**) The four-quadrant plot of differential genes, in which red denotes co-increasing differential genes; blue denotes co-decreasing differential genes; yellow denotes differential genes that rise in PCV2 infection sequencing and fall in PF-treated sequencing; and green denotes differential genes that fall in PCV2 infection sequencing and rise in PF-treated sequencing rising differential genes. (**B**) KEGG bubble map: the bubbles represent the top 20 KEGG pathways enriched in DEGs. The size of the dots indicates the degree of DEG enrichment. The color of the dots indicates the importance of DEG enrichment.

### *3.5. PF Modulates the AKT/mTOR Pathway to Inhibit Autophagy to Suppress PCV2 Replication*

To test whether PF (200 mM) can regulate the AKT/mTOR pathway, we examined the activity of the AKT/mTOR pathway by Western blot and performed a rescue experiment utilizing the autophagy activator rapamycin. The results indicated that the addition of PF probably contributed to the activity enhancement of the AKT/mTOR pathway, while RAPA effectively inhibited this effect (Figures 6A,B and S2). However, the addition of RAPA diminished the efficacy of PF, resulting in a rescue of PCV2's inhibitory effects (Figures 6C,D and S2). To further validate the results regarding the activity of the

AKT/mTOR pathway, we conducted a detection of the autophagy level downstream of this pathway. The samples were divided into four groups: the blank control group, the PF-treated group, the autophagy activator RAPA-treated group, and the group treated with both PF and rapamycin. Western blot was employed to determine the expression levels of the P62 and LC3 proteins. The results demonstrated that in the PF-treated group, the expression of the LC3-II protein was significantly reduced. Conversely, in the rapamycin-treated group, the expression of the LC3-II protein was elevated. When cells were treated with both PF and rapamycin, the effect of PF on decreasing the expression of the LC3-II protein was diminished (Figures 6E and S2). For the purpose of obtaining a more comprehensive understanding, we further explored the impact of PCV2 infection on the level of cellular autophagy. Specifically, we utilized Western blot to analyze the expression levels of P62 and LC3 proteins in four distinct groups: the blank infection group, the PF-treated group after infection, the autophagy activator RAPA-treated group after infection, and the group co-treated with PF and RAPA. The results demonstrated that after PCV2 infection, the expression of the LC3-II protein was significantly enhanced. However, when PF treatment was applied, the expression of the LC3-II protein was markedly reduced. When RAPA was added to the PCV2-infected cells, the increase in LC3-II protein expression was further aggravated. When PCV2-infected cells were treated with both PF and RAPA simultaneously, the inhibitory effect of PF on the expression of the LC3-II protein was neutralized by RAPA (Figures 6F and S2).

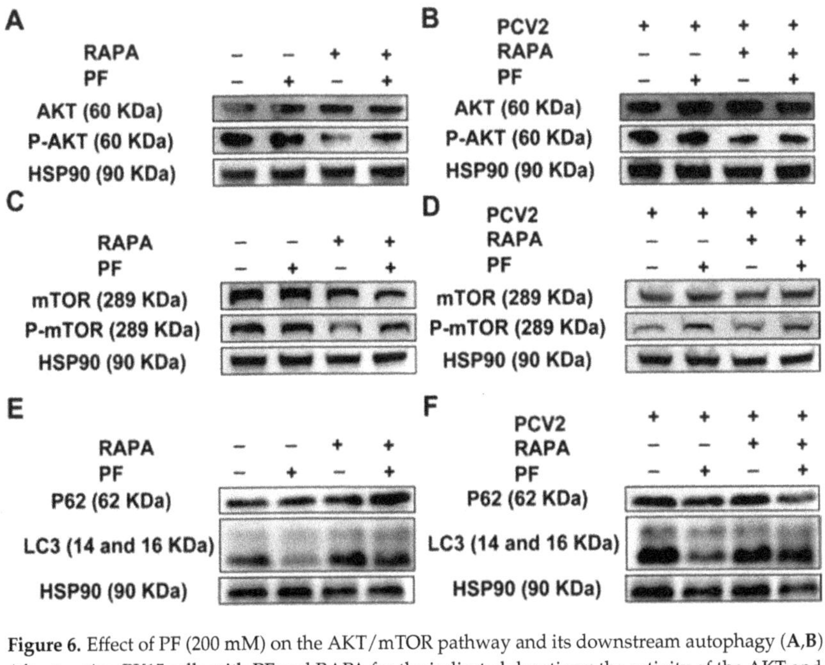


**Figure 6.** Effect of PF (200 mM) on the AKT/mTOR pathway and its downstream autophagy (**A**,**B**) After treating PK15 cells with PF and RAPA for the indicated durations, the activity of the AKT and mTOR pathways was measured. (**C**,**D**) After treating PCV2-infected cells with PF and RAPA for the indicated durations, the activity of the AKT and mTOR pathways was measured. (**E**) After treating PK15 cells with PF and RAPA for the indicated durations, the levels of LC3 and P62 were detected. (**F**) After treating PCV2-infected cells with PF and RAPA for the indicated durations, the levels of LC3 and P62 were detected.

To further validate the aforementioned results, we opted to monitor autophagy flux using the lentiviral LV-mRFP-GFP-LC3 vector. It should be noted that red LC3 spots represent autophagosomes and autolysosomes, while green LC3 spots only represent autophagosomes. The obtained images presented a clear pattern. PF treatment led to a decrease in yellow and red spots. Conversely, autophagy activator RAPA treatment resulted in an increase in yellow and red spots. Notably, when these two treatments were combined, the reduction in yellow and red spots was effectively attenuated (Figures 7 and 8). In conclusion, these results strongly suggest that PF inhibits the production of intracellular autophagy by promoting the phosphorylation of the AKT/mTOR pathway. This finding not only deepens our understanding of the underlying mechanisms but also paves the way for further research and potential therapeutic applications in the context of PCV2 infection and autophagy regulation.

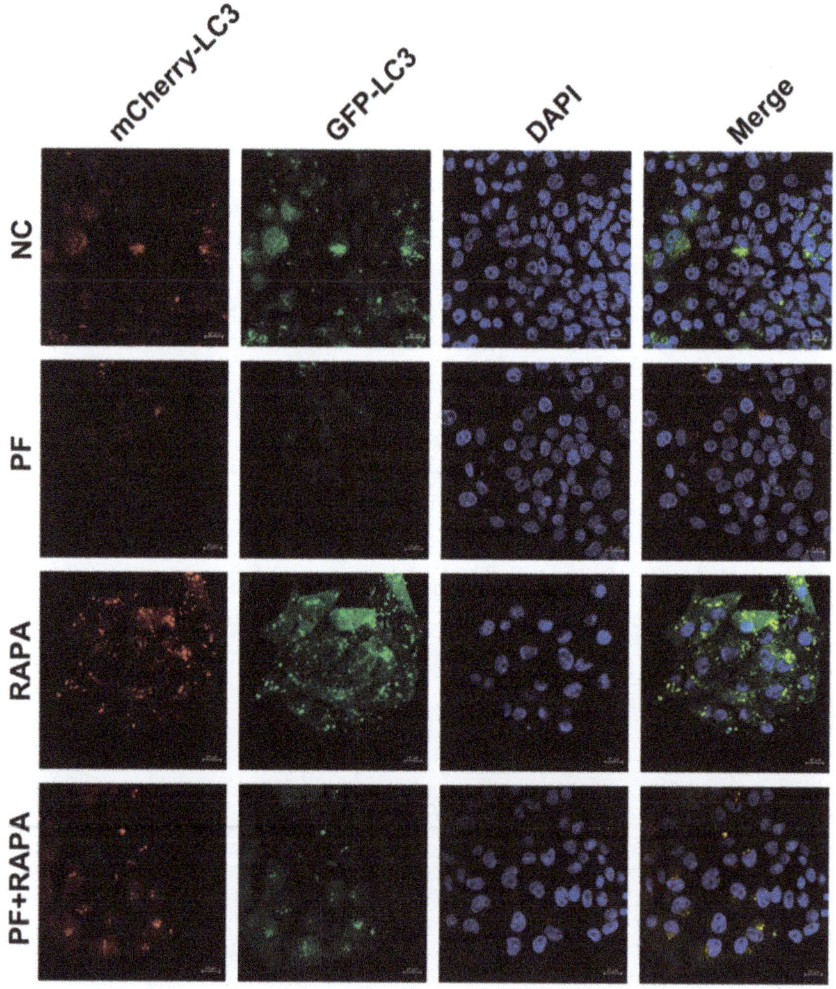


**Figure 7.** After PK15 cells were treated with PF (200 mM) and RAPA for the indicated times, confocal fluorescent images of PK15 cells expressing mRFP-GFP-LC3 were obtained. In the confocal fluorescence image, red LC3 dots represent autophagosomes and autolysosomes, green LC3 dots represent only autophagosomes and blue dots represent the nucleus. The scale of the image is 20 μm.

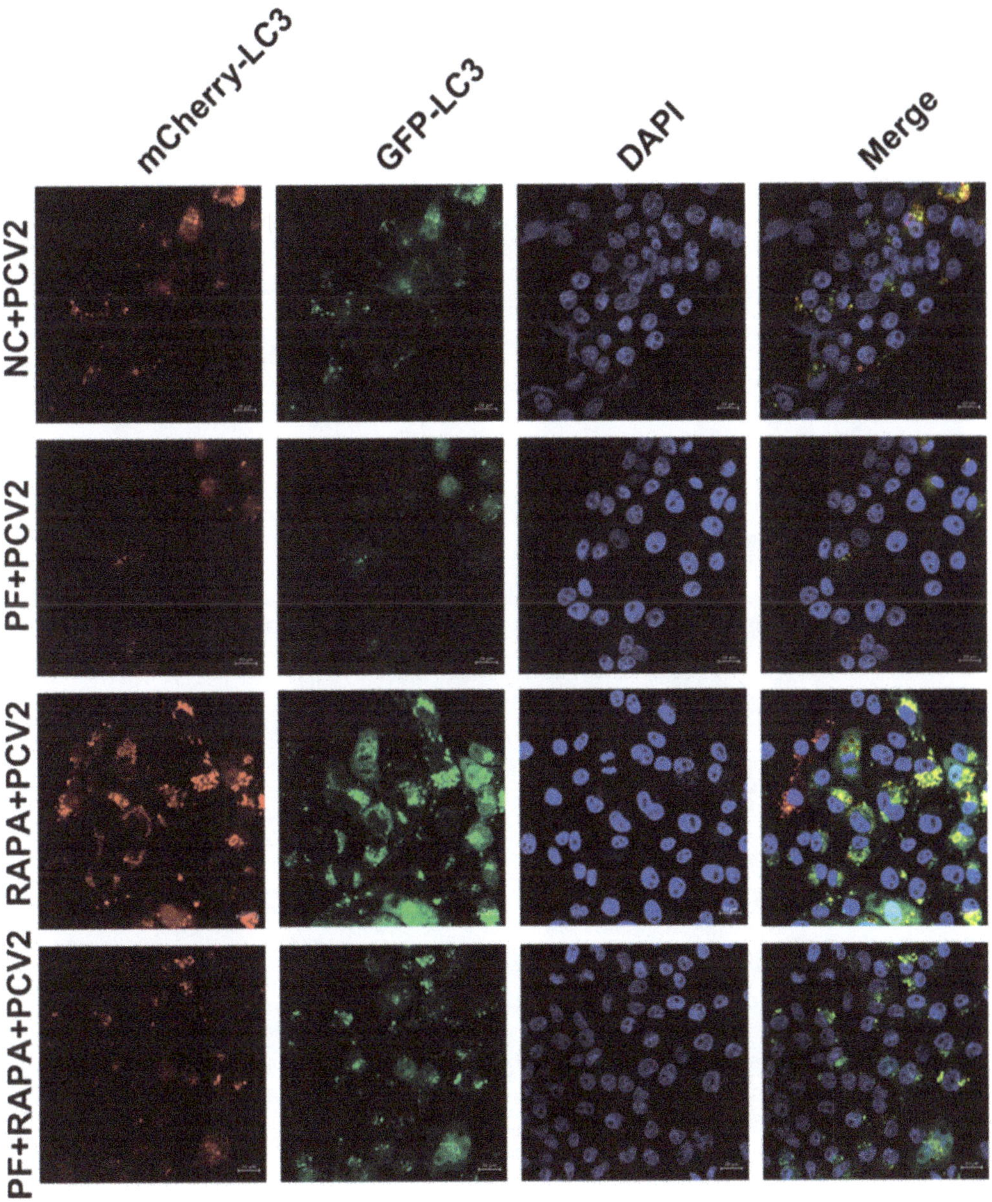


**Figure 8.** After PCV2-infected cells were treated with PF (200 mM) and RAPA for the indicated times, confocal fluorescent images of PK15 cells expressing mRFP-GFP-LC3 were obtained. In the confocal fluorescence images, red LC3 spots signify autophagosomes and autolysosomes, green LC3 spots indicate only autophagosomes and blue spots represent the nucleus. The scale of the image is 20 µm.

In order to verify the above results, we further conducted experiments on PF (200 mM) and autophagy activator RAPA-treated groups and detected the expression level of PCV2 *CAP* by Western blot and qPCR. The results indicated that PF had a significant inhibitory effect on PCV2. When we added the autophagy activator RAPA to the treatment, the results showed that the replication of the virus was promoted when RAPA was added alone, and the inhibitory effect of PF on the virus was attenuated when PF and RAPA were added together. It can be seen that RAPA counteracted the inhibitory effect of PF on virus replication. Taking the above results together, we can clearly speculate that PF reduces the replication of the virus through the inhibition of autophagy (Figure 9A). Meanwhile, the results of indirect immunofluorescence and Western blot also proved this (Figure 9B,C).

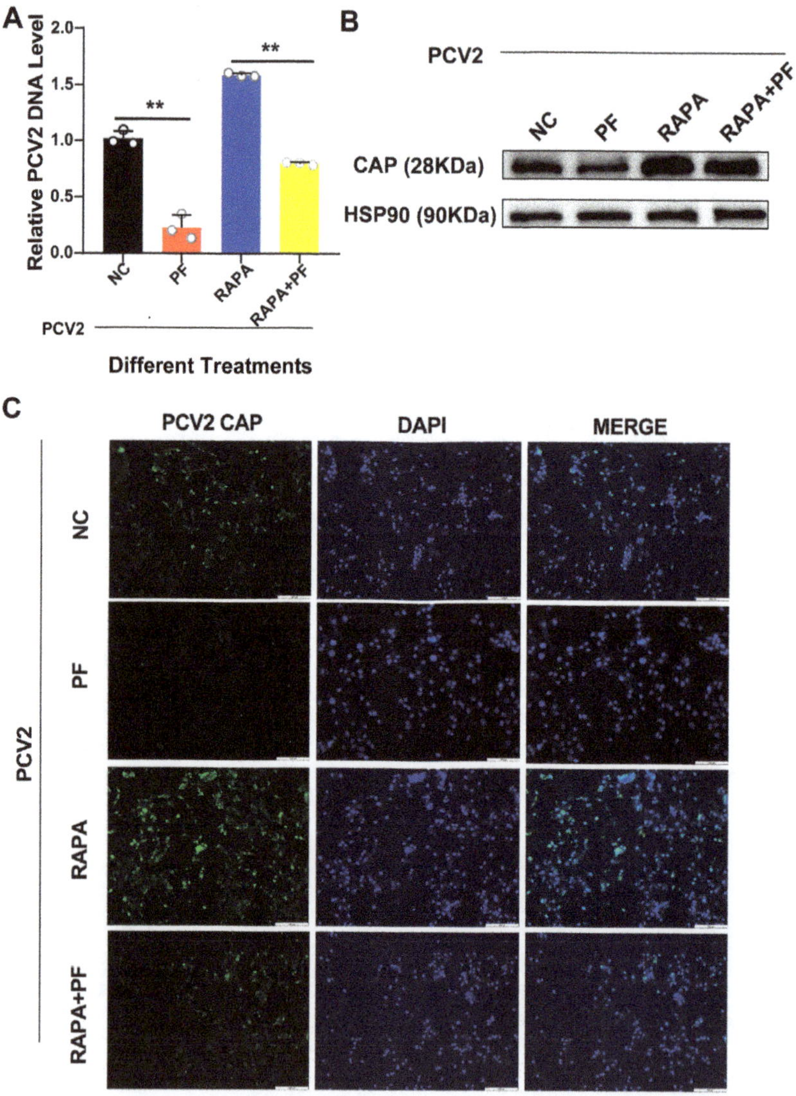


**Figure 9.** The effect of PF (200 mM) on PCV2 replication through autophagy and a rescue experiment with the autophagy activator RAPA. (**A**) The copy number of porcine circovirus type 2 (PCV2) in PK15 cells was counted after the treatment with PF and RAPA. (**B**) The protein expression of the PCV2 capsid protein (CAP) in PK15 cells was examined after PF and RAPA treatment. (**C**) The results of the fluorescent expression of the PCV2 capsid protein in PK15 cells were obtained after the application of PF and RAPA treatment. All data are presented as the mean ± SD, ** $p < 0.01$.

## 4. Discussion

In this research, we investigated the antiviral impact of paeoniflorin on PK15 cells infected with PCV2. In previous studies, paeoniflorin has been shown to exert an inhibitory effect on several human and mammalian viruses, which is assumed to be a broadly active antiviral plant extract [27,28]. The results indicated a dose-dependent inhibitory effect, highlighting a significant relationship between paeoniflorin and PCV2 infection. Upon treating PCV2-infected cells with paeoniflorin, alterations in the AKT/mTOR signaling pathway were observed, implying that this compound may target this pathway. Additional experimental results revealed that the inhibition of the AKT/mTOR pathway by paeoniflorin promotes autophagy, which, in turn, curtails the replication and spread of PCV2 in PK15 cells. These findings underscore the therapeutic potential of paeoniflorin in managing PCV2 infections post-onset and suggest its promising role as a therapeutic agent against PCV2.

To clarify the mechanism by which PF inhibits PCV2 replication, it can be stated more accurately as follows: When PF enters the cell, it targets the AKT/mTOR pathway and promotes this pathway. This promotion leads to an inhibition of autophagy. Since autophagy has been shown to be a critical process that supports PCV2 replication, the enhancement of autophagy by the inhibition of the AKT/mTOR pathway actually works to suppress PCV2 replication. Thus, PF effectively reduces PCV2 replication by inhibiting autophagy through the promotion of the AKT/mTOR pathway (Figure 10).

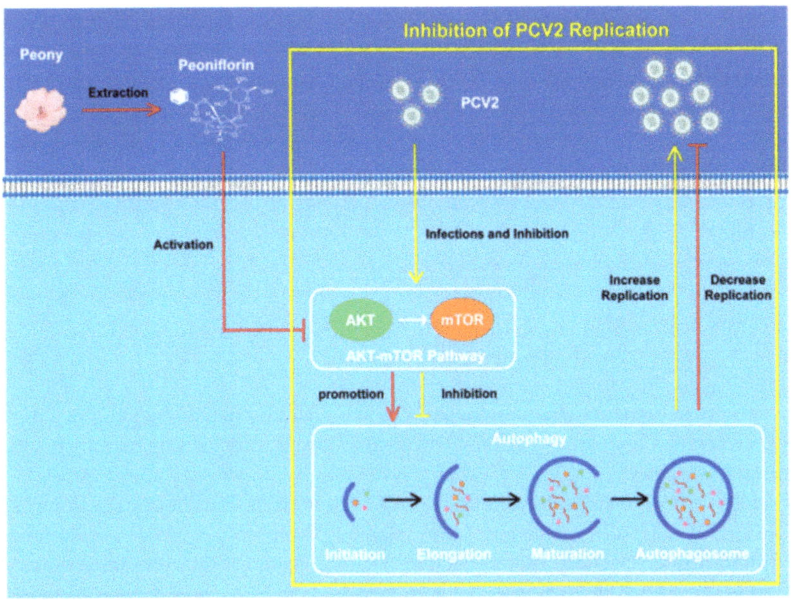


**Figure 10.** Mechanism of PF inhibiting PCV2 replication by influencing autophagy through the AKT/mTOR pathway.

Moreover, there exists a close connection between the AKT/mTOR pathway and viral replication. The AKT/mTOR pathway is a significant signaling pathway that participates in the regulation of cell growth, proliferation, and metabolism [29]. When a virus infects a cell, it utilizes the host cell's signaling pathways to create favorable conditions for its own replication and survival [30]. The AKT/mTOR pathway has a crucial regulatory function in viral replication. Some viruses can activate this pathway to enhance their replication. For instance, by activating the AKT/mTOR pathway, viruses like the hepatitis

B virus (HBV) and human immunodeficiency virus (HIV) can modify the metabolic state of the host cells, thereby providing sufficient energy and raw materials to support viral replication [31]. During the infection with the hepatitis B virus (HBV), the activation of the PI3K/Akt/mTOR signaling pathway is capable of elevating the viral load and the extent of liver fibrosis [32]. In the context of human immunodeficiency virus (HIV) infection, the PI3K/Akt signaling pathway assumes a critical role. It regulates T-cell and mitochondrial functions, promotes HIV replication, re-activates latent HIV, and sustains the viral reservoir [33]. On the contrary, other viruses, such as the influenza A virus and adenovirus, choose to inhibit the AKT/mTOR pathway to take advantage of the host cell's antiviral mechanisms [34]. Adenoviruses utilize the E1A protein to bind to transcriptional co-activators such as p300/CBP. This binding impedes the interaction between these co-activators and the transcription factors associated with the AKT/mTOR pathway. Meanwhile, the E1A protein down-regulates the expression or activity of PI3K, reducing AKT phosphorylation and thereby suppressing the AKT/mTOR pathway [35]. The inhibition of the AKT/mTOR pathway activates cellular autophagy. Adenoviruses take advantage of this to induce moderate autophagy, supplying essential nutrients like amino acids for viral replication and assembly. Moreover, autophagosomes can serve as sites for viral replication, enabling the virus to evade recognition and clearance by the host immune system [35]. Influenza A viruses make use of the interaction between the NS1 protein and the regulatory subunit p85 of PI3K. This prevents the interaction between PI3K and AKT, inhibits AKT phosphorylation and activation, and further suppresses the AKT/mTOR pathway, thus interfering with the host cell's immune response [36,37]. In doing so, these viruses can restrict the growth and proliferation of host cells and maintain their own stable state, thus evading the host's immune attack. In this study, we have shown that the inhibitory effect of paeoniflorin on PCV2 is achieved through activating the AKT/mTOR signaling pathway. It is implied that the PCV2 infection inhibits the AKT/mTOR pathway to utilize the host cell's antiviral mechanisms to assist its own replication. Paeoniflorin, however, can attenuate this inhibition, activating the AKT/mTOR pathway to regulate the replication of PCV2.

Appropriate autophagy is a necessary condition for the host's defense and immune responses [38]. Specifically, autophagy can serve as an effective defense mechanism against various viruses by degrading viral components or particles [38,39]. Meanwhile, numerous studies have demonstrated that multiple viruses can induce autophagy to facilitate their replication [40]. These findings indicate that autophagy has a dual role in the host's antiviral responses. There is growing evidence suggesting that PCV2 infection can trigger autophagic processes in both PK-15 cells and 3D4/21 cells, which promotes PCV2 replication through different mechanisms [41,42]. Consequently, it is of great significance to investigate the mechanism of action of PF in inhibiting PCV2. In this study, it was found that not only is the AKT/mTOR signaling pathway closely related to the inhibition of PCV2 replication, but autophagy, which is downstream of the AKT/mTOR signaling pathway, also plays an important role [43,44]. The research results show that PF reduces the generation of intracellular autophagy by inhibiting the autophagy induced by PCV2, thereby reducing the replication of PCV2, which takes advantage of autophagy. To conclude, the results of this study show that PF significantly inhibits PCV2 replication in PK15 cells. Specifically, PF inhibits the intracellular autophagy induced by PCV2 by activating the AKT/mTOR signaling pathway to suppress virus replication. These research findings suggest that PF could be a promising and effective drug for preventing and treating PCV2 infections.

Although this study has obtained certain achievements in investigating the mechanism of PF against PCV2 infection, it is nonetheless accompanied by several limitations. The current research is principally restricted to the cellular level, which implies a lack of comprehensive verification within a whole animal model. Consequently, this shortcoming renders it extremely challenging to conduct a thorough assessment of the therapeutic efficacy and potential side effects of PF in intricate in vivo environments during PCV2 infection. Furthermore, the detailed molecular mechanisms through which PF activates the AKT/mTOR signaling pathway suppresses autophagy and, subsequently, the impact of PCV2 replication remains inadequately elucidated. In particular, the alterations in the modification of key proteins within the pathway and their interactions with viral proteins demand further in-depth exploration.

In future studies, it is of the utmost importance to establish appropriate animal models. In doing so, a comprehensive evaluation of the pharmacokinetic, pharmacodynamic properties, and safety of PF in vivo can be carried out. This evaluation can, in turn, ensure a more dependable foundation for its clinical application. Additionally, a meticulous analysis of the precise mechanism by which PF acts on the AKT/mTOR signaling pathway and autophagy-related molecules is indispensable [45]. Such an analysis will permit accurate targeting and regulation of virus replication. Concurrently, delving into and capitalizing on the potential of paeoniflorin as an antiviral agent is expected to initiate new trials and introduce novel and efficient strategies for the prevention and treatment of PCV2 infectious diseases.

## 5. Conclusions

In summary, this study shows that paeoniflorin significantly enhances the resistance to PCV2 infection in PK15 cells. In addition, paeoniflorin may increase inhibited PCV2 replication by activating AKT/mTOR signaling. It regulates downstream target proteins through the AKT/mTOR signaling pathway to inhibit the production of intracellular autophagy. These findings indicate that paeoniflorin may be suitable as an effective therapeutic agent for targeting anti-PCV2 infection.

**Supplementary Materials:** The following supporting information can be downloaded at https://www.mdpi.com/article/10.3390/vetsci12020117/s1. Table S1: Primer information for the RT-qPCR; Table S2: Statistics for the sequencing data; Table S3: Differentially expressed genes between PF treatment and negative control groups.; Table S4: List of KEGG pathways obtained by analysis; Table S5: List of KEGG pathways obtained by co-analysis; Figure S1: Measurement of the protein gray value in Figure 3; Figure S2: Measurement of the protein gray value in Figure 6.

**Author Contributions:** Conceptualization, Z.W.; methodology, Z.W. and L.Y.; software, Z.W.; validation, Z.W., L.Y. and Y.H.; formal analysis, Z.W.; investigation, Z.W.; resources, W.B. and S.W.; data curation, Z.W.; writing—original draft preparation, Z.W. and L.Y.; writing—review and editing, S.W.; visualization, Z.W.; supervision, Z.W.; project administration, Z.W.; funding acquisition, S.W. All authors have read and agreed to the published version of the manuscript.

**Funding:** This study was supported by the Key Research and Development Project (Modern Agriculture) of Jiangsu Province (BE2020320), the National Key Research and Development Program of China (2023YFF1000900), the Open Project of International Research Laboratory of Prevention and Control of Important Animal Infectious Diseases and Zoonotic Diseases of Jiangsu Higher Education Institutions, Yangzhou University, Yangzhou, China, and the Priority Academic Program Development of Jiangsu Higher Education Institutions.

**Institutional Review Board Statement:** Not applicable.

**Informed Consent Statement:** Not applicable.

**Data Availability Statement:** The RNA-seq data have been submitted to NCBI's SRA repository under BioProject IDs: PRJNA1173311.

**Conflicts of Interest:** The authors declare that they have no known competing financial interests or personal relationships that could have appeared to influence the work reported in this paper.

## Abbreviations

PCV2: porcine circovirus type 2; PF: paeoniflorin; PCVD: porcine circovirus disease; PRSE: polyphenol-rich sugarcane extract; IAV: influenza a virus; CVB3: coxsackie B3 virus; VMC: viral myocarditis; PK15: porcine kidney cells; FBS: fetal bovine serum; DMEM: dulbecco's modified eagle's medium; RAPA: rapamycin; DMSO: dimethyl sulfoxide; CCK-8: cell counting kit-8; PBS: phosphate-buffered saline; BSA: bovine serum albumin; DAPI: 4′,6-diamidino-2-phenylindole; PVDF: polyvinylidene fluoride; GO: Gene Ontology; KEGG: Kyoto Encyclopedia of Genes and Genomes; SD: standard deviation; IFA: indirect fluorescent antibody.

## References


1. Opriessnig, T.; Karuppannan, A.K.; Castro, A.M.M.G.; Xiao, C.T. Porcine circoviruses: Current status, knowledge gaps and challenges. *Virus Res.* **2020**, *286*, 198044. [CrossRef] [PubMed]
2. Ren, L.; Chen, X.; Ouyang, H. Interactions of porcine circovirus 2 with its hosts. *Virus Genes* **2016**, *52*, 437–444. [CrossRef] [PubMed]
3. Segalés, J. Porcine circovirus type 2 (PCV2) infections: Clinical signs, pathology and laboratory diagnosis. *Virus Res.* **2012**, *164*, 10–19. [CrossRef]
4. Karuppannan, A.K.; Opriessnig, T. Porcine Circovirus Type 2 (PCV2) Vaccines in the Context of Current Molecular Epidemiology. *Viruses* **2017**, *9*, 99. [CrossRef]
5. Lv, Q.Z.; Guo, K.K.; Zhang, Y.M. Current understanding of genomic DNA of porcine circovirus type 2. *Virus Genes* **2014**, *49*, 1–10. [CrossRef] [PubMed]
6. Nawagitgul, P.; Morozov, I.; Bolin, S.R.; Harms, P.A.; Sorden, S.D.; Paul, P.S. Open reading frame 2 of porcine circovirus type 2 encodes a major capsid protein. *J. Gen. Virol.* **2000**, *81*, 2281–2287. [CrossRef] [PubMed]
7. Rose, N.; Opriessnig, T.; Grasland, B.; Jestin, A. Epidemiology and transmission of porcine circovirus type 2 (PCV2). *Virus Res.* **2012**, *164*, 78–89. [CrossRef] [PubMed]
8. Madson, D.M.; Opriessnig, T. Effect of porcine circovirus type 2 (PCV2) infection on reproduction: Disease, vertical transmission, diagnostics and vaccination. *Anim. Health Res. Rev.* **2011**, *12*, 47–65. [CrossRef] [PubMed]
9. Wallgren, P.; Belák, K.; Ehlorsson, C.J.; Bergström, G.; Lindberg, M.; Fossum, C.; Allan, G.M.; Robertsson, J.A. Postweaning multisystemic wasting syndrome (PMWS) in Sweden from an exotic to an endemic disease. *Vet. Q.* **2007**, *29*, 122–137. [CrossRef] [PubMed]
10. Opriessnig, T.; Meng, X.J.; Halbur, P.G. Porcine circovirus type 2 associated disease: Update on current terminology, clinical manifestations, pathogenesis, diagnosis, and intervention strategies. *J. Vet. Diagn. Investig.* **2007**, *19*, 591–615. [CrossRef] [PubMed]
11. Darwich, L.; Segalé, J.; Domingo, M.; Mateu, E. Changes in CD4(+), CD8(+), CD4(+) CD8(+), and immunoglobulin M-positive peripheral blood mononuclear cells of postweaning multisystemic wasting syndrome-affected pigs and age-matched uninfected wasted and healthy pigs correlate with lesions and porcine circovirus type 2 load in lymphoid tissues. *Clin. Diagn. Lab. Immunol.* **2002**, *9*, 236–242. [CrossRef] [PubMed]
12. Grierson, S.S.; King, D.P.; Tucker, A.W.; Donadeu, M.; Mellencamp, M.A.; Haverson, K.; Banks, M.; Bailey, M. Ontogeny of systemic cellular immunity in the neonatal pig: Correlation with the development of post-weaning multisystemic wasting syndrome. *Vet. Immunol. Immunopathol.* **2007**, *119*, 254–268. [CrossRef]
13. Ramamoorthy, S.; Meng, X.J. Porcine circoviruses: A minuscule yet mammoth paradox. *Anim. Health Res. Rev.* **2009**, *10*, 1–20. [CrossRef] [PubMed]
14. Darwich, L.; Mateu, E. Immunology of porcine circovirus type 2 (PCV2). *Virus Res.* **2012**, *164*, 61–67. [CrossRef]
15. Sun, J.; Huang, L.; Wei, Y.; Wang, Y.; Chen, D.; Du, W.; Wu, H.; Liu, C. Prevalence of emerging porcine parvoviruses and their co-infections with porcine circovirus type 2 in China. *Arch. Virol.* **2015**, *160*, 1339–1344. [CrossRef] [PubMed]
16. Baggieri, M.; Gioacchini, S.; Borgonovo, G.; Catinella, G.; Marchi, A.; Picone, P.; Vasto, S.; Fioravanti, R.; Bucci, P.; Kojouri, M.; et al. Antiviral, virucidal and antioxidant properties of Artemisia annua against SARS-CoV-2. *Biomed. Pharmacother.* **2023**, *168*, 115682. [CrossRef] [PubMed]
17. Tang, C.; Carrera Montoya, J.; Fritzlar, S.; Flavel, M.; Londrigan, S.L.; Mackenzie, J.M. Polyphenol rich sugarcane extract (PRSE) has potential antiviral activity against influenza A virus in vitro. *Virology* **2024**, *590*, 109969. [CrossRef] [PubMed]

18. Zhang, L.; Wei, W. Anti-inflammatory and immunoregulatory effects of paeoniflorin and total glucosides of paeony. *Pharmacol. Ther.* **2020**, *207*, 107452. [CrossRef]
19. Hu, Y.; Ge, K.; Du, Y. Paeoniflorin alleviates TGF-β2-mediated extracellular matrix remodeling and oxidative stress in human trabecular meshwork cells. *Int. Ophthalmol.* **2024**, *44*, 229. [CrossRef]
20. Zhou, H.; Bian, D.; Jiao, X.; Wei, Z.; Zhang, H.; Xia, Y.; He, Y.; Dai, Y. Paeoniflorin protects against lipopolysaccharide-induced acute lung injury in mice by alleviating inflammatory cell infiltration and microvascular permeability. *Inflamm. Res.* **2011**, *60*, 981–990. [CrossRef]
21. Zhou, W.; Zuo, H.; Qian, Y.; Miao, W.; Chen, C. Paeoniflorin attenuates particulate matter-induced acute lung injury by inhibiting oxidative stress and NLRP3 inflammasome-mediated pyroptosis through activation of the Nrf2 signaling pathway. *Chem. Biol. Interact.* **2024**, *395*, 111032. [CrossRef]
22. Tao, Y.E.; Wen, Z.; Song, Y.; Wang, H. Paeoniflorin attenuates hepatic ischemia/reperfusion injury via anti-oxidative, anti-inflammatory and anti-apoptotic pathways. *Exp. Ther. Med.* **2016**, *11*, 263–268. [CrossRef]
23. Wang, L.; Lei, J.; Zhao, Z.; Jia, J.; Wang, L. Therapeutic effects of paeoniflorin on irritable bowel syndrome in rats. *J. Vet. Sci.* **2023**, *24*, e23. [CrossRef]
24. Wang, S.; Jia, D.; Lu, H.; Qu, X. Paeoniflorin improves myocardial injury via p38 MAPK/NF-KB p65 inhibition in lipopolysaccharide-induced mouse. *Ann. Transl. Med.* **2021**, *9*, 1449. [CrossRef]
25. Hung, T.C.; Jassey, A.; Lin, C.J.; Liu, C.H.; Lin, C.C.; Yen, M.H.; Lin, L.T. Methanolic Extract of Rhizoma Coptidis Inhibits the Early Viral Entry Steps of Hepatitis C Virus Infection. *Viruses* **2018**, *10*, 669. [CrossRef]
26. Shan, Y.; Xiao, Q.; He, K.; Wu, S.; Bao, W.; Wu, Z. Identification of the Effects of 5-Azacytidine on Porcine Circovirus Type 2 Replication in Porcine Kidney Cells. *Vet. Sci.* **2024**, *11*, 135. [CrossRef] [PubMed]
27. Gassmann, M.; Focher, F.; Buhk, H.J.; Ferrari, E.; Spadari, S.; Hübscher, U. Replication of single-stranded porcine circovirus DNA by DNA polymerases alpha and delta. *Biochim. Biophys. Acta* **1988**, *951*, 280–289. [CrossRef] [PubMed]
28. Cheung, A.K. Rolling-circle replication of an animal circovirus genome in a theta-replicating bacterial plasmid in *Escherichia coli*. *J. Virol.* **2006**, *80*, 8686–8694. [CrossRef]
29. Deretic, V.; Saitoh, T.; Akira, S. Autophagy in infection, inflammation and immunity. *Nat. Rev. Immunol.* **2013**, *13*, 722–737. [CrossRef] [PubMed]
30. Kong, N.; Shan, T.; Wang, H.; Jiao, Y.; Zuo, Y.; Li, L.; Tong, W.; Yu, L.; Jiang, Y.; Zhou, Y.; et al. BST2 suppresses porcine epidemic diarrhea virus replication by targeting and degrading virus nucleocapsid protein with selective autophagy. *Autophagy* **2020**, *16*, 1737–1752. [CrossRef] [PubMed]
31. Zhao, S.; Wang, Y.; Yang, N.; Mu, M.; Wu, Z.; Li, H.; Tang, X.; Zhong, K.; Zhang, Z.; Huang, C.; et al. Genome-scale CRISPR-Cas9 screen reveals novel regulators of B7-H3 in tumor cells. *J. Immunother. Cancer* **2022**, *10*, e004875. [CrossRef] [PubMed]
32. Lee, J.; Kim, J.; Lee, R.; Lee, E.; An, H.I.; Kwon, Y.J.; Jin, H.; Pack, C.G.; Kim, I.; Yoon, Y.I.; et al. SOD1 inhibition enhances sorafenib efficacy in HBV-related hepatocellular carcinoma by modulating PI3K/Akt/mTOR pathway and ROS-mediated cell death. *J. Cell. Mol. Med.* **2024**, *28*, e18533. [CrossRef] [PubMed]
33. Cai, J.F.; Zhou, J.S.; Meng, Z.Y.; Wu, Z.Q.; Zhao, J.C.; Peng, H.X.; Liang, X.Y.; Chen, J.J.; Wang, P.P.; Deng, K. Ripretinib inhibits HIV-1 transcription through modulation of PI3K-AKT-mTOR. *Acta Pharmacol. Sin.* **2024**, *45*, 1632–1643. [CrossRef] [PubMed]
34. Sun, M.X.; Huang, L.; Wang, R.; Yu, Y.L.; Li, C.; Li, P.P.; Hu, X.C.; Hao, H.P.; Ishag, H.A.; Mao, X. Porcine reproductive and respiratory syndrome virus induces autophagy to promote virus replication. *Autophagy* **2012**, *8*, 1434–1447. [CrossRef] [PubMed]
35. Göttig, L.; Jummer, S.; Staehler, L.; Groitl, P.; Karimi, M.; Blanchette, P.; Kosulin, K.; Branton, P.E.; Schreiner, S. The human adenovirus PI3K-Akt activator E4orf1 is targeted by the tumor suppressor p53. *J. Virol.* **2024**, *98*, e0170123. [CrossRef] [PubMed]
36. Ehrhardt, C.; Ludwig, S. A new player in a deadly game: Influenza viruses and the PI3K/Akt signalling pathway. *Cell Microbiol.* **2009**, *11*, 863–871. [CrossRef]
37. Ehrhardt, C.; Wolff, T.; Pleschka, S.; Planz, O.; Beermann, W.; Bode, J.G.; Schmolke, M.; Ludwig, S. Influenza A virus NS1 protein activates the PI3K/Akt pathway to mediate antiapoptotic signaling responses. *J. Virol.* **2007**, *81*, 3058–3067. [CrossRef] [PubMed]
38. Wang, X.; Xu, X.; Wang, W.; Yu, Z.; Wen, L.; He, K.; Fan, H. MicroRNA-30a-5p promotes replication of porcine circovirus type 2 through enhancing autophagy by targeting 14-3-3. *Arch. Virol.* **2017**, *162*, 2643–2654. [CrossRef]
39. Lv, J.; Jiang, Y.; Feng, Q.; Fan, Z.; Sun, Y.; Xu, P.; Hou, Y.; Zhang, X.; Fan, Y.; Xu, X.; et al. Porcine Circovirus Type 2 ORF5 Protein Induces Autophagy to Promote Viral Replication via the PERK-eIF2α-ATF4 and mTOR-ERK1/2-AMPK Signaling Pathways in PK-15 Cells. *Front. Microbiol.* **2020**, *11*, 320. [CrossRef]
40. Hu, Y.; Liu, L.; Shan, L.P.; Chen, J. Natural ingredient paeoniflorin could be a lead compound against white spot syndrome virus infection in *Litopenaeus vannamei*. *J. Fish Dis.* **2022**, *45*, 349–359. [CrossRef] [PubMed]
41. Cao, Y.; Jing, P.; Yu, L.; Wu, Z.; Gao, S.; Bao, W. miR-214-5p/C1QTNF1 axis enhances PCV2 replication through promoting autophagy by targeting AKT/mTOR signaling pathway. *Virus Res.* **2023**, *323*, 198990. [CrossRef] [PubMed]
42. Qian, G.; Liu, D.; Hu, J.; Gan, F.; Hou, L.; Zhai, N.; Chen, X.; Huang, K. SeMet attenuates OTA-induced PCV2 replication promotion by inhibiting autophagy by activating the AKT/mTOR signaling pathway. *Vet. Res.* **2018**, *49*, 15. [CrossRef]

43. Ersahin, T.; Tuncbag, N.; Cetin-Atalay, R. The PI3K/AKT/mTOR interactive pathway. *Mol. Biosyst.* **2015**, *11*, 1946–1954. [CrossRef] [PubMed]
44. Peng, Y.; Wang, Y.; Zhou, C.; Mei, W.; Zeng, C. PI3K/Akt/mTOR Pathway and Its Role in Cancer Therapeutics: Are We Making Headway? *Front. Oncol.* **2022**, *12*, 819128. [CrossRef] [PubMed]
45. Ma, L.; Zhang, R.; Li, D.; Qiao, T.; Guo, X. Fluoride regulates chondrocyte proliferation and autophagy via PI3K/AKT/mTOR signaling pathway. *Chem. Biol. Interact.* **2021**, *349*, 109659. [CrossRef]

*veterinary sciences* MDPI

*Article*

# Development and Efficacy Evaluation of a Novel Nanoparticle-Based Hemagglutination Inhibition Assay for Serological Studies of Porcine Epidemic Diarrhea Virus

**Fengyan Liang [1,†], Wenyue Qiao [1,†], Mengjia Zhang [2], Zhangtiantian Hu [1], Shan Zhao [1], Qigui Yan [1], Wentao Li [2,*] and Yifei Lang [1,*]**

1 College of Veterinary Medicine, Sichuan Agricultural University, Chengdu 611130, China; fdzdyyy@163.com (F.L.)
2 National Key Laboratory of Agricultural Microbiology, Hubei Hongshan Laboratory, College of Veterinary Medicine, Huazhong Agricultural University, Wuhan 430070, China
* Correspondence: wentao@mail.hzau.edu.cn (W.L.); y_langviro@163.com (Y.L.)
† These authors contributed equally to this work.

**Simple Summary:** Porcine epidemic diarrhea virus (PEDV) is a highly contagious enteric pathogen and has caused severe economic losses to the global swine industry. In view of this, the development of rapid and accurate serological diagnostic methods is essential for the monitoring of antibody levels after PEDV infection or vaccine development. In this study, a safe and easy-to-operate method for the detection of PEDV neutralizing antibodies was constructed. The PEDV S1$^{0A}$-Spy recombinant protein containing a sialic acid-binding motif was expressed and then presented onto SpyCatcher-mi3 nanoparticles to establish a novel virus-free hemagglutination inhibition (HI) assay. The HI assay showed good specificity and sensitivity, and the obtained HI titers correlated well with the corresponding virus neutralizing titers. It can be used as an important tool for the monitoring of PEDV neutralizing antibodies and a new way for the serological diagnosis of PEDV.

**Abstract:** Porcine epidemic diarrhea virus (PEDV) is a major pathogen that causes serious economic losses to the swine industry. To aid PEDV clinical diagnosis and vaccine development, sensitive and precise serological methods are demanded for rapid detection of (neutralizing) antibodies. Aiming for the development of a novel virus-free hemagglutination inhibition (HI) assay, the N-terminal region of the PEDV S1 subunit, encompassing the sialic acid-binding motif, was first expressed as an Fc-fusion protein with a C-terminal Spy Tag (S1$^{0A}$-Spy). The S1$^{0A}$-Spy protein was then presented on SpyCatcher-mi3 nanoparticles, forming virus-like particles designated S1$^{0A}$-NPs. Electron microscopy and dynamic light scattering analysis confirmed its topology, and the hemagglutination assay showed that S1$^{0A}$-NPs can efficiently agglutinate red blood cells. The HI assay based on S1$^{0A}$-NPs was then validated with PEDV-positive and -negative samples. The results showed that the HI assay had high specificity for the detection of PEDV antibodies. Next, a total of 253 clinical serum samples were subjected to the HI testing along with virus neutralization (VN) assay. The area under the receiver operating characteristic curve with VN was 0.959, and the *kappa* value was 0.759. Statistical analysis of the results indicated that the HI titers of the samples tested exhibited high consistency with the VN titers. Taken together, a novel virus-free HI assay based on the multivalent display of a chimeric PEDV spike protein upon self-assembling nanoparticles was established, providing a new approach for PEDV serological diagnosis.

**Keywords:** porcine epidemic diarrhea virus; serology; S1$^{0A}$ protein; nanoparticle; hemagglutination inhibition assay

## 1. Introduction

Porcine epidemic diarrhea (PED), a swine epidemic disease caused by porcine epidemic diarrhea virus (PEDV), is characterized by severe watery diarrhea, vomiting, dehydration and high mortality in piglets [1]. PEDV is an enveloped virus with a single-stranded, positive-sense RNA genome approximately 28 kb in length. It contains seven open reading frames, which encode four structural proteins, namely the spike protein (S), envelope protein (E), membrane protein (M) and nucleocapsid protein (N) [2,3]. S is the main surface glycoprotein budded on the virus membrane of PEDV, which can be further divided into S1 and S2 subunits. The S1 subunit contains major neutralizing epitopes and mediates the binding of virus to specific host cell receptors and induces neutralizing antibodies in natural hosts, while the S2 subunit mediates the fusion of the virus and host cell membrane post virus attachment [4–6].

PEDV strains can be classified into two genotypes by phylogenetic analyses based on the S coding sequence: Genotype I (GI) and Genotype II (GII) [7]. The GI CV777 strain was first isolated in 1978 and epidemic in Europe [8]. In 2010, highly virulent GII PEDV strains emerged in China, causing severe PEDV outbreaks that subsequently spread across the country, resulting in a mortality rate of up to 100% in piglets and causing serious economic losses to the swine breeding industry [9,10]. Highly pathogenic PEDV was first detected in the United States in April 2013 and rapidly spread to 31 states, as well as Mexico and Canada [11]. PEDV GII quickly became the dominant strain worldwide and currently circulates in swine farms in Asia, Europe and North America, causing enormous economic losses to the swine industry and posing a potential threat to public health [12]. It has been shown that variations in the S gene, the major virulence factor, can alter the pathogenicity of PEDV, which enlarged the challenge of PEDV prevention and control in China [13,14]. In PEDV-infected pigs, the retention time of S protein antibodies in serum was longer than that of N protein antibodies due to the robust antigenicity of the S protein [15]. Furthermore, the S1 region has been shown to be a suitable target for determining genetic relationships among PEDV isolates and for developing differential diagnostic assays and effective vaccines [16,17]. Therefore, the S protein is the ideal antigen for vaccine development and the establishment of diagnostic methods.

The S1 subunit is comprised of four core domains, $S1^A$ to $S1^D$, whereas many alphacoronaviruses, including PEDV, contain an additional N-terminal domain, $S1^0$ [18,19]. The $S1^B$ domain is known to function as a receptor-binding domain for a variety of coronaviruses, such as transmissible gastroenteritis virus (TGEV) [20], middle east respiratory syndrome coronavirus (MERS-CoV) [21,22] and severe acute respiratory syndrome coronavirus 2 (SARS-CoV-2) [23]. In addition, certain PEDV strains have sialic acid (SA)-binding activity and hemagglutination activity, which were mapped to the amino-terminal 246 residues, constituted mostly by $S1^0$ of the PEDV S1 subunit [6]. Li et al. identified six neutralizing or non-neutralizing epitopes in the S1 subunit of the PEDV spike protein, which further confirmed that the neutralizing epitopes may be mainly located in the region with SA-binding activity (amino acids 19-246, mostly $S1^0$) or the deduced proteinaceous receptor-binding domain ($S1^B$) [24].

Taking into consideration the huge threat that PEDV poses to the swine breeding industry, it is of utmost importance to establish rapid and accurate PEDV detection methods to effectively block PEDV transmission and prevent major outbreaks. Diagnostic methods for PEDV are currently divided into two categories: etiological and serological diagnostic methods. Etiological diagnostic methods are mainly based on the detection of viral nucleic acids or proteins, such as by polymerase chain reaction (PCR). Serological diagnostic methods investigate for antibodies produced by the host immune system post virus infection, such as by the indirect immunofluorescence test (IFA), virus neutralization test (VN) and enzyme-linked immunosorbent assay (ELISA) [25,26]. Serological assays determine previous pathogen exposure by identifying the presence of antibodies. Ferrara et al. used an enzyme-linked immunosorbent assay (ELISA) method to explore the transmission of porcine coronaviruses (including PEDV) and porcine reproductive and respiratory syndrome virus in southern Italy, and the results showed that PEDV was the most common coronavirus circulating in pigs in the region [27]. In the first seroepidemiological study of PEDV in Mexico, it was reported that 61.66% of the serum samples were positive, showing that PEDV is still circulating in the main pig-producing states in Mexico [28]. Serology studies of PEDV can effectively reveal the epidemic status of PEDV in certain areas, which is of great significance for PED prevention and control. The outcome of serological methods can monitor the immune status of the population, which is important for determining the infection status of a specific herd or evaluating the protecting effects post vaccination. One major hurdle for PEDV serological methods is the assessment of neutralizing antibodies elicited by PEDV infection. The VN assay is generally recognized as the "golden standard", but its application requires culturing of contagious PEDV contained within biosafety laboratories. Therefore, the development of safe, easily operatable methods with results congruent to VN is genuinely necessitated for PEDV serology.

As a classic technique in the diagnosis of infectious pathogens, the hemagglutination assay (HA) is operatable and economically efficient and has been widely used in the qualitative or quantitative analysis of viruses with agglutinating ability, such as influenza virus [29]. The derived hemagglutination inhibition (HI) assay was henceforth developed to detect antibodies that block the virus–erythrocyte interaction [30]. Due to its high sensitivity and specificity, the HI assay has now become one of the classic tools of global influenza surveillance, and VN and HI assays are generally recognized as the "gold standards" for detecting influenza virus-neutralizing antibodies [31]. Of note, to avoid culturing of virus particles, designations of de novo HA and HI assays through presenting recombinant viral proteins on platforms such as magnetic beads or nanoparticles have been applied in numerous studies [32,33].

In the present study, we developed a novel HI assay for the detection of PEDV antibody levels through a multimeric exhibition of the chimeric PEDV S protein on self-assembling mi3 nanoparticles. Our data show that the results obtained with the newly developed HI assay are highly compatible with the VN results, which underscores the potential of large-scale application of HI assay in monitoring PEDV-specific antibody response for seroprevalence analyses or the evaluation of vaccine compliance.

## 2. Material and Methods

### 2.1. Viruses, Cell Lines and Clinical Samples

The PEDV/SC/2020 strain was isolated and preserved by the College of Veterinary Medicine, Sichuan Agricultural University. Human embryonic kidney 293 cells (HEK-293T) and African green monkey kidney epithelial (Vero) cells, preserved by the College of Veterinary Medicine, Sichuan Agricultural University, were cultured in Dulbecco's modified Eagle medium (DMEM; Gibco, Grand Island, NY, USA) supplemented with 10% (*w*/*v*) fetal bovine serum (FBS) (ExCell Bio, Suzhou, China), penicillin (100 IU/mL) and streptomycin (100 µg/mL). PEDV-negative sera were collected from specific-pathogen-free (SPF) pigs, while PEDV-positive sera were collected from recovered pigs post PEDV infection from farms and validated by virus neutralization tests in our laboratory. From 2020 to 2022, 253 clinical serum samples were collected across four provinces in China (Sichuan, Fujian, Zhejiang and Chongqing) from pigs of different ages, including nursery pigs, fattening pigs, milking sows, breeding boars, gilts and piglets and stored at −80 °C in our laboratory.

### 2.2. Construct Design, Protein Expression and Purification

The PEDV $S1^{0A}$ domain (amino acids 19 to 504) encoding was cloned from viral cDNA into a pcDNA-3.1 expression plasmid. Briefly, the $S1^{0A}$ gene was cloned in frame with a sequence encoding a CD5 N-terminal signal peptide and the Fc domain of mouse IgG2a (mFc) in the C-terminal for protein purification. In order to produce the $S1^{0A}$ protein ($S1^{0A}$-Spy) to be conjugated to SpyCatcher, the construct of $S1^{0A}$ was designed as described above, with the addition of a sequence encoding a Spy tag (AHIVMVDAYKPTK) at the C-terminal of the construct, which produces a fusion protein, $S1^{0A}$-Fc-Spy. The constructed plasmid was verified by sequencing. The $S1^{0A}$-Spy protein was expressed in HEK-293T cells. In detail, Lipo293$^{TM}$ (Beyotime, Shanghai, China) transfection reagent was used to transfect pcDNA3.1-$S1^{0A}$-Fc-Spy plasmid into HEK-239T cells. At 6 to 7 days post-transfection, cell supernatants containing the soluble $S1^{0A}$ protein were harvested, and the protein was purified by protein A-affinity chromatography (Sino Biological, China) according to the manufacturer's instructions. The purified protein was quantified by NanoDrop spectrophotometry, checked by Western blot and stored at −80 °C until further usage.

### 2.3. Construction, Expression and Purification of SpyCatcher-mi3 Nanoparticles

The gene encoding the trimeric aldolase (mi3) derived from the hyperthermophilic bacterium *Thermotoga maritima* and SpyCatcher sequence from the CnaB2 domain of the FbaB adhesion protein from the common Streptococcus pyogenes (GenBank accession no.MH425515) was synthesized by Sangon (Biotech, Shanghai, China). The gene sequence was codon-optimized for optimized *E. coli* expression and then cloned into pET28a expression plasmid. In addition, a sequence encoding the Strep-tag at the C-terminal for protein purification was introduced into the gene sequence. The pET28a expression plasmid was transformed into *Escherichia coli* strain BL21 (Vazyme, Nanjing, China). A single colony was picked into a 10 mL starter culture of LB medium containing 100 μg/mL ampicillin and incubated for 16 h at 37 °C with shaking at 200 rpm. Then, the culture was expanded in LB liquid medium containing 100 μg/mL ampicillin at 1% inoculum. At $A_{600}$ 0.8, cultures were induced with 0.5 mM IPTG (isopropyl β-D-thiogalactoside) and grown for 16–20 h with shaking at 200 rpm at 22 °C. After induction, the bacteria were collected by centrifugation and sonicated on ice. Then, the lysis mixture was spun down at 14,000× *g* for 30 min at 4 °C. After collecting the supernatant, SpyCatcher-mi3 nanoparticle (NP) was purified using Strep-Tactin Sepharose beads according to the manufacturer's instructions (IBA, Germany). Purified NPs were stored at −80 °C until further use.

### 2.4. Sodium Dodecyl Sulfate-Polyacrylamide Gel Electrophoresis (SDS-PAGE) and Western Blotting (WB)

Purified proteins as described above were mixed with 5 × SDS-PAGE loading buffer (with DTT) (Solarbio, China) and boiled at 95 °C for 10 min. The proteins were separated on SDS-PAGE gel under reduction conditions and stained with Coomassie bright blue. In addition, the $S1^{0A}$-Spy protein was loaded on an SDS-PAGE gel and subsequently electroblotted onto a polyvinylidene (PVDF) difluoride membrane. The membrane was then blocked with 5% powdered skim milk with 0.05% Tween-20 (TBST) at room temperature for 2 h and then reacted directly with the goat anti-mouse IgG HRP-conjugated (ABclonal, Wuhan, China) secondary antibody at a 1:1000 dilution for 1 h at room temperature. Finally, the protein was visualized using Pierce ECL Western blotting substrate (Thermo Fisher, Waltham, MA, USA) according to the manufacturer's protocol.

### 2.5. Transmission Electron Microscopy (TEM)

$S1^{0A}$-Spy was conjugated at 25 °C for 36 h with SpyCatcher-mi3 nanoparticles (NP: $S1^{0A}$, molar ratio 1:1.5) in reaction buffer (25 mM Tris-HCl [pH 8.5], 150 mM NaCl [pH 8.5]).

Subsequently, 100 μL solution was transferred into a 1.5 mL EP tube and sent to Servicebio (Wuhan, China) for TEM observation. The sample (0.2 mg/mL) was applied to freshly glow-discharged carbon 400 mesh copper grids for 2 min and blotted with filter paper. Then, the nanoparticles were stained with 8 μL 1% phosphotungstic acid (PTA) for 1 min, the staining solution was removed by filter paper and the structure of the nanoparticles was observed by transmission electron microscope after drying at room temperature.

### 2.6. Dynamic Light Scattering (DLS)

The particle size distribution of the nanoparticles in this study was measured by DLS using a Zetasizer (Malvern Panalytical, Malvern, UK). The samples were diluted 0.1% (*w/v*) with deionized water with a pH = 7.0; a total of 1 mL of the sample was absorbed and injected into the colorimetric dish followed by triplicate analysis using this apparatus with an angle of 173°. The temperature was 25 °C and equilibration time was set to 120 s. The data were analyzed using the software provided by Malvern Instruments. Particle dH was obtained from the peak of intensity (%) vs. diameter (nm) curve (PSDi).

### 2.7. Preparation of PEDV $S1^{0A}$ Protein Mouse Polyantiserum

The purified PEDV $S1^{0A}$ protein was used as an immunogen, and three six-week-old female BALB/c mice (Dossy, Chengdu, China) were immunized by subcutaneous injection at a dose of 100 μg per mouse. Freund's complete adjuvant (BBI, Shanghai, China) was used for primary immunization, and Freund's incomplete adjuvant (BBI, Shanghai, China) was used for secondary immunization. Serum samples were collected 14 days after the secondary immunization. All the animal use protocols commissioned in this study were reviewed and approved by the Animal Ethics Committee of Sichuan Agricultural University.

### 2.8. Hemagglutination and Hemagglutination Inhibition Assays with $S1^{0A}$-Conjugated SpyCatcher-mi3 Nanoparticles

$S1^{0A}$ was conjugated with SpyCatcher-mi3 nanoparticles to form $S1^{0A}$-NPs, as described above. Subsequently, $S1^{0A}$-NPs were then 2-fold serially diluted and mixed 1:1 with human erythrocytes (0.5% in PBS). Hemagglutination was assessed after incubation on ice for 2 h, and the hemagglutinating units (HAUs) were calculated for each $S1^{0A}$-NP.

To study the HI ability of serum samples, the serum samples were first mixed with 1% human red blood cell (Immocell, Xiamen, China) suspension at a 1:1 ratio, placed at 4 °C for adsorption overnight, centrifuged to collect the supernatant, and then the serum was inactivated at 56 °C for 30 min. The treated serum samples were then 2-fold serially diluted and mixed 1:1 with PBS containing 4 HAUs of $S1^{0A}$-NPs. The mixtures were incubated at room temperature for 30 min and then mixed 1:1 with human erythrocytes (0.5% in PBS). Hemagglutination inhibition was recorded after 2 h of incubation on ice. The maximum dilution factor of completely inhibiting erythrocyte agglutination was used as the coagulation-inhibiting titer of the serum. Each serum sample was analyzed three times independently, and the average HI titer was used. Serum samples with an HI titer higher than 1:40 were considered to be seropositive based on the collective analysis of PEDV reference serum samples.

### 2.9. Virus Neutralization (VN) Test

Serum samples were continuously diluted twice with serum-free DMEM, then added to 96-well plates, then added with the equivalent of 200$TCID_{50}$ PEDV virus and mixed at 37 °C for 1 h. The serum–virus mixture was inoculated onto Vero cells at 37 °C for 2 h, the supernatant was discarded, washed twice with PBS, 200 μL serum-free DMEM (containing 5 μg/mL trypsin) was added and the cytopathic effect (CPE) was observed at each pore for

48–72 h. The virus neutralization titer was defined as the reciprocal of the highest serum dilution for absolute suppression of PEDV infection. Each serum sample was analyzed three times independently, and the average VN titer was used.

### 2.10. Comparison Between HI Assay and VN with Clinical Samples

Clinical serum samples were collected and tested by HI assay and VN test. Results were compared with the VN to assess the performance of HI in terms of relative sensitivity—[(true positive/(true positive + false negative)] × 100%—and relative specificity—[(true negative/(true negative + false positive)] × 100%—and to analyze the coincidence rate (Kappa value) between the HI assay and VN. GraphPad Prism software (version 9.0) and IBM SPSS Statistics 27 software were used for the statistical analyses. Cohen's kappa values were determined as a measure of the overall agreement. *p*-values of less than 0.05 were considered statistically significant.

## 3. Results

### 3.1. Production of $S1^{0A}$-Spy Recombinant Protein and Spycatcher-mi3 Nanoparticles

In order to produce the $S1^{0A}$ protein ($S1^{0A}$-Spy) conjugated with its matching protein SpyCatcher, the Spy Tag sequence was added to the C-terminus of the mFc region (Figure 1A). This chimeric protein was then produced in mammalian cells (HEK-293T) by transfection and purification. The recombinant protein was then verified by WB and shown to migrate to its expected size, indicating the successful expression and purification of the $S1^{0A}$-Spy protein (Figure 1B).

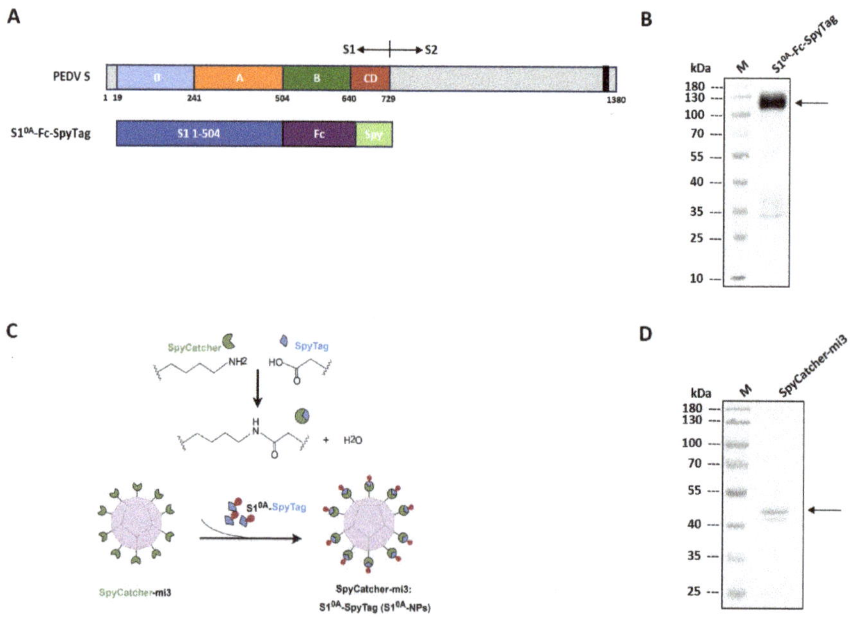


**Figure 1.** Construction and characterization of the recombinant $S1^{0A}$-Spy protein and SpyCatcher-mi3 nanoparticles. (**A**): Schematic representation of the PEDV spike protein sequence (drawn to scale). S1 and S2 subunits of the 1380-aa-long PEDV spike protein are indicated, as well as the five domains of S1 and their respective boundaries: 0 (blue), A (orange), B (green) and CD (red). The S2 subunit is marked in dark gray. The position of the transmembrane domain (black bar) is indicated. Diagram of

the S1 variant used in this study. (**B**): Affinity-purified Fc-tagged S1$^{0A}$-Spy was identified by Western blot. M, protein marker, with numbers on the left side indicating molecular masses (in kilodaltons). (**C**): Cartoon presentation of the Plug-and-Display decoration of S1$^{0A}$-nanoparticles (S1$^{0A}$-NPs) by forming spontaneous isopeptide bonds between S1$^{0A}$-Spy protein and SpyCatcher-mi3 nanoparticles. (**D**): Purified NPs were analyzed by SDS-PAGE. M, protein marker; numbers on the left are molecular masses (in kilodaltons).

To create a plug-and-display nano scaffold, the coding sequence of the trimeric aldolase (mi3) derived from the hyperthermophilic bacterium *Thermotoga maritima* was synthesized, which can self-assemble into a well-organized 60 mer dodecahedral nanoparticle. To allow multivalent presentation of Spy-tagged PEDV S1$^{0A}$, SpyCatcher was displayed on the nanoparticle surface, and the S1$^{0A}$ protein was confirmed to bind SpyCatcher via their Spy Tag (Figure 1C). The SpyCatcher-mi3 nanoparticles (NPs) with N-terminal strep tags were appended for affinity purification. Successful expression and purification of NPs were confirmed with SDS-PAGE analysis (Figure 1D).

### 3.2. PEDV S1$^{0A}$ Protein Can Be Displayed Successfully on mi3 Nanoparticles

The purified recombinant protein S1$^{0A}$-Spy and SpyCatcher-mi3 nanoparticles (molar ratio 1:1.5) were incubated at 25 °C for 36 h to achieve full coupling, thus generating S1$^{0A}$-Spy: Spycatter-mi3 (S1$^{0A}$-NPs). As shown in Figure 2A, SDS-PAGE analysis showed that effective conjugation was achieved between S1$^{0A}$-Spy and SpyCatcher-mi3, with the expected molecular weight migration, which confirmed the occurrence of assembly reaction and high conjugation efficiency. The shape and size of the S1$^{0A}$-NPs were analyzed by transmission electron microscopy (TEM) to explore whether the S1$^{0A}$-NPs were correctly assembled. The results showed that S1$^{0A}$-Spy and SpyCatcher-mi3 were able to assemble to the particle structure of 60 polymers as regular dodecahedrons (Figure 2B). To further verify the multivalent display of the S1$^{0A}$-Spy protein on the surface of Spycatard-mi3 nanoparticles, the particle size distribution of Spycatard-mi3 nanoparticles and S1$^{0A}$-NPs were determined by dynamic light scattering (DLS) analysis. Figure 2C,D shows that the mean diameter of mi3 NPs is about 47.44 nm, and that of S1$^{0A}$-NPs is about 90.11 nm, which confirms the occurrence of the conjugation, indicating that the S1$^{0A}$-Spy protein and SpyCatcher-mi3 nanoparticles were correctly coupled through the isopeptic bond.

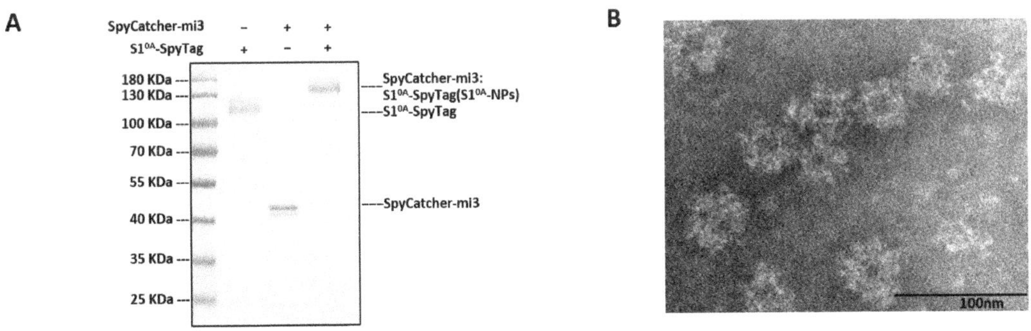


**Figure 2.** *Cont.*

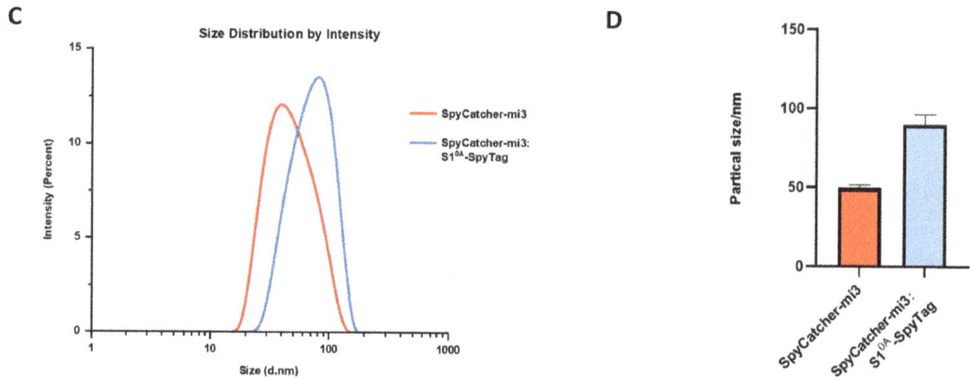


**Figure 2.** S1$^{0A}$-Spy can be efficiently conjugated to SpyCatcher-mi3 NPs. (**A**): SpyCatcher-mi3 was incubated with S1$^{0A}$-Spy at a 1.5:1 molar ratio at 25 °C for 36 h, followed by analysis of S1$^{0A}$-NP formation via reducing SDS-PAGE. (**B**): TEM image of S1$^{0A}$-NPs was negatively stained. Scale bar: 100 nm. (**C**,**D**): Particle sizes of SpyCatcher-mi3 NPs and S1$^{0A}$-NPs were analyzed by DLS.

### *3.3. NP-Based HI Assay Has High Specificity and Is Suitable for Clinical Sample Detection*

We next tested the ability of the SpyCatcher-mi3: S1$^{0A}$-Spy nanoparticles (S1$^{0A}$-NPs) to agglutinate human erythrocytes by using "empty" NPs as a negative control. Strong agglutination of human erythrocytes was indeed observed for NPs displaying S1$^{0A}$ (S1$^{0A}$-NPs), with 2048 hemagglutination units (HAUs) at a concentration of 1 µM, but not for trimeric "empty" NPs (Figure 3A). The above results indicate that S1$^{0A}$-NPs display robust hemagglutination ability and indicate that these NPs can be used for serology screening using the HI assay.

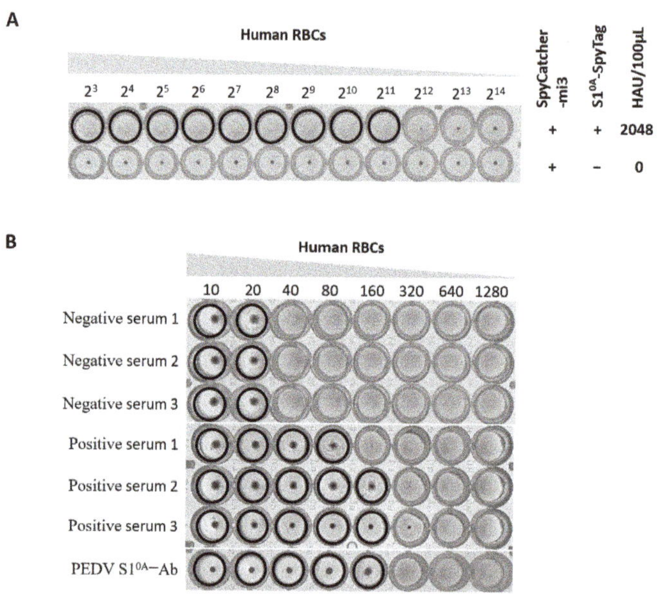


**Figure 3.** NP-based HI assay has high specificity. (**A**) Hemagglutination of human erythrocytes using S1$^{0A}$-NPs. S1$^{0A}$-NPs or empty self-assembling mi3 nanoparticles were serially diluted and tested for the ability to agglutinate human RBCs. (**B**) Specificity of the HI assay for the detection of PEDV reference serum. Serum samples were serially diluted and tested for the ability to block S1$^{0A}$-NP-induced hemagglutination of human RBCs.

Next, we explored the inhibitory properties of virus neutralization assay-validated porcine serum samples with HI assay by using the S1$^{0A}$-NPs. Based on the results of the HA assay, the hemagglutination titer of S1$^{0A}$-NPs was first determined, and a PBS solution containing 4 HAUs of S1$^{0A}$-NPs was prepared as an antigen. The reference sera were then analyzed for HI ability. As shown in Figure 3B, PEDV negative samples ($n = 3$) showed no hemagglutination inhibition (HI titer < 40), while the HA activity of S1$^{0A}$-NPs could be inhibited by PEDV-positive serum samples ($n = 3$, neutralization titer of 1:64) with different HI titers (HI titers of 1:80, 1:160 and 1:160, respectively), indicating that the HI based on S1$^{0A}$-NPs is specifically due to the S1$^{0A}$–antibody interaction. In addition, polyclonal antibodies targeting PEDV S1$^{0A}$ prepared from mice immunized with the PEDV S1$^{0A}$ protein were also tested through VN and HI assay, resulting in a neutralization titer of 1:5 and HI titer of 1:160. VN and HI assay showed consistency in the analysis of this serum, which also indicated that antibodies specific to the PEDV S1$^{0A}$ protein could contribute to the neutralization process of PEDV. Apparently, the S1$^{0A}$-NP-based HI assay exhibits high specificity for the detection of PEDV antibodies, which is a desirable characteristic for accurately identifying and differentiating PEDV infections.

### *3.4. Application of S1$^{0A}$-NP-Based HI to Clinical Samples*

To further validate the S1$^{0A}$-NP-based HI assay in the application of an analysis of PEDV antibody prevalence in clinical samples, a total of 253 porcine clinal samples were examined with the newly developed HI assay side by side with the VN assay. As shown in Table 1 and Figure 4A, the HI assay detected 54 positive samples and 199 negative samples, while the VN detected 36 positive samples and 217 negative samples (titers detailed in Supplementary Table S1). Notably, no HI-negative sample was defined as VN-positive, indicating that the HI analysis was accurate and sensitive. The coincidence rate (*Kappa* value) between the results of the two methods analyzed by SPSS Statistics 27 software was 0.759, which indicates that the two methods have good consistency (*Kappa* $\geq 0.75$). The receiver operating characteristic (ROC) analysis was further utilized to analyze the test results, and the results indicate that the HI assay and VN assay are well-correlated, with the area under the ROC curve (AUC) being 0.959 (Figure 4B). The relative sensitivity of the HI assay was 100.00%, and the relative specificity of the HI assay was 91.71%, which showed high specificity and sensitivity compared with the VN assay.

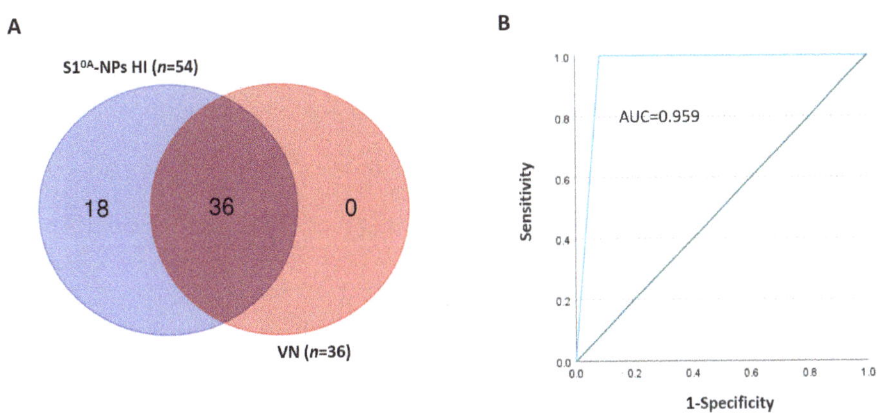


**Figure 4.** Venn diagram and ROC curve analysis of HI and VN detection results of pig serum samples. (**A**): Using Venn diagrams, the overlap between the number of positive samples in serum samples tested for HI and VN was shown. (**B**): The correlation of HI and VN to serum detection results was analyzed by the ROC curve.

**Table 1.** Detection of anti-PEDV antibodies in pig serum samples.

| Method | | VN | | Total |
|---|---|---|---|---|
| | | **Positive Sample No.** | **Negative Sample No.** | |
| **HI assay** | **Positive sample no.** | 36 | 18 | 54 |
| | **Negative sample no.** | 0 | 199 | 199 |
| Total | | 36 | 217 | 253 |
| *Kappa* value | | | | 0.759 |

## 4. Discussion

Since 2010, a large-scale PEDV outbreak occurred in China, and highly virulent Genotype II (GII) PEDV variants emerged in various regions, resulting in mortality up to 100% in piglets [9]. Therefore, effective PEDV detection and antibody monitoring methods provide an important basis for the prevention and control of PED. Currently, PEDV detection methods include etiological and serological diagnosis, among which serological detection can be used for herd immunity status monitoring and evaluation of the protective effects post vaccination for a long period of time [26]. Normally, serology methods, including indirect immunofluorescence (IFA) assay, virus neutralization (VN) assay and traditional enzyme-linked immunosorbent assay (ELISA), need to culture infectious viruses using certain facilities. Therefore, the assays using recombinant proteins as antigens are easier to operate and standardize, which is more suitable for regular laboratory diagnosis. A Bac-to-Bac baculovirus expression system was applied previously to express a truncated PEDV S1 protein, and it established an indirect ELISA method for detecting PEDV IgA antibodies, which was used for the examination of herd immunity and the evaluation of antibody levels post vaccination [34].

The classic hemagglutination inhibition (HI) assay is the derivation of hemagglutination assay (HA) for serological analysis of erythrocyte-agglutinating viruses, such as influenza virus and Newcastle disease virus [35,36]. The essence of HA and HI is the multivalent display of viral antigens, which previously could only be achieved by using live or inactivated virus particles. Here in the present study, we designed a chimeric PEDV spike protein carrying a C-terminal Spy tag ($S1^{0A}$-Spy), which forms isopeptide covalent bonds with the SpyCatcher protein fused to the N-terminal of the self-assembling trimeric aldolase (mi3) nanoparticle (NP). The polyhedral configuration of the generated virus-like particle, designated $S1^{0A}$-NP, was confirmed with transmission electron microscopy (TEM) and dynamic light scattering (DLS). The HA assay showed that the $S1^{0A}$-NPs had a decent hemagglutinating activity, with an HAU of 2048 at a concentration of 1 μM. The "empty" NPs do not have an HA ability, indicating that the PEDV $S1^{0A}$ protein is key to erythrocyte agglutination. The following HI analysis showed the agglutination could be effectively inhibited through PEDV $S1^{0A}$–antibody interaction. Utilization of this approach allows the conduction of HA or HI assays without virus propagation, which collectively minimalizes the absolute biosafety requirements. Similar HI assays were also designed for Middle East respiratory syndrome coronavirus (MERS-CoV) and influenza virus (IAV), both of which have an HA ability [32,33]. The results described in these two studies suggest that the NP-based HI assay is highly akin to HI assays performed with virus particles and could be used to differentiate antibody responses between virus genotypes.

Notably, despite being vastly reactive to antibodies, those "virus mimics" are also antigenic as potential vaccine candidates. Recent advances have shown that the artificial virus-like particles (VLPs) based on a rational design of viral proteins can elicit rapid and effective immune responses in vivo post vaccination with VLPs or VLP-coding mRNAs [37–39]. Therefore, the scheme for antigen exhibition on NPs used in this study could also be considered for the future "plug in and display" design of vaccine candidates.

The S1 subunit of the S protein has the highest immunogenicity among coronaviral proteins [40]. In the case of PEDV, the $S1^B$ domain, also known as the collagenase equivalent (COE) domain, was first identified as the receptor-binding domain (RBD) and the main target of (neutralizing) antibodies [41,42]. Meanwhile, researchers had identified that the N-terminal 249 amino acid residues of the S1 subunit of GII PEDV, mainly comprising the $S1^0$ domain, promote virus entry through adhesion to sialic acid (SA) moieties [6]. It was then confirmed that the $S1^0$ domain is also a major target of neutralizing antibodies, which function through blocking the SA attachment [24]. Therefore, the antibodies detected by HI can intuitively reflect the level of $S1^0$-targeting neutralizing antibodies. Recent epidemiological studies indicate that the $S1^0$ domain of PEDV strains circulating in China undergoes rapid evolution, which might be associated with changes in virulence [43]. Consequently, serological analysis of the coordinative antibody responses shall be implemented, where the HI assay could serve as a versatile tool.

In the present study, we stabilized the PEDV $S1^0$ by Fc-fused expression with the adjacent $S1^A$ and a C-terminal Spy tag, resulting in the chimeric protein $S1^{0A}$-Spy. This protein elicited neutralizing antibodies when used as an immunogen, and the newly developed HI assay based on this protein could distinguish PEDV-negative and -positive serum samples. In the analysis of 253 clinical serum samples, the results obtained by the HI assay are highly consistent with VN, as shown by statistical analysis. No VN-positive sample was negative for HI, suggesting that the HI assay is specific in neutralizing antibody detection. Meanwhile, one-third of the HI-positive samples (18 out of 54) is negative in VN analysis. Considering the overall results, it is unlikely that those samples contain only PEDV $S1^0$ antibodies, but rather the VN assay is less sensitive than the HI, and the neutralizing antibody titers in those samples are below the detection limit. It is possible that with the rapid evolution of PEDV $S1^0$, the dominant epitopes that elicit neutralizing immune responses had gradually shift from $S1^B$ to $S1^0$, while future epitope mapping analysis will be implemented with newly emerging PEDV strains. Taken together, due to the high specificity and sensitivity, the newly developed HI assay could be used as an alternative or complementary serology method for VN, especially when a large number of samples need to be screened within a short period of time.

## 5. Conclusions

We successfully developed and validated an HI assay for detecting neutralizing antibody levels in porcine serum samples. The results indicated that the nanoparticle-based, virus-free HI assay had good specificity and sensitivity, and the obtained HI titers have good relevancy with the corresponding VN titers. With the advantages in efficiency and biosafety, the newly established HI assay will find future utility in PEDV serology, such as in the evaluation of herd immunity and monitoring of vaccine compliance.

**Supplementary Materials:** The following supporting information can be downloaded at: https://www.mdpi.com/article/10.3390/vetsci12020101/s1, Figure S1: Figure 1B. Affinity-purified Fc-tagged $S1^{0A}$-Spy was identified by Western blot; Figure S2: Figure 1D. Purified NPs were analyzed by SDS-PAGE; Figure S3: Figure 2B. TEM image of $S1^{0A}$-NPs was negatively stained; Figure S4: Figure 2A. SpyCatcher-mi3 was incubated with $S1^{0A}$-SpyTag at a 1.5,1 molar ratio at 25°C for 36 h, followed by analysis of $S1^{0A}$-NP formation via reducing SDS-PAGE; Table S1: HI and VN titer of 54 positive samples detected by HI assay.

**Author Contributions:** Conceptualization, S.Z. and Y.L.; Methodology, Y.L.; Software, F.L.; Validation, W.Q. and Z.H.; Formal Analysis, M.Z.; Investigation, W.L.; Resources, Q.Y.; Data Curation, Y.L.; Writing—Original Draft Preparation, F.L.; Writing—Review and Editing, W.L. and Y.L.; Visualization, S.Z.; Supervision, W.L. and Y.L.; Project Administration, S.Z.; Funding Acquisition, W.L. and Y.L. All authors have read and agreed to the published version of the manuscript.

**Funding:** This work was supported by the National Natural Science Foundation of China (32473007) and the Joint Funds of Science Technology and Education of Sichuan Province (2024NSFSC2063).

**Institutional Review Board Statement:** The procedures for the care and use of animals in this animal study were approved by the Ethics Committee of the Sichuan Agricultural University Institutional Animal Care and Use Committee and were performed according to animal ethics guidelines and approved protocols (approval code 20240327).

**Informed Consent Statement:** Informed consent was obtained from all subjects involved in the study.

**Data Availability Statement:** All data analyzed during this study are available from the corresponding author upon reasonable request.

**Conflicts of Interest:** The authors declare no conflicts of interest.

## References

1. Wood, E.N. An apparently new syndrome of porcine epidemic diarrhoea. *Vet. Rec.* **1977**, *100*, 243–244. [CrossRef] [PubMed]
2. Chen, J.F.; Liu, X.Z.; Shi, D.; Shi, H.Y.; Zhang, X.; Feng, L. Complete genome sequence of a porcine epidemic diarrhea virus variant. *J. Virol.* **2012**, *86*, 3408. [CrossRef] [PubMed]
3. Morris, D.R.; Geballe, A.P. Upstream open reading frames as regulators of mRNA translation. *Mol. Cell. Biol.* **2020**, *20*, 8635–8642. [CrossRef]
4. Ji, C.M.; Wang, B.; Zhou, J.Y.; Huang, Y.W. Aminopeptidase-N-independent entry of porcine epidemic diarrhea virus into Vero or porcine small intestine epithelial cells. *Virology* **2018**, *517*, 16–23. [CrossRef] [PubMed]
5. Sternberg, A.; Naujokat, C. Structural features of coronavirus SARS-CoV-2 spike protein: Targets for vaccination. *Life Sci.* **2020**, *257*, 257–263. [CrossRef] [PubMed]
6. Li, W.T.; Van Kuppeveld, F.J.M.; He, Q.G.; Rottier, P.J.M.; Bosch, B.J. Cellular entry of the porcine epidemic diarrhea virus. *Virus Res.* **2016**, *226*, 117–127. [CrossRef]
7. Zhang, Y.H.; Chen, Y.J.; Yuan, W.F.; Peng, Q.; Zhang, F.F.; Ye, Y.; Huang, D.Y.; Ding, Z.; Lin, L.H.; He, H.J.; et al. Evaluation of Cross-Protection between G1a- and G2a-Genotype Porcine Epidemic Diarrhea Viruses in Suckling Piglets. *Animals* **2020**, *10*, 1674. [CrossRef] [PubMed]
8. Pensaert, M.B.; De Bouck, P. A new coronavirus-like particle associated with diarrhea in swine. *Adv. Virol.* **1978**, *58*, 243–247. [CrossRef]
9. Li, W.T.; Li, H.; Liu, Y.B.; Pan, Y.F.; Deng, F.; Song, Y.H.; Tang, X.B.; He, Q.G. New Variants of Porcine Epidemic Diarrhea Virus, China, 2011. *Emerg. Infect. Dis.* **2012**, *18*, 1350–1353. [CrossRef] [PubMed]
10. Sun, R.Q.; Cai, R.J.; Chen, Y.Q.; Liang, P.S.; Chen, D.K.; Song, C.X. Outbreak of porcine epidemic diarrhea in suckling piglets, China. *Emerg. Infect. Dis.* **2012**, *18*, 161–163. [CrossRef] [PubMed]
11. Oka, T.; Saif, L.J.; Marthaler, D.; Esseili, M.A.; Meulia, T.; Lin, C.M.; Vlasova, A.N.; Jung, K.; Zhang, Y.; Wang, Q. Cell culture isolation and sequence analysis of genetically diverse US porcine epidemic diarrhea virus strains including a novel strain with a large deletion in the spike gene. *Vet. Microbiol.* **2014**, *173*, 258–269. [CrossRef]
12. Huang, Y.W.; Dickerman, A.W.; Piñeyro, P.; Li, L.; Fang, L.; Kiehne, R.; Opriessnig, T.; Meng, X.J. Origin, evolution, and genotyping of emergent porcine epidemic diarrhea virus strains in the United States. *mBio* **2013**, *4*, e00737-13. [CrossRef]
13. Deng, F.; Ye, G.; Liu, Q.Q.; Navid, M.T.; Zhong, X.l.; Li, Y.W.; Wan, C.Y.; Xiao, S.B.; He, Q.G.; Fu, Z.F.; et al. Identification and Comparison of Receptor Binding Characteristics of the Spike Protein of Two Porcine Epidemic Diarrhea Virus Strains. *Viruses* **2016**, *8*, 55. [CrossRef] [PubMed]
14. Lin, C.M.; Saif, L.J.; Marthaler, D.; Wang, Q.H. Evolution, antigenicity and pathogenicity of global porcine epidemic diarrhea virus strains. *Virus Res.* **2016**, *226*, 20–39. [CrossRef] [PubMed]
15. Lin, H.X.; Zhou, H.; Gao, L.; Li, B.; He, K.W.; Fan, H.J. Development and application of an indirect ELISA for the detection of antibodies to porcine epidemic diarrhea virus based on a recombinant spike protein. *BMC Vet. Res.* **2018**, *14*, 243–250. [CrossRef]
16. Oh, J.S.; Lee, K.W.; Choi, H.W.; Lee, C. Immunogenicity and protective efficacy of recombinant S1 domain of the porcine epidemic diarrhea virus spike protein. *Arch. Virol.* **2014**, *159*, 2977–2987. [CrossRef]

17. Lee, D.K.; Park, C.K.; Kim, S.H.; Lee, C. Heterogeneity in spike protein genes of porcine epidemic diarrhea viruses isolated in Korea. *Virus Res.* **2010**, *149*, 175–182. [CrossRef] [PubMed]
18. Kirchdoerfer, R.N.; Bhandari, M.; Martini, O.; Sewall, L.M.; Bangaru, S.; Yoon, K.J.; Ward, A.B. Structure and immune recognition of the porcine epidemic diarrhea virus spike protein. *Structure* **2021**, *29*, 385–392. [CrossRef] [PubMed]
19. Luque, D.; Caston, J.R. Cryo-electron microscopy for the study of virus assembly. *Nat. Chem. Biol.* **2020**, *16*, 231–239. [CrossRef]
20. Reguera, J.; Ordoño, D.; Santiago, C.; Enjuanes, L.; Casasnovas, J.M. Antigenic modules in the N-terminal S1 region of the transmissible gastroenteritis virus spike protein. *J. Gen. Virol.* **2011**, *92*, 1117–1126. [CrossRef] [PubMed]
21. Raj, V.S.; Mou, H.; Smits, S.L.; Dekkers, D.H.; Müller, M.A.; Dijkman, R.; Muth, D.; Demmers, J.A.; Zaki, A.; Fouchier, R.A.; et al. Dipeptidyl peptidase 4 is a functional receptor for the emerging human coronavirus-EMC. *Nature* **2013**, *495*, 251–254. [CrossRef] [PubMed]
22. Mou, H.; Raj, V.S.; Van Kuppeveld, F.J.; Rottier, P.J.; Haagmans, B.L.; Bosch, B.J. The receptor binding domain of the new Middle East respiratory syndrome coronavirus maps to a 231-residue region in the spike protein that efficiently elicits neutralizing antibodies. *J. Virol.* **2013**, *87*, 9379–9383. [CrossRef] [PubMed]
23. Wang, C.Y.; Li, W.T.; Drabek, D.; Okba, N.M.A.; van Haperen, R.; Osterhaus, A.D.M.E.; van Kuppeveld, F.J.M.; Haagmans, B.L.; Grosveld, F.; Bosch, B.J. A human monoclonal antibody blocking SARS-CoV-2 infection. *Nat. Commun.* **2020**, *11*, 2251–2256. [CrossRef] [PubMed]
24. Li, C.H.; Li, W.T.; Lucio De Esesarte, E.; Guo, H.B.; van den Elzen, P.; Aarts, E.; van den Born, E.; Rottier, P.J.M.; Bosch, B.J. Cell Attachment Domains of the Porcine Epidemic Diarrhea Virus Spike Protein Are Key Targets of Neutralizing Antibodies. *J. Virol.* **2017**, *91*, e00273-17. [CrossRef]
25. Olech, M. Current State of Molecular and Serological Methods for Detection of Porcine Epidemic Diarrhea Virus. *Pathogens* **2022**, *11*, 1074. [CrossRef] [PubMed]
26. Diel, D.G.; Lawson, S.; Okda, F.; Singrey, A.; Clement, T.; Fernandes, M.H.V.; Christopher-Hennings, J.; Nelson, E.A. Porcine epidemic diarrhea virus: An overview of current virological and serological diagnostic methods. *Virus Res.* **2016**, *226*, 60–70. [CrossRef] [PubMed]
27. Ferrara, G.; D'Anza, E.; Rossi, A.; Improda, E.; Iovane, V.; Pagnini, U.; Iovane, G.; Montagnaro, S. A Serological Investigation of Porcine Reproductive and Respiratory Syndrome and Three Coronaviruses in the Campania Region, Southern Italy. *Viruses* **2023**, *15*, 300. [CrossRef] [PubMed]
28. García-González, E.; Cerriteño-Sánchez, J.L.; Cuevas-Romero, J.S.; García-Cambrón, J.B.; Castañeda-Montes, F.J.; Villaseñor-Ortega, F. Seroepidemiology Study of Porcine Epidemic Diarrhea Virus in Mexico by Indirect Enzyme-Linked Immunosorbent Assay Based on a Recombinant Fragment of N-Terminus Domain Spike Protein. *Microorganisms* **2023**, *11*, 1843. [CrossRef] [PubMed]
29. Killian, M.L. Hemagglutination Assay for Influenza Virus. *Methods Mol. Biol.* **2020**, *2123*, 3–10. [PubMed]
30. Anderson, T.C.; Crawford, P.C.; Katz, J.M.; Dubovi, E.J.; Landolt, G.; Gibbs, E.P. Diagnostic performance of the canine Influenza A Virus subtype H3N8 hemagglutination inhibition assay. *J. Vet. Diagn. Investig.* **2012**, *24*, 499–508. [CrossRef]
31. Morrison, B.J.; Martin, N.J.; Rehman, T.; Ewing, D.; Dewar, R.L.; Metcalf, J.; Sun, P.; Beigel, J.; Luke, T.C.; Raviprakash, K. Influence of sample collection tube method, anticoagulant-containing plasma versus serum, on influenza virus hemagglutination inhibition titer and microneutralization titer serological assays. *BMC Health Serv. Res.* **2018**, *18*, 651–661. [CrossRef] [PubMed]
32. Zhao, S.; Schuurman, N.; Tieke, M.; Quist, B.; Zwinkels, S.; van Kuppeveld, F.J.M.; de Haan, C.A.M.; Egberink, H. Serological Screening of Influenza A Virus Antibodies in Cats and Dogs Indicates Frequent Infection with Different Subtypes. *J. Clin. Microbiol.* **2020**, *58*, e01689-20. [CrossRef] [PubMed]
33. Okba, N.M.A.; Widjaja, I.; Li, W.T.; GeurtsvanKessel, C.H.; Farag, E.A.B.A.; Al-Hajri, M.; Park, W.B.; Oh, M.D.; Reusken, C.B.E.M.; Koopmans, M.P.G.; et al. Serologic Detection of Middle East Respiratory Syndrome Coronavirus Functional Antibodies. *Emerg. Infect. Dis.* **2020**, *26*, 1024–1027. [CrossRef]
34. Shan, Y.; Gao, Q.; Mao, J.Y.; Zheng, J.Y.; Xu, X.H.; Zhang, C.N.; Huang, X.J.; Xu, J.D.; Shi, F.S.; Yue, M.; et al. Establishment of enzyme-linked immunosorbent assays based on recombinant S1 and its truncated proteins for detection of PEDV IgA antibody. *BMC Vet. Res.* **2022**, *67*, 2589–2601. [CrossRef] [PubMed]
35. Wu, W.L.; Chen, Y.; Wang, P.; Song, W.J.; Lau, S.Y.; Rayner, J.M.; Smith, G.J.; Webster, R.G.; Peiris, J.S.; Lin, T.W.; et al. Antigenic Profile of Avian H5N1 Viruses in Asia from 2002 to 2007. *J. Virol.* **2008**, *82*, 1798–1807. [CrossRef] [PubMed]
36. Cattoli, G.; Susta, L.; Terregino, C.; Brown, C. Newcastle disease: A review of field recognition and current methods of laboratory detection. *J. Vet. Diagn. Investig.* **2011**, *23*, 637–656. [CrossRef] [PubMed]
37. Li, M.H.; Wang, Y.; Wang, Y.N.; Li, R.Q.; Wang, S.Q.; Ding, P.Y.; Zhang, G.P. Accurate location of two conserved linear epitopes of PEDV utilizing monoclonal antibodies induced by S1 protein nanoparticles. *Int. J. Biol. Macromol.* **2023**, *253*, 127276–127287. [CrossRef]

38. Hutchinson, G.B.; Abiona, O.M.; Ziwawo, C.T.; Werner, A.P.; Ellis, D.; Tsybovsky, Y.; Leist, S.R.; Palandjian, C.; West, A.; Fritch, E.J.; et al. Nanoparticle display of prefusion coronavirus spike elicits S1-focused cross-reactive antibody response against diverse coronavirus subgenera. *Nat. Commun.* **2023**, *14*, 6195–6205. [CrossRef] [PubMed]
39. Liu, J.; Sun, J.; Ding, X.; Liu, W.H.; Wang, Y.P.; Wang, Z.H.; Peng, H.Y.; Zhang, Y.; Su, W.H.; Jiang, C.I. A nucleoside-modified mRNA vaccine forming rabies virus-like particle elicits strong cellular and humoral immune responses against rabies virus infection in mice. *Emerg. Microbes Infect.* **2024**, *13*, 2389115–2389131. [CrossRef] [PubMed]
40. Yang, Y.L.; Qin, P.; Wang, B.; Liu, Y.; Xu, G.H.; Peng, L.; Zhou, J.; Zhu, S.J.; Huang, Y.W. Broad Cross-Species Infection of Cultured Cells by Bat HKU2-Related Swine Acute Diarrhea Syndrome Coronavirus and Identification of Its Replication in Murine Dendritic Cells In Vivo Highlight Its Potential for Diverse Interspecies Transmission. *J. Virol.* **2019**, *93*, e01448-19. [CrossRef]
41. Chang, C.Y.; Cheng, I.C.; Chang, Y.C.; Tsai, P.S.; Lai, S.Y.; Huang, Y.L.; Jeng, C.R.; Pang, V.F.; Chang, H.W. Identification of Neutralizing Monoclonal Antibodies Targeting Novel Conformational Epitopes of the Porcine Epidemic Diarrhoea Virus Spike Protein. *Sci. Rep.* **2019**, *9*, 2529–2539. [CrossRef] [PubMed]
42. Chang, S.H.; Bae, J.L.; Kang, T.J.; Kim, J.; Chung, G.H.; Lim, C.W.; Laude, H.; Yang, M.S.; Jang, Y.S. Identification of the Epitope Region Capable of Inducing Neutralizing Antibodies against the Porcine Epidemic Diarrhea Virus. *Mol. Cells* **2002**, *14*, 295–299. [CrossRef] [PubMed]
43. Su, M.J.; Li, C.Q.; Qi, S.S.; Yang, D.; Jiang, N.; Yin, B.S.; Guo, D.H.; Kong, F.Z.; Yuan, D.W.; Feng, L.; et al. A molecular epidemiological investigation of PEDV in China: Characterization of co-infection and genetic diversity of S1-based genes. *Transbound. Emerg. Dis.* **2019**, *67*, 1129–1140. [CrossRef] [PubMed]

veterinary sciences

MDPI

*Opinion*

# Getah Virus: A New Contaminant in Veterinary Vaccines

**Pin-Pin Chu [1,†], Sheng-Nan Chen [2,3,†], Xia Zhou [1], Zu-Zhang Wei [2,*] and Shao-Lun Zhai [1,*]**

1 Department of Swine Diseases, Institute of Animal Health, Guangdong Academy of Agricultural Sciences, Guangdong Provincial Key Laboratory of Livestock Disease Prevention, Scientific Observation and Experiment Station of Veterinary Drugs and Diagnostic Techniques of Guangdong Province, Ministry of Agriculture and Rural Affairs, Guangzhou 510640, China; cpp1900@163.com (P.-P.C.); xiazhou698868@163.com (X.Z.)

2 Laboratory of Animal infectious Diseases and Molecular Immunology, College of Animal Science and Technology, Guangxi University, Nanning 530005, China; chen.shengnan@sslab.com.cn (S.-N.C.)

3 Guangzhou Sino-Science Gene Testing Service Co., Ltd., Guangzhou 510700, China

* Correspondence: zuzhangwei@gxu.edu.cn (Z.-Z.W.); zhaishaolun@163.com (S.-L.Z.)

† These authors contributed equally to this work.

**Simple Summary:** Vaccines are essential for the prevention and control of infectious diseases in livestock farming. Among these, live veterinary vaccines play an important role. The production of live vaccines requires high-level biosafety, toxicity and potential contaminants should be closely monitored. Unfortunately, other viral contaminants in commercial live-attenuated vaccines against a multitude of viruses have been discovered, which are difficult to detect and can cause huge losses. Similar situations have occurred in commercial live-attenuated vaccines against porcine reproductive and respiratory syndrome virus (PRRSV), which arouses our considerable interest.

**Abstract:** Mycoplasma, reticuloendotheliosis virus (REV), avian leukosis virus (ALV), chicken infectious anemia virus (CIAV), bovine polyomavirus (BPV), bovine viral diarrhea virus (BVDV), and porcine circovirus (PCV) are considered common contaminants in live veterinary vaccines against Newcastle disease virus (NDV), fowlpox virus (FPV), infectious bursal disease virus (IBDV), classical swine fever virus (CSFV), pseudorabies virus (PRV), and porcine reproductive and respiratory syndrome virus (PRRSV). In the past five years, Getah virus (GETV), an arbovirus affecting many farming mammals, was reported as a new contaminant in live PRRSV vaccines in two previous studies, which arouses our considerable interest. Therefore, in this paper, we aim to analyze and discuss the source, biological hazard, and genomic characteristics of these contaminating GETV strains further.

**Keywords:** porcine reproductive and respiratory syndrome virus; Getah virus; live vaccine contaminant; vaccination infection; exogenous pathogens

## 1. Introduction

Veterinary vaccines are important and economic measures for controlling animal infectious diseases. Safety and efficacy are important criteria for evaluating the quality of vaccines. The contamination of exogenous pathogens in live vaccines can result in the spread of exogenous pathogens and severe diseases in vaccinated animals [1,2]. Therefore, surveillance of exogenous pathogens is crucial during the development of vaccines.

Previously, mycoplasma, REV, ALV, CIAV, BVDV, FERV, and PCV have been considered common contaminants in live veterinary vaccines against NDV, FPV, IBDV, CPV, and CSFV [3–20].

 https://doi.org/10.3390/vetsci12020082

PRRSV is a very important pathogen in the world swine industry and can infect pigs of all ages and cause serious respiratory and reproductive disorders, delayed growth, and even death. Moreover, PRRSV infection often causes immunosuppression, which provides opportunities for co-infection or secondary infection with other pathogens. PRRSV, with high variability in the genome, undergoes rapid evolution through mutation and recombination. PRRSV variants, including highly pathogenic PRRSV-like PRRSV, NADC30-like PRRSV, and NADC34-like PRRSV, are widely distributed around the world. All of these make prevention and control difficult [21–23]. There are currently two main kinds of vaccines against PRRSV available. One is an inactivated vaccine, and another is an attenuated live vaccine. Although the attenuated live vaccine has a better immune effect than the inactivated vaccine, the attenuated live vaccine can easily be contaminated by exogenous pathogens, including BVDV [17]. This might result in outbreaks of BVDV via vaccination with the live PRRSV vaccine in pig herds [17].

GETV, an arbovirus, can infect many kinds of animals and even humans, resulting in significant economic losses and death [24–26]. The original GETV strain MM2021 was isolated from mosquitoes in Malaysia in 1955, and GETV infection was first reported in diseased Japanese racehorses (16.48%, 70/455) in 1978. Seven years later, in 1985, a GETV outbreak occurred in a pig farm in Japan and led to death in newborn piglets. Since then, GETV infection has occurred in pigs and other animals such as cattle, foxes, and red pandas in different regions and countries, leading to substantial economic losses [27–29]. Equally, serological surveys have revealed the presence of GETV antibodies in humans. Expansion of the host range and geographical distribution of GETV has posed potential threats to public health. Although GETV can cause reproductive failure in sows and high mortality in neonatal piglets and result in huge economic losses in pig breeding, it is easily neglected because its symptoms are similar to those of infections caused by other viruses, such as PRRSV. In the past five years, there have been two studies to report the contamination of GETV in a live PRRSV vaccine [30,31]. GETV, as a new contaminant, should be given considerable attention. In this paper, we aim to analyze and discuss the source, hazard, and genomic characteristics of the contaminating GETV strains further.

## 2. Evidence of Contamination with GETV in Live PRRSV Vaccines

The first evidence of contamination by GETV in a live PRRSV vaccine was recorded by a Chinese research team in 2020 [30]. In fact, the collection time of the live PRRSV vaccines was the year of 2017. According to the description in the reference above, after immunization with the live PRRSV vaccine, abortions in pregnant sows in a pig farm increased. The author suspected that there may be other potential pathogenic pathogens in the vaccine. Firstly, Marc-145 cells were used to isolate potential contaminant pathogens, and the results showed that the cytopathic effect (CPE) significantly differed from that caused by PRRSV strains. More cell lines, including Vero, PK-15, BHK-21, and human hepatocellular carcinoma cells (HepG-2), were used, showing a similar phenomenon. And then, a specific polymerase chain reaction (PCR) using specific primers (the forward primer 5′-ACCGAAGAAGCCGAAGAA-3′ and reverse primer 5′-GCACTCRAGGTCATACTTG-3′) identified the presence of GETV nucleic acid in the vaccine samples. In addition, an immunofluorescence assay (IFA), transmission electron microscopy (TEM), and genome sequencing (of the isolate of GETV-V1) were used to confirm GETV as the contaminant in the live PRRSV vaccines. Further serological investigations using a self-designed enzyme-linked immunosorbent assay (ELISA) method based on the GETV Cap protein showed that a 100% antibody-positive rate occurred in the sow farms vaccinated with the live PRRSV vaccine, and it was higher than that (96.7%) in sow farms without vaccination with the live PRRSV vaccines. This result showed that GETV infection might have occurred on all of the sow farms at some points in the past. To further assess the side effect of GETV as a contaminant in the live PRRSV vaccine, animal experiments should be conducted in two groups of animals: one group should receive the live PRRSV vaccine, and the other should be served as a control [30].

In 2023, another Chinese research team recorded the second piece of evidence of contamination by GETV in live PRRSV vaccines [31]. One of several commercially modified live PRRSV vaccines from the same batch was found to be GETV-positive via specific RT-PCR. Subsequently, a new GETV isolate, named BJ0304, was isolated in ST cells. Animal experiment showed that the BJ0304 strain did not cause clinical signs and obvious histopathological changes in mice. Only testis and kidney samples from GETV-inoculated mice were positive by 7 and 14 days post-inoculation (dpi). This reveals the low pathogenicity of the BJ0304 strain [31].

## 3. Genomic Comparison Between Vaccine-Contaminated GETV Strains and Other GETV Strains

### *3.1. Genomic Comparison of GETV-V1 and Other GETV Strains*

The genome sequences of two above-mentioned vaccine-contaminated GETV strains were collected from the GenBank database. Meanwhile, 88 complete or nearly complete genomic sequences from the GenBank database as of 21 May 2024 were also downloaded for sequence analysis. Among these, the GETV-V1 strain had the highest full-length genome nucleotide similarity with a porcine-origin strain AH9192 (GenBank no. MG865965) isolated in 2017 (Table 1). A phylogenetic analysis based on the complete genome indicated that GETV-V1 belonged to group III (Figure 1). Its 5′ untranslated region (UTR) is relatively conservative, having 98.7% nucleotide similarity with most reference 5′ UTR sequences (Table 1). In terms of the non-structural polyprotein gene, the GETV-V1 strain had the highest nucleotide (99.5%) and amino acid (99.6%) similarity with a porcine-origin strain JS18 (GenBank no. MT210319) isolated in 2018. In terms of the structural polyprotein gene, it had the highest nucleotide (99.4%) similarity with the AH9192 and JS18 strains, while it had the highest amino acid (99.7%) similarity with the JS18 strain. Moreover, its 3′ UTR had the highest nucleotide similarity (99.5%) with that from AH9192. In short, the GETV-V1 strain was highly related to these porcine-origin GETV strains.

**Table 1.** Similarity of two vaccine-contaminated GETV strains with 88 reference strains.

| Strains | GETV-V1 | | | | | | | BJ0304 | | | | | | |
|---|---|---|---|---|---|---|---|---|---|---|---|---|---|---|
| | Complete Genome | 5′ UTR | Non-Structural Polyprotein | | Structural Polyprotein | | 3′ UTR | Complete Genome | 5′ UTR | Non-Structural Polyprotein | | Structural Polyprotein | | 3′ UTR |
| | nt | nt | nt | aa | nt | aa | nt | nt | nt | nt | aa | nt | aa | nt |
| MM2021 | 95.2 | 97.4 | 95.4 | 98.7 | 95.1 | 98.1 | 93.5 | 95.3 | 98.7 | 95.5 | 98.5 | 95.2 | 98 | 94.2 |
| Sagiyamavirus | 97.1 | 98.7 | 97.4 | 99 | 97 | 98.4 | 94.6 | 97.3 | 100 | 97.6 | 99 | 97 | 98.4 | 95.3 |
| M1 | 98 | 98.7 | 98.2 | 99 | 97.9 | 98.8 | 97.2 | 98.2 | 100 | 98.3 | 99 | 98 | 98.8 | 98 |
| 16-I-676 | 98.8 | 98.7 | 98.9 | 99.4 | 98.8 | 99.6 | 98.2 | 98.9 | 100 | 98.9 | 99.2 | 98.9 | 99.6 | 99 |
| 15-I-752 | 98.8 | 98.7 | 99 | 99.5 | 98.8 | 99.6 | 98.2 | 98.9 | 100 | 98.9 | 99.3 | 98.9 | 99.6 | 99 |
| 14-I-605-C2 | 98.8 | 98.7 | 99 | 99.5 | 98.8 | 99.6 | 98 | 98.9 | 100 | 99 | 99.3 | 98.9 | 99.6 | 98.7 |
| 12IH26 | 98.9 | 98.7 | 99 | 99.6 | 98.9 | 99.6 | 98.2 | 99 | 100 | 99 | 99.4 | 98.9 | 99.6 | 99 |
| GETV-GDFS2-2018 | 98.7 | 98.7 | 98.8 | 99.4 | 98.7 | 99.4 | 98.2 | 98.8 | 100 | 98.8 | 99.2 | 98.8 | 99.4 | 99 |
| HNJZ-S2 | 98.8 | 98.7 | 98.9 | 99.6 | 98.9 | 99.6 | 98.5 | 98.9 | 100 | 98.9 | 99.4 | 99 | 99.6 | 99.2 |
| HNPDS-2 | 98.9 | 98.7 | 99.1 | 99.5 | 98.8 | 99.7 | 98 | 99.1 | 100 | 99.2 | 99.5 | 98.9 | 99.7 | 98.7 |
| HNNY-2 | 98.9 | 98.7 | 99.1 | 99.5 | 98.9 | 99.7 | 98.2 | 99.1 | 100 | 99.1 | 99.5 | 99 | 99.7 | 99 |
| JL17/08 | 98.9 | 98.7 | 99.1 | 99.5 | 98.9 | 99.6 | 98 | 99 | 100 | 99.1 | 99.4 | 98.9 | 99.6 | 98.7 |
| HNJZ-S1 | 98.9 | 98.7 | 99.1 | 99.4 | 98.9 | 99.7 | 98.2 | 99.1 | 100 | 99.1 | 99.2 | 99 | 99.7 | 99 |
| SC266 | 98.5 | 98.7 | 98.7 | 99.1 | 98.5 | 99.4 | 97.5 | 98.7 | 100 | 98.7 | 99 | 98.6 | 99.4 | 98.2 |
| SC483 | 98.5 | 98.7 | 98.7 | 99.2 | 98.4 | 99.4 | 97.5 | 98.7 | 100 | 98.8 | 99.1 | 98.5 | 99.4 | 98.2 |
| SC201807 | 98.7 | 97.4 | 98.9 | 99.5 | 98.7 | 99.6 | 95.5 | 98.8 | 98.7 | 98.9 | 99.4 | 98.8 | 99.6 | 96.2 |

**Table 1.** *Cont.*

| Strains | GETV-V1 | | | | | | | BJ0304 | | | | | | |
|---|---|---|---|---|---|---|---|---|---|---|---|---|---|---|
| | Complete Genome | 5′ UTR | Non-Structural Polyprotein | | Structural Polyprotein | | 3′ UTR | Complete Genome | 5′ UTR | Non-Structural Polyprotein | | Structural Polyprotein | | 3′ UTR |
| | nt | nt | nt | aa | nt | aa | nt | nt | nt | nt | aa | nt | aa | nt |
| JL1707 | 98.9 | 98.7 | 99.1 | 99.5 | 98.8 | 99.3 | 98.5 | 99.1 | 100 | 99.1 | 99.3 | 98.9 | 99.3 | 99.2 |
| HB0234 | 98.9 | 98.7 | 99.1 | 99.3 | 98.8 | 99.3 | 98.5 | 99.1 | 100 | 99.2 | 99.2 | 98.9 | 99.3 | 99.2 |
| JS18 | / | / | 99.5 | 99.6 | 99.4 | 99.7 | / | 96.6 | / | 99 | 99.3 | 98.6 | 99.2 | / |
| AH9192 | 99.4 | 98.7 | 99.3 | 99.3 | 99.4 | 99.6 | 99.5 | 98.7 | 100 | 98.9 | 99.1 | 98.6 | 99.2 | 98.7 |
| GETV-V1 | 100 | 100 | 100 | 100 | 100 | 100 | 100 | 99 | 98.7 | 99.1 | 99.3 | 98.9 | 99.5 | 98.2 |
| YN12042 | 98.4 | 98.7 | 98.8 | 99.3 | 98.9 | 99.6 | 88.3 | 98.5 | 100 | 98.8 | 99.2 | 99 | 99.7 | 88.6 |
| SC1210 | 98.7 | 98.7 | 98.9 | 99.4 | 98.6 | 99.5 | 96.7 | 98.8 | 100 | 98.9 | 99.3 | 98.7 | 99.5 | 97.5 |
| YN0540 | 98.8 | 98.7 | 99 | 99.4 | 98.9 | 99.6 | 96.7 | 99 | 100 | 99 | 99.3 | 99 | 99.6 | 97.5 |
| South Korea | 98.2 | 98.7 | 99.2 | 99.5 | 99.2 | 99.6 | 70.2 | 98.4 | 100 | 99.3 | 99.4 | 99.3 | 99.8 | 70.6 |
| GX201808 | 97.3 | 97.4 | 97.3 | 99 | 97.3 | 99 | 98.5 | 97.5 | 98.7 | 97.4 | 99 | 97.5 | 99 | 98.2 |
| HuN1 | 97.5 | 98.7 | 97.4 | 99.2 | 97.8 | 99.2 | 97.7 | 97.7 | 100 | 97.5 | 99.1 | 97.9 | 99.2 | 98 |
| SD17/09 | 97.7 | 98.7 | 97.7 | 99.3 | 97.6 | 99.2 | 98.5 | 97.8 | 100 | 97.8 | 99.3 | 97.7 | 99.2 | 98.7 |
| JL1808 | 97.7 | 98.7 | 97.8 | 99.3 | 97.7 | 99.3 | 98.2 | 97.9 | 100 | 97.9 | 99.3 | 97.7 | 99.3 | 98.5 |
| LEIV 17741 MPR | 98.5 | 98.7 | 98.5 | 99.3 | 98.7 | 99.7 | 98.2 | 98.7 | 100 | 98.6 | 99.2 | 98.8 | 99.7 | 99 |
| MI-110-C1 | 98.6 | 98.7 | 98.7 | 99.5 | 98.6 | 99.6 | 98 | 98.7 | 100 | 98.7 | 99.3 | 98.7 | 99.6 | 98.7 |
| B254 | / | / | 96.3 | 98.4 | 96.3 | 98.4 | / | / | / | 96.4 | 98.2 | 96.1 | 98 | / |
| YN12031 | 96.2 | 97.4 | 96.3 | 98.6 | 96.4 | 98.8 | 95.5 | 96.1 | 98.7 | 96.3 | 98.5 | 96.3 | 98.4 | 96 |
| GETV/SW/Thailand/2017 | 96 | 98.7 | 96.3 | 98.8 | 95.8 | 98.6 | 94.7 | 95.9 | 100 | 96.3 | 98.7 | 95.6 | 98.3 | 95.2 |
| LEIV 16275 Mag | 97.4 | 97.4 | 97.6 | 99.2 | 97.5 | 99.2 | 96 | 97.4 | 98.7 | 97.7 | 99.1 | 97.6 | 99.2 | 96.7 |
| GETV-JX-CHN-22 | 98.6 | 97.4 | 98.8 | 99.5 | 98.7 | 99.5 | 98 | 98.8 | 98.7 | 98.8 | 99.3 | 98.7 | 99.5 | 98.7 |
| GETV-YL | 98.7 | 98.7 | 98.9 | 99.5 | 98.8 | 99.7 | 98.2 | 98.8 | 100 | 98.9 | 99.3 | 98.8 | 99.6 | 99 |
| GETV-XJ-2019-07 | 98.3 | 98.7 | 98.5 | 99.3 | 98.4 | 99.5 | 97.5 | 98.5 | 100 | 98.5 | 99.2 | 98.4 | 99.5 | 98.2 |
| GETV/SCrph328 | / | / | 98.9 | 99.6 | 98.5 | 99.6 | / | / | / | 98.9 | 99.5 | 98.7 | 99.6 | / |
| GS11-155 | 98.9 | 98.7 | 99.1 | 99.5 | 98.8 | 99.7 | 98.2 | 99.1 | 100 | 99.2 | 99.4 | 98.9 | 99.7 | 98.5 |
| dog202206 | 0 | 98.7 | 98.8 | 99.5 | 98.5 | 99.3 | 0 | 0 | 100 | 98.9 | 99.4 | 98.5 | 99.2 | 0 |
| Rbsq202206 | 96 | 98.7 | 96.3 | 98.7 | 95.9 | 98.5 | 94.7 | 95.9 | 100 | 96.4 | 98.6 | 95.7 | 98.2 | 95.2 |
| GDQY2022 | 98.8 | 98.7 | 99 | 99.6 | 98.9 | 99.7 | 98 | 98.9 | 100 | 99 | 99.4 | 98.9 | 99.7 | 98.7 |
| GDJM2022 | 98.5 | 97.4 | 98.7 | 99.5 | 98.5 | 99.3 | 98.2 | 98.6 | 98.7 | 98.6 | 99.3 | 98.6 | 99.3 | 99 |
| SCZY202010 | 98.6 | 98.7 | 98.6 | 99.1 | 98.8 | 99.6 | 97.5 | 98.7 | 100 | 98.7 | 98.9 | 98.8 | 99.6 | 98.2 |
| HeN2021 | 98.7 | 98.7 | 98.9 | 99.5 | 98.7 | 99.5 | 98 | 98.9 | 100 | 98.9 | 99.3 | 98.8 | 99.5 | 98.7 |
| GD201907-1 | / | / | 97.2 | 98.7 | 97.5 | 99.1 | / | / | / | 97.3 | 98.7 | 97.6 | 99.1 | / |
| GX201909 | / | / | 99.4 | 99.6 | 99.4 | 99.7 | 0 | / | / | 98.9 | 99.3 | 98.6 | 99.2 | / |
| HeB201707 | / | / | 98.6 | 99 | 98.4 | 99.4 | 97.5 | / | / | 98.6 | 98.9 | 98.5 | 99.4 | 98.2 |
| HeN201907 | / | / | 98.8 | 99.5 | 98.7 | 99.6 | 0 | / | / | 98.9 | 99.4 | 98.7 | 99.4 | / |
| BJ0304 | 99 | 98.7 | 99.1 | 99.3 | 98.9 | 99.5 | 98.2 | 100 | 100 | 100 | 100 | 100 | 100 | 100 |
| 19-703 | / | / | 98.6 | 99.4 | 98.5 | 99.3 | 0 | / | / | 98.6 | 99.2 | 98.6 | 99.3 | / |
| NMDK1813-1 | 98.3 | 98.7 | 98.8 | 99.6 | 98.5 | 99.6 | 85.8 | 98.4 | 100 | 98.9 | 99.5 | 98.7 | 99.6 | 86.5 |
| GZ201808 | / | / | 99.4 | 99.5 | 99.4 | 99.6 | / | / | / | 99 | 99.3 | 98.6 | 99.2 | / |
| GDHYLC23 | 98.3 | 97.4 | 98.8 | 99.5 | 98.5 | 99.6 | 89.4 | 98.5 | 98.7 | 98.8 | 99.4 | 98.6 | 99.6 | 90 |
| GD2202/mosquito/China/2/2022 | 98.9 | 98.7 | 99.1 | 99.6 | 98.8 | 99.6 | 98 | 99 | 100 | 99.1 | 99.3 | 99 | 99.7 | 98.7 |
| GETV-GDFS9-2018 | 98.7 | 98.7 | 98.8 | 99.4 | 98.7 | 99.4 | 98.2 | 98.8 | 100 | 98.8 | 99.2 | 98.8 | 99.4 | 99 |
| HNPDS-1 | 98.9 | 98.7 | 99.1 | 99.5 | 98.9 | 99.7 | 97.5 | 99.1 | 100 | 99.2 | 99.5 | 98.9 | 99.7 | 98.2 |
| HNNY-1 | 98.9 | 98.7 | 74.3 | 74.5 | 98.9 | 99.7 | 98 | 99.1 | 100 | 74.3 | 74.5 | 99 | 99.7 | 98.7 |
| 15-I-1105 | 98.8 | 98.7 | 98.9 | 99.4 | 98.8 | 99.6 | 98 | 98.9 | 100 | 98.9 | 99.2 | 98.9 | 99.6 | 98.7 |
| 16-I-674 | 98.8 | 98.7 | 98.9 | 99.4 | 98.8 | 99.6 | 98.2 | 98.9 | 100 | 98.9 | 99.2 | 98.9 | 99.6 | 99 |
| 16-I-599 | 98.8 | 98.7 | 98.9 | 99.4 | 98.8 | 99.6 | 98 | 98.9 | 100 | 98.9 | 99.2 | 98.9 | 99.6 | 98.7 |
| 14-I-605-C1 | 98.8 | 98.7 | 99 | 99.5 | 98.8 | 99.6 | 98.2 | 98.9 | 100 | 99 | 99.3 | 98.9 | 99.6 | 99 |
| MI-110-C2 | 98.6 | 98.7 | 98.7 | 99.5 | 98.6 | 99.7 | 98 | 98.7 | 100 | 98.7 | 99.4 | 98.7 | 99.7 | 98.7 |
| Kochi/01/2005 | 97.8 | 98.7 | 97.9 | 99.3 | 97.8 | 99.3 | 98 | 97.9 | 100 | 97.9 | 99.3 | 97.9 | 99.3 | 98.2 |
| SC202010 | 98.4 | 98.7 | 98.6 | 99.1 | 98.4 | 99 | 97.5 | 98.6 | 100 | 98.6 | 98.9 | 98.5 | 99.2 | 98.2 |
| GD201909 | / | 98.7 | 99.3 | 99.5 | 99.4 | 99.7 | / | / | 100 | 98.8 | 99.3 | 98.6 | 99.2 | / |
| GX202005 | / | / | 97.3 | 98.9 | 97.6 | 98.8 | / | / | / | 97.4 | 98.8 | 97.6 | 98.7 | / |
| JS201809-2 | / | / | 98.8 | 99.4 | 98.7 | 99.6 | / | / | / | 98.8 | 99.3 | 98.7 | 99.4 | / |
| JX202004 | / | / | 97.2 | 98.9 | 97.4 | 99.1 | 98.5 | / | / | 97.3 | 98.9 | 97.5 | 99.1 | 98.7 |
| SX201809 | / | / | 99 | 99.5 | 98.7 | 99.7 | / | / | / | 99 | 99.3 | 98.8 | 99.7 | / |
| FJ202005-2 | / | / | 99.4 | 99.5 | 99.1 | 99.1 | / | / | / | 98.9 | 99.3 | 98.2 | 98.6 | / |
| HuB201905 | / | / | 99.1 | 99.6 | 98.9 | 99.7 | / | / | / | 99.1 | 99.4 | 99 | 99.7 | / |
| SD201910 | / | / | 98.9 | 99.4 | 98.9 | 99.7 | 97.7 | / | / | 98.9 | 99.2 | 99 | 99.7 | 98.5 |
| SC202009 | / | / | 98.8 | 99.4 | 98.5 | 99.5 | 97.5 | / | / | 98.8 | 99.3 | 98.6 | 99.6 | 98.2 |
| FJ201807-1 | / | / | 99.5 | 99.6 | 99.4 | 99.7 | / | / | / | 99 | 99.3 | 98.6 | 99.2 | / |
| HeN202009-2 | 98.8 | 98.7 | 98.9 | 99.5 | 98.9 | 99.7 | 98 | 98.9 | 100 | 98.9 | 99.3 | 98.9 | 99.7 | 98.7 |
| GX1 | / | / | 98.9 | 99.5 | 98.7 | 99.6 | / | / | / | 98.9 | 99.4 | 98.7 | 99.4 | / |

**Table 1.** *Cont.*

| Strains | GETV-V1 | | | | | | | BJ0304 | | | | | | |
|---|---|---|---|---|---|---|---|---|---|---|---|---|---|---|
| | Complete Genome | 5′ UTR | Non-Structural Polyprotein | | Structural Polyprotein | | 3′ UTR | Complete Genome | 5′ UTR | Non-Structural Polyprotein | | Structural Polyprotein | | 3′ UTR |
| | nt | nt | nt | aa | nt | aa | nt | nt | nt | nt | aa | nt | aa | nt |
| GETV/SCrph129/2020 | / | / | 98.8 | 99.5 | 98.5 | 99.6 | / | / | / | 98.8 | 99.4 | 98.6 | 99.6 | / |
| NM,JA_F2_18-8L-NH-Cxp-Y-1-1 | 98.6 | 98.7 | 98.8 | 99.5 | 98.4 | 99.6 | 97.5 | 98.8 | 100 | 98.9 | 99.4 | 98.5 | 99.6 | 98.2 |
| Getahvirus-1 | 98.2 | 98.7 | 99.2 | 99.5 | 99.2 | 99.6 | 70.2 | 98.4 | 100 | 99.3 | 99.4 | 99.3 | 99.8 | 70.6 |
| 16-0810-26 | / | / | 99 | 99.4 | 98.8 | 99.6 | / | / | / | 98.9 | 99.2 | 98.9 | 99.6 | / |
| GETV-China/GX2020 | / | 96.1 | 95.6 | 98.3 | 95.1 | 98 | / | / | 97.4 | 95.6 | 98.2 | 95 | 97.7 | / |
| YN2305/cattle/China/5/2023 | 97.7 | 98.7 | 73.3 | 74.3 | 97.7 | 99.3 | 98.5 | 97.9 | 100 | 73.3 | 74.3 | 97.7 | 99.3 | 98.7 |
| JLy1 | 98.8 | 98.7 | 98.9 | 99.4 | 98.7 | 99.3 | 97.2 | 98.9 | 100 | 99 | 99.3 | 98.9 | 99.5 | 98 |
| SD2206 | 98.9 | 98.7 | 99.1 | 99.6 | 98.9 | 99.7 | 98 | 99.1 | 100 | 99.1 | 99.4 | 99 | 99.7 | 98.7 |
| HN-QH23-As-10 | / | / | 98.6 | 99.4 | 98.3 | 99.4 | / | / | / | 98.6 | 99.3 | 98.3 | 99.4 | / |
| HN-QH23-As-12 | / | / | 98.5 | 99.4 | 98.3 | 99.5 | / | / | / | 98.6 | 99.3 | 98.4 | 99.5 | / |
| HN-QH23-As-14 | / | / | 98.6 | 99.3 | 98.3 | 99.4 | / | / | / | 98.6 | 99.3 | 98.3 | 99.4 | / |
| GETV-JX-CHN-22-P7 | 98.6 | 97.4 | 98.7 | 99.4 | 98.8 | 99.7 | 98 | 98.8 | 98.7 | 98.7 | 99.3 | 98.8 | 99.7 | 98.7 |

"/" indicates there being no available data due to incomplete sequences.

### *3.2. Genomic Comparison of BJ0304 and Other GETV Strains*

A whole genome sequence analysis showed that BJ0304 had the highest similarity (99.1%) to six porcine-origin strains (HNJZ-S1 from 2011, GenBank no. KY363862; HNNY-1 and HNNY-2 from 2016, GenBank nos. MG865966 and MG865967; HNPDS-1 and HNPDS-2 from 2017, GenBank nos. MG865968 and MG865969; SD2206 from 2022, GenBank no. PP623164) and three mosquito-origin strains (JL1707 from 2017, GenBank no. MH722255; HB0234 from 2002, GenBank no. EU015062; and GS11-155 from 2011, GenBank no. ON828424) (Table 1). A phylogenetic analysis based on the complete genome indicated that BJ0304 belonged to group III (Figure 1). Its 5′ UTR is quite conservative, having 100% nucleotide similarity with most reference 5′ UTR sequences. In terms of the non-structural polyprotein gene, BJ0304 had the highest nucleotide (99.3%) similarity with two porcine-origin strains (GenBank nos. AY702913 and NC_006558) isolated from South Korea, while at the amino acid level, it shared the highest similarity (99.5%) with three porcine-origin strains (HNNY-2 from 2016; HNPDS-1 and HNPDS-2 from 2017), one red-panda-origin strain (SCrph328 from 2018, GenBank no. MZ357111), and one mosquito-origin strain (NMDK1813-1 from 2018, GenBank no. MW512827). In the structural polyprotein gene, it had the highest nucleotide (99.3%) and amino acid (99.8%) similarity with two porcine-origin strains (GenBank nos. AY702913 and NC_006558) isolated from South Korea. In addition, its 3′ UTR had the highest nucleotide similarity (99.2%) with that in HNJZ-S2 (a porcine-origin strain, GenBank no. KY363863); and JL1707 and HB0234 (two mosquito-origin strains). In general, BJ0304 was also highly close to these porcine-origin GETV strains.

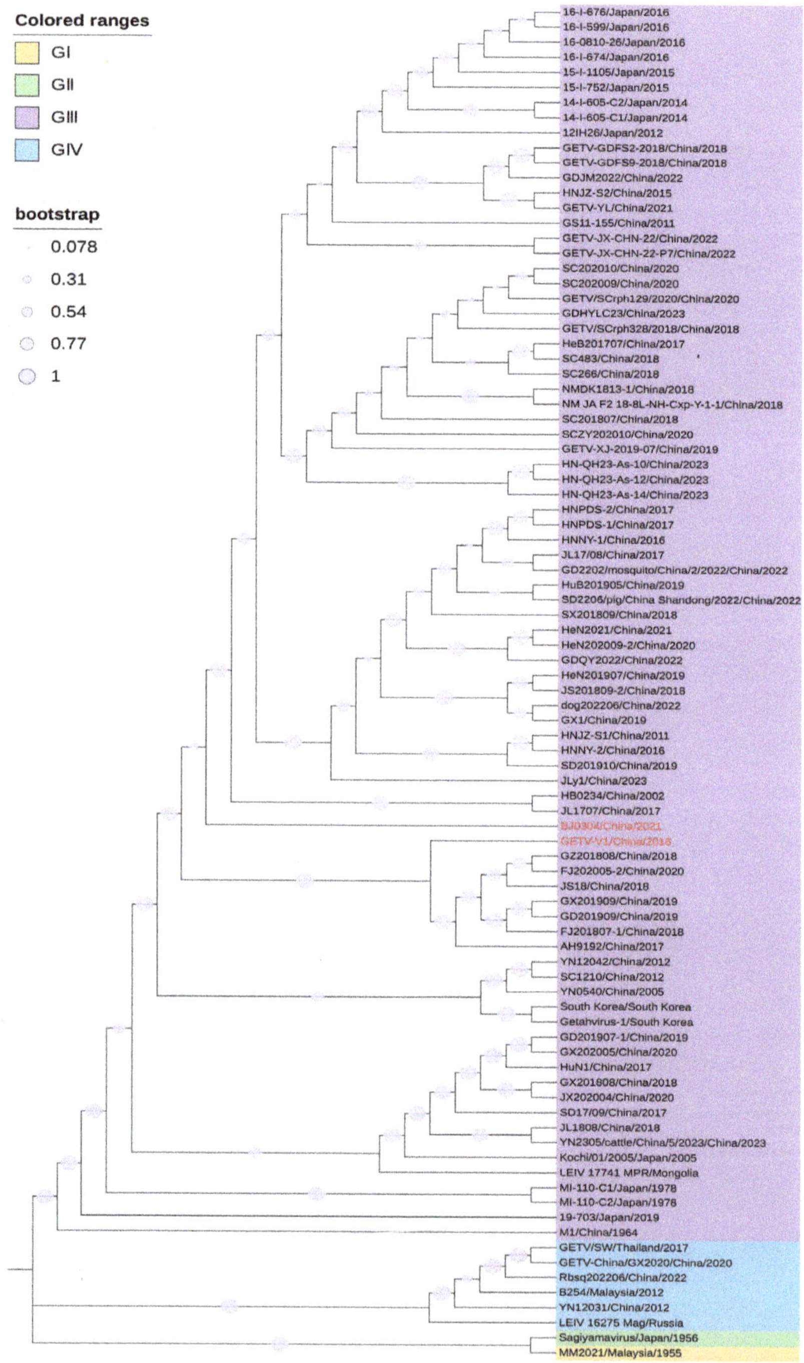


**Figure 1.** Phylogenetic analysis of GETVs. A phylogenetic tree based on complete genomic sequences was constructed using the maximum likelihood method with 1000 replicates and the GTR+G model in MEGA (version 11.0.13).

## 4. Conclusions

To summarize, this study provides a detailed analysis of the sequences of two PRRSV vaccine-derived GETVs, revealing their high relevance to porcine-origin GETV strains. The contaminant of GETV in the live PRRSV vaccine may have come from raw and auxiliary materials of animal origin, such as trypsin and fetal bovine serum, used during the vaccine production process. There are several live pig vaccines against Japanese encephalitis virus (JEV), CSFV, PRV, and PRRSV that are widely used in Chinese swine herds. GETV exists in the PRRSV vaccine; this may just be the tip of the iceberg. In recent years, GETV cases have increased in Chinese pig herds [27,32–37], and these vaccines contaminated with GETVs might play an important role in its transmission. In the future, GETV should be included in live veterinary vaccine quarantines, and the detection of GETV in the raw and auxiliary materials and final vaccine products should be strengthened during vaccine manufacturing. Moreover, surveillance and vaccine development for GETV are also important in pig herds.

**Author Contributions:** Conceptualization: P.-P.C. and S.-L.Z. Methodology: P.-P.C. Software: P.-P.C. Formal analysis: P.-P.C., S.-N.C. and S.-L.Z. Data curation: P.-P.C., S.-N.C. and X.Z. Writing—original draft preparation: P.-P.C. and S.-L.Z. Writing—review and editing: S.-L.Z. and Z.-Z.W. Supervision: Z.-Z.W. and S.-L.Z. Project administration: S.-L.Z. Funding acquisition: S.-L.Z. All authors have read and agreed to the published version of the manuscript.

**Funding:** This research was funded by Guangdong provincial key research and development program, grant number 2023B0202150001; the Introduction of Scientific and Technological Talents of the Guangdong Academy of Agricultural Sciences Special Fund program, grant number R2024YJ-YB3016; Guangdong Provincial Key Laboratory of Livestock Disease Prevention program, grant number 2023B1212060040; and Guangxi Natural Science Foundation (No. 2023GXNSFAA026494).

**Institutional Review Board Statement:** Not applicable.

**Informed Consent Statement:** Not applicable.

**Data Availability Statement:** The original contributions presented in this study are included in the article. Further inquiries can be directed to the corresponding authors.

**Conflicts of Interest:** Sheng-Nan Chen is a veterinarian and technical supporter in Guangzhou Sino-Science Gene Testing Service Co., Ltd. and her company did not provide the grant support. So the authors declare no conflicts of interest.

## References

1. Barkema, H.W.; Bartels, C.J.; van Wuijckhuise, L.; Hesselink, J.W.; Holzhauer, M.; Weber, M.F.; Franken, P.; Kock, P.A.; Bruschke, C.J.; Zimmer, G.M. Outbreak of Bovine virus diarrhea on Dutch dairy farms induced by a bovine herpesvirus 1 marker vaccine contaminated with bovine virus diarrhea virus type 2. *Tijdschr. Diergeneeskd.* **2001**, *126*, 158–165. [PubMed]
2. Su, Q.; Li, Y.; Zhang, Y.W.; Zhang, Z.H.; Meng, F.F.; Cui, Z.Z.; Chang, S.; Zhao, P. Newcastle disease virus-attenuated vaccine LaSota played a key role in the pathogenicity of contaminated exogenous virus. *Vet. Res.* **2018**, *49*, 80. [CrossRef]
3. Thornton, D.H. A survey of mycoplasma detection in veterinary vaccines. *Vaccine* **1986**, *4*, 237–240. [CrossRef] [PubMed]
4. Sasaki, T.; Harasawa, R.; Shintani, M.; Fujiwara, H.; Sasaki, Y.; Horino, A.; Kenri, T.; Asada, K.; Kato, I.; Chino, F. Application of PCR for detection of mycoplasma DNA and pestivirus RNA in human live viral vaccines. *Biologicals* **1996**, *24*, 371–375. [CrossRef]
5. Kojima, A.; Takahashi, T.; Kijima, M.; Ogikubo, Y.; Nishimura, M.; Nishimura, S.; Harasawa, R.; Tamura, Y. Detection of Mycoplasma in avian live virus vaccines by polymerase chain reaction. *Biologicals* **1997**, *25*, 365–371. [CrossRef]
6. Kojima, A.; Takahashi, T.; Kijima, M.; Ogikubo, Y.; Tamura, Y.; Harasawa, R. Detection of mycoplasma DNA in veterinary live virus vaccines by the polymerase chain reaction. *J. Vet. Med. Sci.* **1996**, *58*, 1045–1048. [CrossRef]
7. Awad, A.M.; Abd El-Hamid, H.S.; Abou Rawash, A.A.; Ibrahim, H.H. Detection of reticuloendotheliosis virus as a contaminant of fowl pox vaccines. *Poult. Sci.* **2010**, *89*, 2389–2395. [CrossRef] [PubMed]
8. Li, J.P.; Dong, X.; Yang, C.H.; Li, Q.H.; Cui, Z.Z.; Chang, S.; Zhao, P.; Yu, K.Z.; Yang, H.C. Isolation, identification, and whole genome sequencing of reticuloendotheliosis virus from a vaccine against Marek's disease. *Poult. Sci.* **2015**, *94*, 643–649. [CrossRef]

9. Li, Y.; Cui, S.; Cui, Z.Z.; Chang, S.; Zhao, P. Genome analysis and pathogenicity of reticuloendotheliosis virus isolated from a contaminated vaccine seed against infectious bursal disease virus: First report in China. *J. Gen. Virol.* **2016**, *97*, 2809–2815. [CrossRef] [PubMed]
10. Zavala, G.; Cheng, S. Detection and characterization of avian leukosis virus in Marek's disease vaccines. *Avian Dis.* **2006**, *50*, 209–215. [CrossRef] [PubMed]
11. Shittu, I.; Adedeji, A.J.; Luka, P.D.; Asala, O.O.; Sati, N.M.; Nwagbo, I.O.; Chinyere, C.N.; Arowolo, O.O.; Adole, J.A.; Emennaa, P.; et al. Avian leukosis virus subgroup-J as a contaminant in live commercially available poultry vaccines distributed in Nigeria. *Biologicals* **2019**, *57*, 29–33. [CrossRef]
12. Mao, Y.Q.; Su, Q.; Li, J.P.; Jiang, T.Z.; Wang, Y.X. Avian leukosis virus contamination in live vaccines: A retrospective investigation in China. *Vet. Microbiol.* **2020**, *246*, 108712. [CrossRef] [PubMed]
13. Li, Y.; Hu, Y.; Cui, S.; Fu, J.Y.; Wang, Y.X.; Cui, Z.Z.; Fang, L.C.; Chang, S.; Zhao, P. Molecular characterization of chicken infectious anemia virus from contaminated live-virus vaccines. *Poult. Sci.* **2017**, *96*, 1045–1051. [CrossRef]
14. Li, Y.; Wang, J.J.; Chen, L.F.; Wang, Q.; Zhou, M.; Zhao, H.; Chi, Z.N.; Wang, Y.X.; Chang, S.; Zhao, P. Genomic Characterization of CIAV Detected in Contaminated Attenuated NDV Vaccine: Epidemiological Evidence of Source and Vertical Transmission From SPF Chicken Embryos in China. *Front. Vet. Sci.* **2022**, *9*, 930887. [CrossRef] [PubMed]
15. Gómez-Romero, N.; Velazquez-Salinas, L.; Ridpath, J.F.; Verdugo-Rodríguez, A.; Basurto-Alcántara, F.J. Detection and genotyping of bovine viral diarrhea virus found contaminating commercial veterinary vaccines, cell lines, and fetal bovine serum lots originating in Mexico. *Arch. Virol.* **2021**, *166*, 1999–2003. [CrossRef]
16. Harasawa, R. Adventitious pestivirus RNA in live virus vaccines against bovine and swine diseases. *Vaccine* **1995**, *13*, 100–103. [CrossRef]
17. Zhang, Z.; Zhang, M.J.; Dong, Y.Q.; Wang, J.; Wu, F.X.; Liu, S.; Shao, W.X.; Li, X.C.; Wang, S.S. Detection and Analysis of Bovine Viral Diarrhea Virus Contamination in Classical Swine Fever and Porcine Reproductive and Respiratory Syndrome Vaccines. *China Anim. Health Insp.* **2015**, *32*, 76–79.
18. Yoshikawa, R.; Shimode, S.; Sakaguchi, S.; Miyazawa, T. Contamination of live attenuated vaccines with an infectious feline endogenous retrovirus (RD-114 virus). *Arch. Virol.* **2014**, *159*, 399–404. [CrossRef]
19. Yoshikawa, R.; Sato, E.; Miyazawa, T. Contamination of infectious RD-114 virus in vaccines produced using non-feline cell lines. *Biologicals* **2011**, *39*, 33–37. [CrossRef]
20. Zhai, S.L.; Lu, S.S.; Wei, W.K.; Lv, D.H.; Wen, X.H.; Zhai, Q.; Chen, Q.L.; Sun, Y.W.; Xi, Y. Reservoirs of Porcine Circoviruses: A Mini Review. *Front. Vet. Sci.* **2019**, *19*, 319. [CrossRef]
21. Cui, X.Y.; Xia, D.S.; Luo, L.Z.; An, T.Q. Recombination of Porcine Reproductive and Respiratory Syndrome Virus: Features, Possible Mechanisms, and Future Directions. *Viruses* **2024**, *16*, 929. [CrossRef]
22. Zhou, L.; Yu, J.F.; Zhou, J.; Long, Y.P.; Xiao, L.; Fan, Y.D.; Yang, D.J.; Zhang, B.; Zhang, Z.D.; Liu, L. A novel NADC34-like porcine reproductive and respiratory syndrome virus 2 with complex genome recombination is highly pathogenic to piglets. *Infect. Genet. Evol.* **2023**, *112*, 105436. [CrossRef]
23. Zhao, D.S.; Yang, B.; Yuna, X.G.; Shen, C.C.; Zhang, D.J.; Shi, X.J.; Zhang, T.; Cui, H.M.; Yang, J.K.; Chen, X.H.; et al. Advanced Research in Porcine Reproductive and Respiratory Syndrome Virus Co-infection With Other Pathogens in Swine. *Front. Vet. Sci.* **2021**, *8*, 699561. [CrossRef]
24. Fukunaga, Y.; Kumanomido, T.; Kamada, M. Getah virus as an equine pathogen. *Vet. Clin. N. Am. Equine Pract.* **2000**, *16*, 605–617. [CrossRef]
25. Lu, G.; Chen, R.A.; Shao, R.; Dong, N.; Liu, W.Q.; Li, S.J. Getah virus: An increasing threat in China. *J. Infect.* **2020**, *80*, 350–371. [CrossRef]
26. Li, B.; Wang, H.Y.; Liang, G.D. Getah Virus (Alphavirus): An Emerging, Spreading Zoonotic Virus. *Pathogens* **2022**, *11*, 945. [CrossRef]
27. Chu, P.P.; Guo, H.Y.; Zhou, X.; Chen, S.N.; Sun, X.R.; Tian, S.C.; Zou, Y.G.; Li, C.L.; Zhang, R.; Zhai, S.L. Emergence of a novel GIII Getah virus variant in pigs in Guangdong, China, 2023. *Microbiol. Spectrum* **2024**, *12*, e0048324. [CrossRef]
28. Yang, T.; Li, R.; Hu, Y.; Yang, L.; Zhao, D.; Du, L.; Li, J.; Ge, M.; Yu, X. An outbreak of Getah virus infection among pigs in China, 2017. *Transbound. Emerg. Dis.* **2018**, *65*, 632–637. [CrossRef]
29. Wu, Y.; Gao, X.P.; Kuang, Z.P.; Lin, L.M.; Zhang, H.; Yin, L.J.; Hong, J.B.; Ren, B.H.; Li, Q.H.; Wang, L.X. Isolation and pathogenicity of a highly virulent group III porcine Getah virus in China. *Front. Cell. Infect. Microbiol.* **2024**, *14*, 1494654. [CrossRef]
30. Zhou, F.; Wang, A.J.; Chen, L.; Wang, X.G.; Cui, D.D.; Chang, H.T.; Wang, C.Q. Isolation and phylogenetic analysis of Getah virus from a commercial modified live vaccine against porcine reproductive and respiratory syndrome virus. *Mol. Cell. Probes* **2020**, *53*, 101650. [CrossRef]
31. Gao, X.T.; Li, J.L.; Wu, T.; Dou, J.P.; Zhang, W.R.; Jia, H.; Zhang, Z.F.; Liu, X.J.; Li, Y.N. The Isolation and Characterization of a Novel Group III-Classified Getah Virus from a Commercial Modified Live Vaccine against PRRSV. *Viruses* **2023**, *15*, 2090. [CrossRef] [PubMed]

32. Shi, N.; Qiu, X.S.; Cao, X.Y.; Mai, Z.H.; Zhu, X.Y.; Li, N.; Zhang, H.; Zhang, J.Y.; Li, Z.X.; Shaya, N.; et al. Molecular and serological surveillance of Getah virus in the Xinjiang Uygur Autonomous Region, China, 2017–2020. *Virol. Sin.* **2022**, *37*, 229–237. [CrossRef]
33. Sun, Q.; Xie, Y.X.; Guan, Z.X.; Zhang, Y.; Li, Y.H.; Yang, Y.; Zhang, J.J.; Li, Z.J.; Qiu, Y.F.; Li, B.B.; et al. Seroprevalence of Getah virus in Pigs in Eastern China Determined with a Recombinant E2 Protein-Based Indirect ELISA. *Viruses* **2022**, *14*, 2173. [CrossRef] [PubMed]
34. Pan, J.H.; Zhang, H.Q.; Chen, X.Q.; Zeng, M.Y.; Han, H.; Guo, Y.J.; Li, J.M.; Luo, S.C.; Yan, G.Z.; Chen, S.N.; et al. Evolutionary characterization and pathogenicity of Getah virus from pigs in Guangdong Province of China. *Arch. Virol.* **2023**, *168*, 258. [CrossRef] [PubMed]
35. Zhao, J.; Dellicour, S.; Yan, Z.Q.; Veit, M.; Gill, M.S.; He, W.T.; Zhai, X.F.; Ji, X.; Suchard, M.A.; Lemey, P.; et al. Early Genomic Surveillance and Phylogeographic Analysis of Getah Virus, a Reemerging Arbovirus, in Livestock in China. *J. Virol.* **2023**, *97*, e0109122. [CrossRef] [PubMed]
36. Sun, X.Y.; Zhang, Q.Y.; Shan, H.; Cao, Z.; Huang, J. Genome characteristics of atypical porcine pestivirus from abortion cases in Shandong Province, China. *Virol. J.* **2023**, *20*, 282. [CrossRef]
37. Lan, J.H.; Fang, M.T.; Duan, L.L.; Liu, Z.; Wang, G.G.; Wu, Q.; Fan, K.; Huang, D.Y.; Ye, Y.; Wan, G.; et al. Novel Porcine Getah Virus from Diarrheal Piglets in Jiangxi Province, China: Prevalence, Genome Sequence, and Pathogenicity. *Animals* **2024**, *14*, 2980. [CrossRef]

*veterinary sciences*

MDPI

*Article*

# Development and Evaluation of an Immunochromatographic Strip and a Magnetic Chemiluminescence Immunoassay for Detection of *Porcine Circovirus Type* 2 Antigen

**Sirui Tao [1,†], Yu Duan [1,†], Yinhe Zha [1,2], Xiaxia Tong [2], Yulong He [1,3], Huapeng Feng [1] and Jianhong Shu [1,2,*]**

[1] College of Life Sciences and Medicine, Zhejiang Sci-Tech University, Hangzhou 310018, China; taosirui2019@163.com (S.T.); 13625267597@163.com (Y.D.); zhayinhe2006@163.com (Y.Z.); heyl79@zstu.edu.cn (Y.H.); fenghuapeng@zstu.edu.cn (H.F.)

[2] Zhejiang Hom-Sun Biosciences Co., Ltd., Shaoxing 312000, China; hzhssw@163.com

[3] Research Center of Animal Vaccines and Diagnostic Reagents, Zhejiang Sci-Tech University Shaoxing Academy of Biomedicine, Shaoxing 312090, China

* Correspondence: shujianhong@zstu.edu.cn

† These authors contributed equally to this work.

**Simple Summary:** This article describes the development and testing of PCV2 (*Porcine circovirus type* 2) diagnostic tools, including the expression and purification of recombinant proteins, the production of antibodies, and the validation of immunoassays. The team successfully expressed the Cap protein of PCV2 in Escherichia coli and obtained a high-purity protein by affinity chromatography purification. Subsequently, the protein was used to immunize rabbits, and polyclonal antibodies were successfully obtained and purified. The study also developed an immunochromatographic test strip and tested its sensitivity and specificity to different PCV2 strains. The results showed that the test strip showed high sensitivity to PCV and good specificity. In addition, this article also compared the immunochromatographic test strip with the traditional ELISA detection method, and the results showed that the two were highly consistent (Kappa coefficient was 0.866). Finally, the study also verified the linear range and detection limit of the magnetic chemiluminescence immunoassay, further improving the accuracy and sensitivity of PCV2 detection.

**Abstract:** *Porcine circovirus type 2* (PCV2) is the main and primary causative agent of Postweaning Multisystemic Wasting Syndrome (PMWS). To date, immunoperoxidase monolayer assay (IPMA), indirect immunofluorescent assay (IFA), and enzyme linked immunosorbent assay (ELISA) are the most commonly diagnostic methods for detecting PCV2 antigens. However, these methods require specialized equipment and technical expertise and are suitable for laboratory use only. This study aims to develop an immunochromatographic strip and a magnetic chemiluminescence immunoassay for the detection of PCV2 antigens. The recombinant protein was constructed using a prokaryotic expression system, and the polyclonal antibody was obtained by animal experiments. Polystyrene microspheres are used as solid phase carriers to covalently bind to the amino groups of proteins to form immunoprobes. Monodisperse beads are covalently bound to antigens or antibodies as solid phases to bind antibodies or antigens in the liquid phase in a superior manner, thereby capturing and separating antigens and antibodies in the liquid phase. The immunochromatographic strip is qualitative detection method, this method can detect PCV2a strain, PCV2b strain, and PCV2d strain. The immunochromatographic strip had minimum detection limits of $10^{2.89}TCID_{50}/0.1$ mL, $10^{3.19}TCID_{50}/0.1$ mL, and $10^{3.49}TCID_{50}/0.1$ mL for PCV2a/LG, PCV2b/SH, and PCV2d/JH. The results of testing PEDV (CV777 strain), PRV (HB2000 strain), CSFV (WH-09 strain), PRRS (JXA1-R strain),

PPV (WH-1 strain), and ASFV (SD strain) were negative. The agreement between the immunochromatographic strip and the ELISA kit was 93.33% (140/150) and the Kappa was 0.866 (Kappa > 0.81). On the premise of ensuring sensitivity, the most important feature of the immunochromatographic strip is that this method can save time when testing; results can be obtained within 5 to 10 min. Magnetic chemiluminescence immunoassay is quantitative detection method; this method can detect PCV2 Cap proteins in swine serum, the linear range of this method was 0.25 ng/mL to 32 ng/mL and $R^2$ of the standard curve was 0.9993. The limit of detection (LOD) is 0.051 ng/mL and the limit of quantitation (LOQ) is 0.068 ng/mL. The agreement between the magnetic chemiluminescence immunoassay and the ELISA kit (test PCV2 Cap proteins) was 97.14% (68/70). This method took less than 30 min to achieve results, which is less than the ELISA kit. The results of this study show that immunochromatographic strip and magnetic chemiluminescence immunoassay for PCV2 antigens had great sensitivity and specificity, which lays the foundation for PCV2 clinical detection.

**Keywords:** prokaryotic expression; immunochromatographic strip; magnetic chemiluminescence immunoassay; *porcine circovirus 2*

## 1. Introduction

*Porcine circovirus type 2* (PCV2) is considered to be the smallest single stranded circular DNA virus known to directly infect mammals [1–3]. The genome consists of 1766 nucleotides and 11 specific open reading frames (ORF) [4]. The main specific ORF is ORF1 and ORF2 [1,5,6]. ORF1 encodes two replication-related proteins (Rep and Rep') and the capsid protein (Cap) encoded by ORF2 is the only structural protein of PCV2 [7]. At present, the Cap protein is considered to be the most immunogenic main protein of PCV2, with a molecular weight of about 27.8 kDa, consisting of 233–234 amino acids, and is closely related to PCV2 infection [8].

PCV2 has caused great harm to the pig industry. With the deepening of research [9–11], it was found that PCV2 not only caused Postweaning Multisystemic Wasting Syndrome (PMWS), but also caused porcine dermatitis and nephropathy syndrome (PDNS), porcine respiratory syndrome (PRDC), reproductive failure (RF), and congenital tremors (CT) [12–14].

PCV3 was first discovered in pig herds in North America and Asia in 2016, and later confirmed in European countries. Over time, it spread to South America and Africa. The virus can be isolated from pig serum, lymph nodes, lungs, oral fluids, and nasal swabs. In China, it has been confirmed that PCV3 genetic material exists in the serum of aborted fetuses, semen, and pigs with symptoms of reproductive disorders [15–18].

Although PCV4 was discovered in Hunan Province, China in 2019, retrospective studies have shown that PCV4 DNA has been detected in pig samples since 2012, indicating that PCV4 has been circulating in pigs for at least ten years. Since 2019, PCV4 has been reported in domestic pigs in multiple provinces of China, including Hunan, Henan, Jiangsu, Anhui, Shanxi, Guangxi, and in Inner Mongolia, as well as in South Korea. PCV4 DNA was not detected in pig samples (serum and tissue) from Europe (Italy and Spain), indicating that PCV4 may have limited geographical distribution. Therefore, further research is needed on the distribution of PCV4 in other geographical regions [19–22].

To date, polymerase chain reaction (PCR), indirect fluorescence assay (IFA), immunoperoxidase monolayer assay (IPMA), in situ hybridization (ISH), immunohistochemistry (IHC), and enzyme-linked immunosorbent assay (ELISA) had been widely used to detect PCV2 antibodies and antigens [23–25]. These methods should only be used in laboratories which are equipped with specialized equipment [26–29]. Therefore, a simple, rapid, and sensitive detection method is needed for antigen detection.

Compared with traditional testing, the immunochromatographic strip and magnetic chemiluminescence immunoassay have several advantages, such as being simple, rapid, and low-cost, with no need for professional equipment [30]. The immunochromatographic strip combines polystyrene microspheres and immunochromatography. Carboxyl can be attached to the polystyrene microspheres, and polystyrene microspheres can be linked to the aminos of antigens or antibodies by carboxyl [30–33]. The monoclonal antibodies were fixed at the test line and goat anti-rabbit IgG secondary antibodies were fixed at the control line on NC membrane [32,34]. The appearance of color depends on the antigen–antibody reaction. Polyclonal antibody-labeled polystyrene microspheres can catch antigens in a sample, monoclonal antibodies, which are in test line, can also catch antigens; therefore, these components compose a sandwich structure—monoclonal antibodies link the nitrocellulose membrane (NC membrane) by hydrophobic interaction. Finally, this sandwich structure stays at test line and a blue-colored band will appear at the test line. The goat anti-rabbit IgG secondary antibodies which were fixed at the control line can catch polyclonal antibody-labeled polystyrene microspheres which are redundant, and a blue-colored band will appear at the control line. The sample is positive when two blue lines appear on the strips. The sample is negative when one blue line is at the control line. No blue line and one blue at the test line show that the result is invalid. Figure 1 shows the fabrication process of both methods.

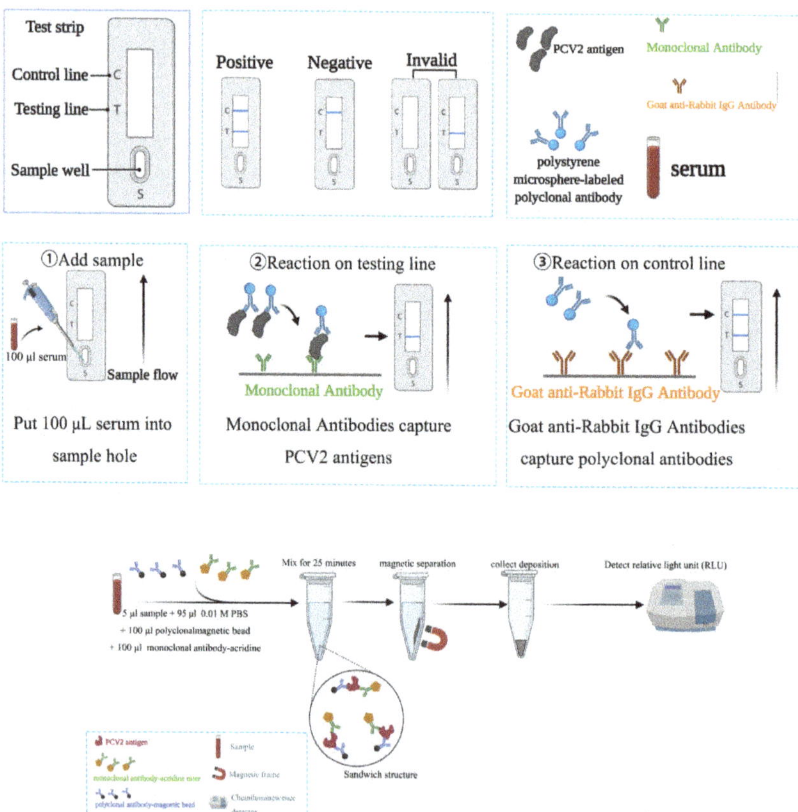


**Figure 1.** Processes of two detection methods.

Now, raw material suppliers can provide polychrome polystyrene microspheres; the red polystyrene microsphere and bule polystyrene microsphere are widely used in immunochromatographic strips. The appearance of color depends on the antigen–antibody reaction; so, we do not need to add any substrates or enzymes to the strip. Instead of polystyrene microspheres, many other alternatives can be chosen to develop strip, such as colloidal gold or fluorescent microspheres. Their sensitivity and specificity will change when we change label because these labels have different physical and chemical properties.

A magnetic chemiluminescence immunoassay combines magnetic particles and chemiluminescent immunoassay (CLIA). Briefly, the double antibody sandwich exists when antibody-magnetic particles and an antibody-acridine ester bind to a PCV2 antigen. After magnetic particle separation, acridine esters which were on monoclonal antibodies can generate an optical signal by buffer A and B; the detection machine will record the relative light unit (RLU). RLU is proportional to the PCV2 antigen [35].

The cores of magnetic particles are iron oxides; magnetic particles have superparamagnetic effect. These magnetic particles can quickly gather when there is an external magnetic field. Meanwhile, magnetic particles can be separated quickly when the external magnetic field is removed [36–38]. Magnetic particles also have a larger specific surface area; magnetic particles provide a more adequate reaction environment for immune response. Carboxyl can be attached to the surface of magnetic particles, and magnetic particles can be linked to the aminos of antigen or antibody by carboxyl [39–41].

The acridine ester compound is the core of the magnetic chemiluminescence immunoassay (CLIA). Compared with other chemiluminescence systems, acridine ester compounds have low luminescence interference and low background and do not need enzymes or catalysts [42–44]. They are suitable for all-automatic and kinetic detection. In addition, this type of compound also has very good stability; many advantages have resulted in it gradually occupying the main position in actual analysis and detection applications [45–47]. In the future, the acridine ester compound will become the chemiluminescent substance with the greatest research and application value [48–50].

Currently, the ELISA kit is the most common method for detecting *porcine circovirus type 2* infection, but it takes several hours to obtain results. Our team hope to develop one or two methods to detect *porcine circovirus type 2* infection within 30 min. So, we chose the immunochromatographic strip and the magnetic chemiluminescence immunoassay. Immunochromatographic strip is a qualitative analysis method. Its greatest feature is that the results can be obtained in 15 min. Meanwhile, magnetic chemiluminescence immunoassay takes less than 30 min; moreover, we can accurately know the number of antigens in serum by using magnetic chemiluminescence immunoassay.

In this study, immunochromatographic strip and magnetic chemiluminescence immunoassay were successfully developed and applied to test for PCV2 antigen. Immunochromatographic strip can achieve accurate results within 10 min and magnetic chemiluminescence immunoassay can achieve accurate results within 25 min. In summary, two methods were used to improve the speed and simplicity of PCV2 diagnosis without compromising sensitivity and specificity. Immunochromatographic strip and magnetic chemiluminescence immunoassay provide new ideas for the detection of PCV2.

## 2. Materials and Methods

### 2.1. Clinical Samples, Cells, Viruses, and Enzyme

One hundred and fifty serum specimens from PCV2-suspected pigs were preserved in the College of Life Sciences and Medicine, Zhejiang Sci-Tech University, Hangzhou, China. All serum specimens were stored at −80 °C.

PK15 cells (ATCC® CCL-33), Dulbecco's modified eagle medium (DMEM, Gibco, Waltham, MA, USA), fetal bovine serum (FBS, Gibco, MA, USA).

XohI restriction enzyme (TKARA, Tokyo, Japan), NcoI restriction enzyme (TKARA, Tokyo, Japan).

Urea (National pharmaceutical reagent, Shanghai, China), Bovine Serum Albumin (BSA, Solarbio, Hangzhou, China), Ni-NTA (Solarbio, Hangzhou, China), SDS gel electrophoresis (Solarbio, Hangzhou, China), BCA assay (Solarbio, Hangzhou, China), Protein A kit (Solarbio, Hangzhou, China), EDC (RHAWN, Hangzhou, China), polystyrene microsphere (Huge Biotechnology, Shanghai, China), magnetic beads (Huge Biotechnology, Shanghai, China), Matrix SephadexTM G-25 resin (Solarbio, Hangzhou, China).

Goat anti-Rabbit IgG (H+L) Highly Cross-Adsorbed Secondary Antibody was supplied by Thermo Fisher Scientific. Mouse monoclonal antibody against PCV2-Cap was supplied by Hangzhou DaYao Biotechnology Co., Ltd. (Hangzhou, 310018, China).

Strain origins genotypes and accession numbers are shown in Table 1.

**Table 1.** Origins of the PCV2 strains used in this study.

| Name | Isolate Region | Isolate Time | Genotype | Genotype | Genome Length | Accession Numbers |
|---|---|---|---|---|---|---|
| LG | China | 2008 | 8839.70 | PCV2a | 1767 bp | HM038034 |
| SZ | China | 2016 | | PCV2a | 1767 bp | KX814351 |
| SH | China | 2006 | | PCV2b | 1767 bp | HM038027 |
| JH | China | 2017 | | PCV2d | 1767 bp | MG245867 |
| WZ | China | 2016 | | PCV2d | 1767 bp | MK604498 |

### 2.2. Antibodies and Microsphere

Goat anti-Rabbit IgG (H+L) Highly Cross-Adsorbed Secondary Antibody was supplied by Thermo Fisher Scientific. Mouse monoclonal antibody against PCV2-Cap was supplied by Hangzhou DaYao Biotechnology Co., Ltd. Polystyrene microsphere and magnetic microspheres were produced by Shanghai Huge Biotechnology Co., Ltd. (Shanghai, 200000, China).

### 2.3. Preparation of Recombinant Cap Protein of PCV2

In this study, the PCV2 ORF2 (GenBank DQ231512.1) complete coding DNA sequences (CDS) and primer sequences were both synthesized by GENEWIZ Biotechnology Co., Ltd. (Suzhou, 215000, China). The forward primer sequence is 5′-GCCCATGGGCAAAAATGGCATCTTCAAC-3′ and the reverse primer is 5′-GCCTCGAG-AGGGTTAAGTGGGGGGTCTT-3′.

XohI and NcoI were used to digest the PCR product, then the purified PCR product was cloned into the pET-28a vector. DNA sequencing was used to confirm the construct (GENEWIZ, Suzhou, China). Finally, recombinant plasmids were transformed into *E. coli* BL21-competent cells.

An optical density of 0.8 at 600 nm was measured using a positive clone of recombinant protein grown at 37 °C in lysogeny broth supplemented with 50 μg/mL kanamycin. In the next step, 0.5 mM of IPTG was added to lysogeny broth. Approximately 15 h after induction at 30 °C, cells were harvested by centrifugation (12,000 rpm for 15 min) and resuspended in 10 mL of 10 mM PBS (pH 7.2). The ice-cold cells were broken down by sonication and centrifuged at 12,000 rpm for 25 min. The precipitation was collected; the protein was purified using Ni-NTA resin as specified by the manufacturer.

The protein solution was added into a pre-treated dialysis bag. Firstly, buffer I (8 mol/L urea, 0.01 M PBS based solution) was used to dialysis protein solution.

Then, buffer II (6 mol/L urea, 0.01 M PBS based solution), buffer III (4.5 mol/L urea, 0.01 M PBS based solution), buffer IV (3.5 mol/L urea, 0.01 M PBS based solution), buffer V (2.5 mol/L urea, 0.01 M PBS based solution), buffer VI (1.5 mol/L urea, 0.01 M PBS based solution), and buffer VII (0.5 mol/L urea, 0.01 M PBS based solution) were used to perform dialysis of the protein solution in turn; the liquid changing interval was 12 h.

Finally, 0.01 M PBS was used to perform dialysis of the protein solution.

High concentrations of urea (8 mol/L) can denature PCV2 Cap protein completely; we reduced the concentration of urea gradually so that PCV2 Cap protein will be refolded gradually.

Before storing at −80 °C, the protein was tested by SDS PAGE.

### 2.4. Preparation of Polyclonal Antibody

Two healthy female rabbits (6 to 8 weeks of age) were provided by Zhejiang University for the purposes of this study. Freund's complete adjuvant (FCA) was used to emulsify 100 μL Cap solution (1 μg/μL) before injecting it into the rabbit on day 0. For secondary and booster injections, Freund's incomplete adjuvant (FIA) was used instead of FCA to prevent inflammation and lesions due to the formation of granulomas on day 14 and day 28. Vaccination of the last immunization was performed without adjutant on day 42. At the same time, a PBS negative control group was set up. An enzyme-linked immunosorbent assay kit (ELISA kit) was used to analyze the antibody level.

According to aseptic purification strategies, specific polyclonal antibodies can be isolated from serum using a Protein A kit. The polyclonal antibody also be tested using sodium dodecyl sulfate-polyacrylamide gel electrophoresis and the BCA protein determination method.

The ethics committee of Zhejiang Sci-Tech University's College of Life Sciences has reviewed and approved the study on mice, and this study was conducted according to its guidelines (Code: 20180401, Date: 2 April 2018).

### 2.5. Conjugation of Polyclonal Antibody with Polystyrene Microsphere

A measure of 0.9 mL of 0.02 M sodium borate buffer was incubated with 0.1 mL of polystyrene microsphere solution with rapid stirring (220 rpm) for 5 min; then, 0.2 mL 1-(3-Dimethylaminopropyl)-3-ethylcarbodiimide hydrochloride solution (EDC, 1 mg/mL) and 0.1 mL of polyclonal antibody was added to the stirred solution. After the mixture was stirred for 15 min at 25 °C, 1 mL of BSA solution (2 mg/mL) was added to the stirred solution and mixture was continuously stirred for 60 min. Next, the stirred solution was harvested by centrifugation (12,000 rpm for 15 min) and the precipitation is the polystyrene microsphere-labeled polyclonal antibody. As a final step, the precipitation was added to 1 mL of 0.02 M sodium borate buffer (containing 0.1% Proclin300, 1% BSA); then, the solution was stored at 4 °C. This solution can be used as the immunoprobe.

Before the assembly of the test strips, 500 μL of immunoprobe is dropped onto the front of the sample pad. The treated sample pad was placed in an oven to dry.

### 2.6. Immobilization of Test Line and Control Line

Monoclonal antibody solution (2 mg/mL) and goat anti-Rabbit IgG (2 mg/mL) were injected into pressure pumps in platform-sensing dispenser. Two lines (testing line and control line) were draw on the nitrocellulose (NC) membrane, and there was approximately 0.5 cm between the 2 lines; NC membranes must be stored in the oven for 6 h [32].

### 2.7. Assembly of Immunochromatographic Strip

Each strip contained a sample pad, NC membrane, and absorbent pad (Figure 2). An adhesive plastic backing had the NC membrane pasted in the center. Overlapping 1 mm on the top and bottom of the NC membrane, the absorbent pad and sample pad were then pasted together. Finally, the boards were cut into 3.0 mm wide strips [51].

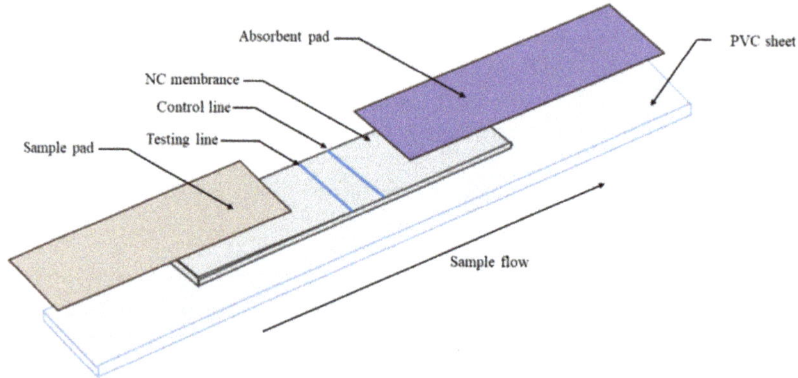


**Figure 2.** Structure schematic of test trip.

### *2.8. Evaluation of Immunochromatographic Strip*

#### 2.8.1. Preparation of Sample

An immunochromatographic strip can be used to detect PCV2 strain in cell culture media (PK-15 cell) or swine serum. Tissue culture infectious dose ($TCID_{50}$) assay was used to detect the virus titer of different genotype PCV2 strains. Add 100 μL cell culture media to the sample pad and wait 5–10 min to observe the result. After centrifugation, swine serum was obtained from whole blood. Add 100 μL swine serum to the sample pad and wait 5–10 min to observe the result.

#### 2.8.2. Sensitivity Test of Immunochromatographic Strip

We collected 3 strains. They are PCV2a/LG ($10^5$ $TCID_{50}$/0.1 mL), PCV2b/SH ($10^5$ $TCID_{50}$/0.1 mL), and PCV2d/JH ($10^5$ $TCID_{50}$/0.1 mL). Three PCV2 strains were used to evaluate the sensitivity of the immunochromatographic strip. Three viruses were serially diluted tenfold from $10^5$ $TCID_{50}$/0.1 mL to $10^{2.29}$ $TCID_{50}$/0.1 mL. A PK-15 cell culture was used for blank control. Immunochromatographic strips were used to test these virus diluent solutions; each diluent solution should be tested 3 times.

#### 2.8.3. Evaluation of the Detection Range of Immunochromatographic Strip

We collected 5 PCV2 strains. They are PCV2a/LG ($10^5$ $TCID_{50}$/0.1 mL), PCV2a/SZ ($10^5$ $TCID_{50}$/0.1 mL), PCV2b/SH ($10^5$ $TCID_{50}$/0.1 mL), PCV2d/JH ($10^5$ $TCID_{50}$/0.1 mL), and PCV2d/WZ ($10^5$ $TCID_{50}$/0.1 mL). These different PCV2 genotype strains were used to evaluate the detection range of the immunochromatographic strip.

#### 2.8.4. Specificity Test of Immunochromatographic Strip

We prepared 20 PCV2-positive serum samples, 10 PCV2-negative serum samples, 5 PEDV-positive serum samples, 5 PRV-positive serum samples, 5 CSFV-positive serum samples, 5 PRRS-positive serum samples, 5 PPV-positive serum samples, ASFV PCV1-positive serum samples, and 1 PCV1 strain. We used immunochromatographic strips to test serum samples and strain; each serum sample or strain was tested 3 times.

We prepared cell culture mediums of PEDV (CV777 strain), PRV (HB2000 strain), CSFV (WH-09 strain), PRRS (JXA1-R strain), PPV (WH-1 strain), and ASFV (SD strain). A measure of 100 μL cell culture medium was added onto the sample pad; each sample should be tested 3 times.

### 2.9. Immunochromatographic Strip Compared with ELISA

In total, 150 serum samples were collected from three swine herds in Zhejiang province and the immunochromatographic strip and Porcinecircovirus 2 Cap ELISA kit (antigen testing, Beijing Median Biotek Co., Ltd., (Beijing, 100000, China) were used to test the 150 serum samples.

The immunochromatographic strip is qualitative detection method, Porcinecircovirus 2 Cap ELISA kit is quantitative detection method. Therefore, we used a Kappa test to analyze the differences between the two sets of data.

The Kappa consistency test is a statistical method used to evaluate the consistency level of two or more evaluators on categorical data. It not only tells us whether there is consistency between evaluators, but also quantifies the degree of consistency through a Kappa value.

Kappa values between 0.61 and 0.80 were considered to indicate 'high' consistency, while those between 0.81 and 1.00 indicated 'extremely strong' consistency.

### 2.10. Development of Magnetic Chemiluminescence Immunoassay

#### 2.10.1. Conjugation of Polyclonal Antibody with Magnetic Bead

A measure of 0.9 mL of 0.02 M sodium borate buffer was incubated with 0.1 mL of magnetic bead solution with rapid stirring (220 rpm) for 5 min; then, 0.2 mL 1-(3-Dimethylaminopropyl)-3-ethylcarbodiimide hydrochloride solution (EDC, 1 mg/mL) and 0.1 mL of anti-PCV2-Cap polyclonal antibody (1 mg/mL) was added to the stirred solution. After the mixture was stirred for 15 min at 25 °C, 1 mL of BSA solution (2 mg/mL) was added to the stirred solution and the mixture was continuously stirred for 60 min. Next, the stirred solution was harvested by centrifugation (12,000 rpm for 15 min) and the precipitation was the magnetic beads-labeled polyclonal antibody. As a final step, the precipitation was added to 1 mL of 0.02 M sodium borate buffer; then, the solution was stored at 4 °C.

#### 2.10.2. Conjugation of Monoclonal Antibody with Acridine Ester

Measures of 0.1 mL of monoclonal antibody (1 mg/mL) and 7.87 μL acridine ester solution (0.5 mg/mL) were added to 0.8 mL of 0.02 M sodium borate buffer; the mixture was stirred (220 rpm) for 3 h. Then, conjugation was ended by the addition of 10 μL lysine solution (10% lysine). Finally, the stirred solution was purified using a Matrix Sephadex™ G-25 Fine.

### 2.11. Evaluation of Magnetic Chemiluminescence Immunoassay

#### 2.11.1. Preparation of Sample Solution

In the testing process of magnetic chemiluminescence immunoassay, place 5 μL of serum sample and 95 μL of sample diluent (0.01 M PBS) into a reaction tube and mix them together. Place 100 μL of probe solution and 100 μL of acridine ester-labeled antibody into a reaction tube and mix them together. The reaction tube should be stored at 37 °C for 25 min. Place the reaction tube on a magnetic rack for magnetic separation; the reaction solution should be discarded and 1000 μL of washing buffer should be added into tube for cleaning. Put the reaction tube into the automatic chemistry system; the system will inject 100 μL buffer A and 100 μL buffer B into the tube. Finally, the system will detect the relative light unit (RLU).

2.11.2. Evaluation of the Linear Range of the Magnetic Chemiluminescence Immunoassay

In this study, we prepared PCV2 Cap solutions of different concentrations (0.25 ng/mL, 0.5 ng/mL, 1 ng/mL, 2 ng/mL, 4 ng/mL, 8 ng/mL, 16 ng/mL, 32 ng/mL). Each sample corresponds to one relative light unit (RLU). Then, the standard curve for detecting PCV2 antigen was developed; meanwhile, the coefficients of determination ($R^2$) of standard curve are also recorded to show the linear relationship ($R^2 > 0.9$).

2.11.3. LOD and LQD of Magnetic Chemiluminescence Immunoassay

According to the Clinical and Laboratory Standards Institute (NCCLS) guidelines, one negative sample and one low-concentration sample were selected to manufacture a standard curve. The negative sample was tested 20 times and, calculating average (Mean) and standard deviation (SD), Mean + 2SD and Mean + 10SD were substituted into the standard curve; the result of Mean + 2SD was LOD and result of Mean + 10SD was LOQ.

2.11.4. Specificity Test of Magnetic Chemiluminescence Immunoassay

We prepared 5 PCV2-positive serum samples, 5 PCV2-negative serum samples, 5 PRV-positive serum samples, 5 PRRSV-positive serum samples, 5 PEDV-positive serum samples, and 5 CSFV-positive serum samples; each serum sample was tested 3 times.

*2.12. Magnetic Chemiluminescence Immunoassay Compared with ELISA*

Magnetic chemiluminescence immunoassay and the Porcinecircovirus 2 Cap ELISA kit (antigen testing, Beijing Median Biotek Co., Ltd.) were used to test 70 serum samples. A Bland–Altman diagram was used to analyze the results between the magnetic chemiluminescence immunoassay and Porcinecircovirus 2 Cap ELISA kit. The most important reason for choosing a Bland–Altman diagram is that the magnetic chemiluminescence immunoassay and Porcinecircovirus 2 Cap ELISA kit both are quantitative analysis methods. We hope to analyze the difference between them for the same serum.

The Bland–Altman diagram is a visual display method for consistency measurement. After the relevant calculation of the measured data, the scatter points are displayed. If the scatter points are within the confidence interval (generally within the range of 1.96 standard deviations of the difference), then the data have a good consistency level.

## 3. Results

*3.1. Expression and Purification of the Cap Protein in E. coli and Western Blot Identification*

The recombinant Cap protein was expressed and purified from *E. coli* (BL21). An inserted 6 × histidine tag is attached to this approximately 27.2 kDa protein. After sonication treatment, SDS-PAGE detection showed that the expressed target protein was presented mainly in the form of inclusion body. BCA protein assay was used to test the concentration of the Cap protein purified by affinity chromatography after renaturation. Purification resulted in protein with over 90% purity according to SDS-PAGE (Figure 3).

*3.2. Purification of Polyclonal Antibody*

After four immunizations, polyclonal antibodies were produced and existed in the serum of the rabbits. Blood was collected from the heart and serum was obtained by centrifuge (3000 rpm). Polyclonal antibodies were purified using saturated ammonium sulfate. Purification resulted in polyclonal antibody according to SDS-PAGE (Figure 4). Polyclonal antibodies were composed of two chains; light chain is 25 kDa and heavy chain is 55 kDa.

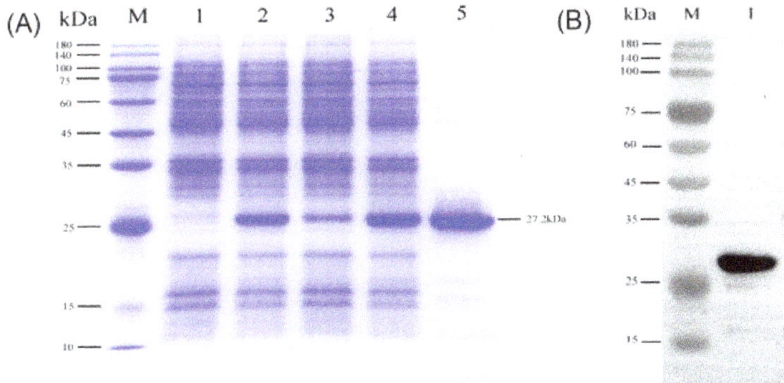


**Figure 3.** Purification of the Cap protein and western blot. (**A**) Identification of purified recombinant proteins by SDS-PAGE: M—Protein molecular weight standard; 1—Not induced; 2—After induction; 3—Upper clear; 4—Precipitation; 5—Purified recombinant Cap protein. (**B**) Western blot analysis of purified recombinant Cap protein: M—Protein molecular weight standard; 1—Purified recombinant Cap protein.

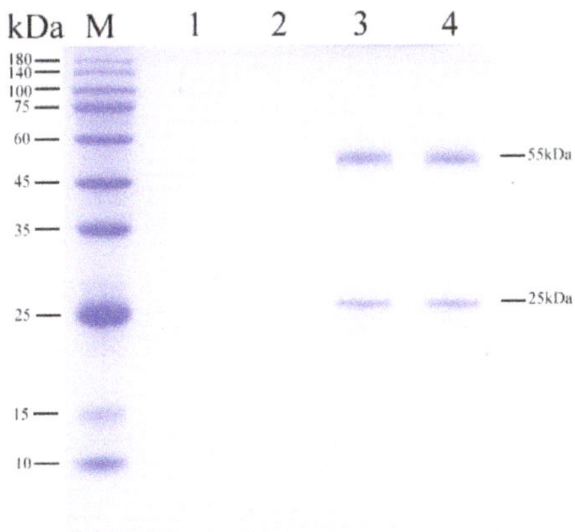


**Figure 4.** Purification of the polyclonal antibodies. M—maker; 1, 2—unimmunized serum; 3, 4—immunized serum.

### 3.3. *Sensitivity of Immunochromatographic Strip*

In this study, PCV2a/LG ($10^5$ $TCID_{50}$/0.1 mL), PCV2b/SH ($10^5$ $TCID_{50}$/0.1 mL) and mL, $10^{3.19}$ $TCID_{50}$/0.1 mL, $10^{3.49}$ $TCID_{50}$/0.1 mL (Figure 5A–C).

We can draw a conclusion that the immunochromatographic strip was more sensitive to PCV2a strains and PCV2b strains than PCV2d strains.

### 3.4. *Detection Range of Immunochromatographic Strip*

We collected five PCV2 strains. They are PCV2a/LG ($10^5$ $TCID_{50}$/0.1 mL), PCV2a/SZ ($10^5$ $TCID_{50}$/0.1 mL), PCV2b/SH ($10^5$ $TCID_{50}$/0.1 mL), PCV2d/JH ($10^5$ $TCID_{50}$/0.1 mL), and PCV2d/WZ ($10^5$ $TCID_{50}$/0.1 mL). The results of all experimental groups were positive. We can reach a conclusion that immunochromatographic strips can be used to test PCV2a, PCV2b, and PCV2d (Figure 6).

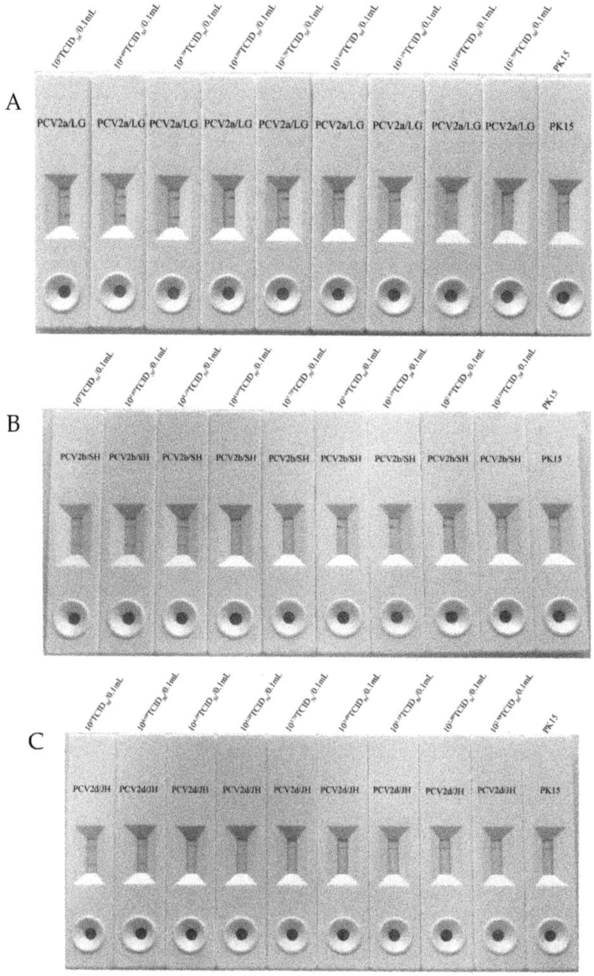


**Figure 5.** Sensitivity and specificity of the test strip. (**A**) PCV2a/LG; (**B**) PCV2b/SH; (**C**) PCV2d/JH.

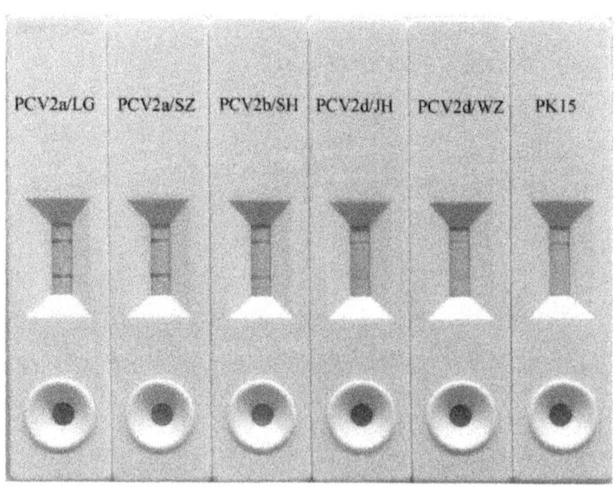


**Figure 6.** Detection range of immunochromatographic strips: PCV2a/LG ($10^5$ $TCID_{50}$/0.1 mL), PCV2a/SZ ($10^5$ $TCID_{50}$/0.1 mL), PCV2b/SH ($10^5$ $TCID_{50}$/0.1 mL), PCV2d/JH ($10^5$ $TCID_{50}$/0.1 mL), and PCV2d/WZ ($10^5$ $TCID_{50}$/0.1 mL). PK-15 cell culture is the blank control.

### 3.5. Specificity of Immunochromatographic Strip

PCV2-positive serum samples, PCV2-negative serum samples, PEDV-positive serum samples, PRV-positive serum samples, CSFV-positive serum samples, PRRS-positive serum samples, PPV-positive serum samples, and an ASFV-positive serum sample, and PCV1 strain solution were tested using the immunochromatographic strip; the results showed that positive results only appeared in PCV2-positive serum samples. Other experimental groups showed negative results (Figure 7A).

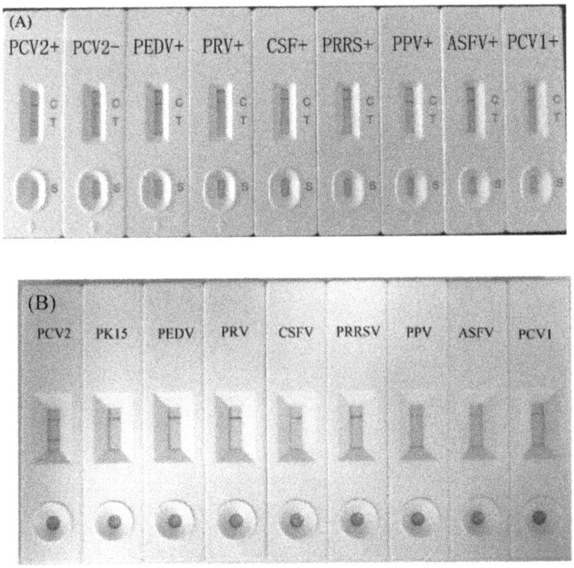


**Figure 7.** Sensitivity and specificity of the test strip. (**A**) The results of serum sample. (**B**) The results of cell culture medium.

Moreover, we collect cell culture medium of PEDV (CV777 strain), PRV (HB2000 strain), CSFV (WH-09 strain), PRRS (JXA1-R strain), PPV (WH-1 strain), ASFV (SD strain), and PCV1 (D strain). All results were negative (Figure 7B).

Based on the results, immunochromatographic strips can test serum samples and strain samples.

### 3.6. Comparison of ELISA Kit with Immunochromatographic Strip

We prepared 150 serum samples and used the developed immunochromatographic strip and ELISA kit to test serum samples (Table 2).

**Table 2.** Comparison of the developed immunochromatographic strip with the enzyme-linked immunosorbent assay (ELISA) kit.

| Strip | ELISA Kit | | | Kappa [1] | $p$ |
|---|---|---|---|---|---|
| | Positive | Negative | Total | | |
| Positive | 77 | 6 | 83 | 0.866 | <0.01 |
| Negative | 4 | 63 | 67 | | |
| Total | 81 | 69 | 150 | | |

[1] Kappa is divided into five levels. Kappa = 0.0–0.20, slight agreement; Kappa = 0.21–0.40, fair agreement; Kappa = 0.41–0.60, moderate agreement; Kappa = 0.61–0.80, substantial agreement; Kappa = 0.81–1.00, almost perfect agreement.

The Kappa statistic was used to prove that the immunochromatographic strip showed a high level of consistency with the enzyme-linked immunosorbent assay (ELISA) kit. The agreement between the immunochromatographic strip and ELISA kit was 93.33% (140/150). The Kappa coefficient between immunochromatographic strip and ELISA kit was 0.866 (Kappa > 0.810); the two methods have almost perfect agreement (Table 2).

### 3.7. Linear Range of the Magnetic Chemiluminescence Immunoassay

PCV2 Cap solutions of different concentrations (0.25 ng/mL, 0.5 ng/mL, 1 ng/mL, 2 ng/mL, 4 ng/mL, 8 ng/mL, 16 ng/mL, 32 ng/mL) were prepared; each sample corresponds to one relative light unit (RLU) and standard curves were established using SPSS. The $R^2$ of standard curves is 0.9993; there is a great correlation between PCV2 antigen and RLU when the concentration of PCV2 antigen is 0.25 ng/mL to 32 ng/mL (Figure 8).

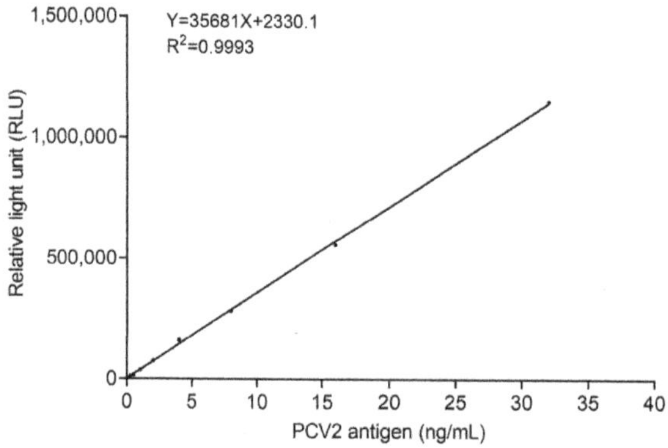


**Figure 8.** Standard curves of magnetic chemiluminescence immunoassay (0.25 ng/mL, 0.5 ng/mL, 1 ng/mL, 2 ng/mL, 4 ng/mL, 8 ng/mL, 16 ng/mL, 32 ng/mL).

### 3.8. The Limit of Detection (LOD) and the Limit of Quantitation (LOQ)

According to the Clinical and Laboratory Standards Institute (NCCLS) guidelines, one negative sample and one low-concentration sample were selected to manufacture a standard curve. The negative sample was tested 20 times and, calculating average (Mean) and standard deviation (SD), Mean + 2SD and Mean + 10SD were substituted into the standard curve; the result of Mean + 2SD was LOD and result of Mean + 10SD was LOQ (Table 3).

**Table 3.** The limit of detection (LOD) and the limit of quantitation (LOQ).

| Mean | SD | Mean + 2SD | Mean + 10SD | LOD | LOQ |
|---|---|---|---|---|---|
| 5283.92 | 355.58 | 5995.08 | 8839.70 | 0.051 ng/mL | 0.068 ng/mL |

### 3.9. Specificity of Magnetic Chemiluminescence Immunoassay

In this study, we collected five PCV2-positive serums, five PCV2-negative serums, five PRV-positive serums, five PRRSV-positive serums, five PEDV-positive serums, and five CSFV-positive serums. These serums were tested by magnetic chemiluminescence immunoassay. The result show that positive results only appeared when we tested PCV2-positive serums. Meanwhile, the results of other experimental groups were negative (Table 4).

**Table 4.** The results of swine serum.

| Serum Name | Result | Judgment | Serum Name | Result | Judgment |
|---|---|---|---|---|---|
| PCV2-positive serum | 727,198.53 | + | PRRSV-positive serum | 2351.51 | - |
| PCV2-positive serum | 682,285.52 | + | PRRSV-positive serum | 2556.29 | - |
| PCV2-positive serum | 926,893.28 | + | PRRSV-positive serum | 3165.25 | - |
| PCV2-positive serum | 652,288.21 | + | PRRSV-positive serum | 2885.76 | - |
| PCV2-positive serum | 715,582.62 | + | PRRSV-positive serum | 3515.19 | - |
| PCV2-negative serum | 3052.91 | - | PEDV-positive serum | 2927.51 | - |
| PCV2-negativeserum | 3251.82 | - | PEDV-positive serum | 3992.31 | - |
| PCV2-negative serum | 2752.51 | - | PEDV-positive serum | 3812.15 | - |
| PCV2-negative serum | 3192.35 | - | PEDV-positive serum | 3165.23 | - |
| PCV2-negative serum | 2951.59 | - | PEDV-positive serum | 3079.25 | - |
| PRV-positive serum | 3515.36 | - | CSFV-positive serum | 5018.27 | - |
| PRV-positive serum | 2656.17 | - | CSFV-positive serum | 3312.82 | - |
| PRV-positive serum | 3586.62 | - | CSFV-positive serum | 2980.65 | - |
| PRV-positive serum | 2588.66 | - | CSFV-positive serum | 3796.61 | - |
| PRV-positive serum | 3815.63 | - | CSFV-positive serum | 3188.56 | - |

We collected cell culture mediums of PCV2a/LG ($10^5$ $TCID_{50}$/0.1 mL), PCV2a/SZ ($10^5$ $TCID_{50}$/0.1 mL), PCV2b/SH ($10^5$ $TCID_{50}$/0.1 mL), PCV2d/JH ($10^5$ $TCID_{50}$/0.1 mL), and PCV2d/WZ ($10^5$ $TCID_{50}$/0.1 mL). The PK-15 cell culture was a blank control; the results are shown (Table 5).

**Table 5.** The results of strains.

| Serum Name | Result | Judgment | Serum Name | Result | Judgment |
|---|---|---|---|---|---|
| PCV2a/LG | 866,125.93 | + | PEDV (CV777 strain) | 3641.83 | - |
| PCV2a/SZ | 796,321.65 | + | PRV (HB2000 strain) | 2306.58 | - |
| PCV2b/SH | 621,432.98 | + | CSFV (WH-09 strain) | 3023.68 | - |
| PCV2d/JH | 468,391.27 | + | PRRS (JXA1-R strain) | 3790.53 | - |
| PCV2d/WZ | 428,625.21 | + | PPV (WH-1 strain) | 2142.19 | - |
| PK-15 cell culture medium | 2816.65 | - | | | |

We tested the cell culture medium of PEDV (CV777 strain), PRV (HB2000 strain), CSFV (WH-09 strain), PRRS (JXA1-R strain), and PPV (WH-1 strain); the results are shown (Table 5).

### *3.10. Comparison of ELISA Kit and Magnetic Chemiluminescence Immunoassay*

In total, 70 serum samples were tested by magnetic chemiluminescence immunoassay and ELISA kit. The Bland–Altman test was used to obtain the agreement between the magnetic chemiluminescence immunoassay and ELISA kit. For the same sample, there was a high level of consistency between the immunochromatographic strip and the ELISA kit when the difference in result was between +1.96SD and −1.96SD (Figure 9). Only two samples had a substantial difference. The agreement between the magnetic chemiluminescence immunoassay and ELISA kit was 97.14% (68/70). The results showed that two measures had high consistency.

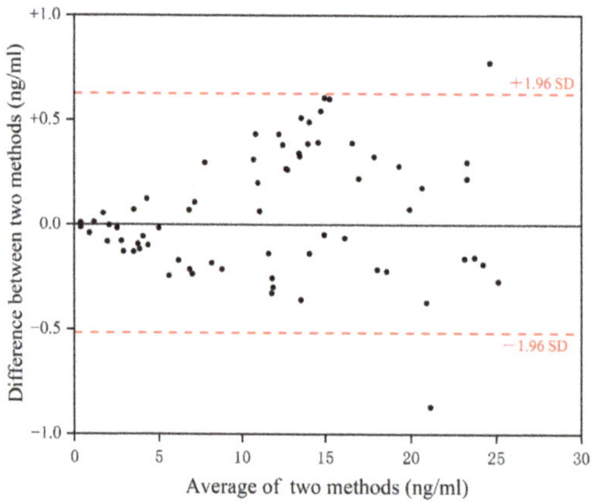


**Figure 9.** Agreement between immunochromatographic strip and ELISA kit.

## 4. Discussion

*Porcine circovirus 2* (PCV2) is the main and primary causative agent of Postweaning Multisystemic Wasting Syndrome (PMWS), which has caused significant economic losses in the swine industry. Several solution techniques have been applied to test for *Porcine circovirus 2*. Immunoperoxidase monolayer assay (IPMA), indirect immunofluorescent assay (IFA), and enzyme-linked immunosorbent assay (ELISA) are the most common diagnostic methods for detecting PCV2 antibodies. However, these methods require specialized equipment and technical expertise and are suitable for laboratory use only.

A technology was needed that has several advantages over traditional immunoassays, such as simplicity of procedure, rapid operation, immediate results, low cost, and no requirements for technical expertise or specialized equipment. In this study, immunochromatographic (detection time 5 min) and magnetic chemiluminescence immunoassay (detection time 20 min) possess high sensitivity and specificity and may be useful for clinical laboratories and rapid diagnosis in the field. In this study, the recombinant Cap protein was expressed and SDS PAGE showed a new protein with a molecular weight of 27.8 KD. Then, the rabbits were immunized with the purified Cap protein to produce multiclonal antibodies.

A rapid (less than 10 min) immunochromatographic strip, using polystyrene microspheres, was successfully developed and applied in the detection of the porcine circovirus 2 (PCV2) antigen. The polyclonal antibody and goat anti-Rabbit IgG secondary antibody were blotted on the nitrocellulose membrane for the test line and control line. The specificity of the strip was 100% (65/65). Regarding sensitivity, the immunochromatographic strip could test for a concentration of Cap antigen as low as 0.25 μg/mL. In total, 150 clinical swine serum samples were detected both by strip and the commercial ELISA kit. The agreement of immunochromatographic strip and ELISA kit was 93.33%. It showed that this strip possesses high sensitivity and specificity and may be useful for clinical laboratories and rapid diagnosis in the field.

We created a magnetic chemiluminescence immunoassay; there are two important features. First, the polyclonal antibody is conjugated with a magnetic bead. Second, the monoclonal antibody is conjugated with an acridine ester.

The magnetic chemiluminescence immunoassay had a high correlation and consistency. This method had a good linear relationship, between 0.25 ng/mL and 25 ng/mL; the $R^2$ is 0.9993. The limit of detection is 0.051 ng/mL and the limit of quantitation is 0.058 ng/mL. In the contrast experiment of 70 serum specimens, results showed that the agreement between the magnetic chemiluminescence immunoassay and ELISA kit was 97.14% (68/70). The magnetic chemiluminescence immunoassay and ELISA kit had high correlation and consistency.

In summary, we established immunochromatographic and magnetic chemiluminescence immunoassay to detect the antigen against PCV2. The assays showed good performance in terms of sensitivity and specificity and have certain application value, which could potentially be used for monitoring the immune responses in pigs to PCV2 infections and vaccination as well.

## 5. Conclusions

PCR, ELISA, and gene chip detection are the main methods for the detection of PCV2, but each has its own limitations, such as the probability of false positives in PCR detection [52], the complexity and low-sensitivity of ELISA process [5], and the high cost and difficulty of gene-chip detection [53]. This study has successfully developed a solution to this problem, developed and applied an immunochromatographic strip and magnetic chemiluminescence immunoassay to test for PCV2 antigen. The immunochromatographic strip can achieve accurate results within 10 min; magnetic chemiluminescence immunoassay can achieve accurate results within 25 min. The two methods were used to improve the speed and simplicity of PCV2 diagnosis without compromising sensitivity and specificity. The immunochromatographic strip and magnetic chemiluminescence immunoassay provide new ideas for the detection of PCV2.

**Author Contributions:** S.T., Y.Z., and J.S. conceived the experiments; S.T., X.T., and Y.Z. designed the experiments; S.T. performed the experiments and analyzed the data; S.T., and Y.D. wrote the paper; Y.H., H.F., and J.S. interpreted data and critically revised the manuscript. All authors have read and agreed to the published version of the manuscript.

**Funding:** This research was supported by "Pioneer" and "Leading Goose" R&D Program of Zhejiang (No. 2023C02047, 2023C02023); Science and Technology Program of the Xinjiang Production and Construction Corps (No. 2023AB009-03).

**Institutional Review Board Statement:** Not applicable.

**Informed Consent Statement:** Not applicable.

**Data Availability Statement:** All data are contained within this manuscript.

**Conflicts of Interest:** Yulong He is an employee of Zhejiang Sci-Tech University. The rest of the authors do not have conflicts of interests to declare.

## References

1. Patterson, A.R.; Opriessnig, T. Epidemiology and Horizontal Transmission of Porcine Circovirus Type 2 (Pcv2). *Anim. Health Res. Rev.* **2010**, *11*, 217–234. [CrossRef] [PubMed]
2. Opriessnig, T.; Karuppannan, A.K.; Castro, A.M.M.G.; Xiao, C.-T. Porcine Circoviruses: Current Status, Knowledge Gaps and Challenges. *Virus Res.* **2020**, *286*, 198044. [CrossRef]
3. Grau-Roma, L.; Fraile, L.; Segalés, J. Recent Advances in the Epidemiology, Diagnosis and Control of Diseases Caused by Porcine Circovirus Type 2. *Vet. J.* **2011**, *187*, 23–32. [CrossRef]
4. Lv, Q.; Wang, T.; Deng, J.; Chen, Y.; Yan, Q.; Wang, D.; Zhu, Y. Genomic Analysis of Porcine Circovirus Type 2 from Southern China. *Vet. Med. Sci.* **2020**, *6*, 875–889. [CrossRef]
5. Karuppannan, A.K.; Opriessnig, T. Porcine Circovirus Type 2 (Pcv2) Vaccines in the Context of Current Molecular Epidemiology. *Viruses* **2017**, *9*, 99. [CrossRef]
6. Meng, X.-J. Porcine Circovirus Type 2 (Pcv2): Pathogenesis and Interaction with the Immune System. *Annu. Rev. Anim. Biosci.* **2013**, *1*, 43–64. [CrossRef]
7. Segalés, J. Porcine Circovirus Type 2 (Pcv2) Infections: Clinical Signs, Pathology and Laboratory Diagnosis. *Virus Res.* **2012**, *164*, 10–19. [CrossRef]

8. Zhang, Z.; Luo, Y.; Zhang, Y.; Guo, K. Enhanced Protective Immune Response to Pcv2 Adenovirus Vaccine by Fusion Expression of Cap Protein with Invc in Pigs. *J. Vet. Sci.* **2019**, *20*, e35. [CrossRef]
9. Ouyang, T.; Zhang, X.; Liu, X.; Ren, L. Co-Infection of Swine with Porcine Circovirus Type 2 and Other Swine Viruses. *Viruses* **2019**, *11*, 185. [CrossRef]
10. Guo, Z.; Ruan, H.; Qiao, S.; Deng, R.; Zhang, G. Co-Infection Status of Porcine Circoviruses (Pcv2 and Pcv3) and Porcine Epidemic Diarrhea Virus (Pedv) in Pigs with Watery Diarrhea in Henan Province, Central China. *Microb. Pathog.* **2020**, *142*, 104047. [CrossRef]
11. Hao, J.; Wang, F.; Xing, G.; Liu, Y.; Deng, R.; Zhang, H.; Cheng, A.; Zhang, G. Design and Preliminary Application of Affinity Peptide Based on the Structure of the Porcine Circovirus Type Ii Capsid (Pcv2 Cap). *PeerJ* **2019**, *7*, e8132. [CrossRef]
12. Chen, Q.; Rong, J.; Li, G.; Xu, B.; Wang, X.; Hu, J.; Rong, M.; Li, H. Establishment of a Rep' Protein Antibody Detection Method to Distinguish Natural Infection with Pcv2 from Subunit Vaccine Immunization. *J. Med. Microbiol.* **2020**, *69*, 1183–1196. [CrossRef]
13. Li, D.; Du, Q.; Wu, B.; Li, J.; Chang, L.; Zhao, X.; Huang, Y.; Tong, D. Immunogenicity of Adenovirus Vaccines Expressing the Pcv2 Capsid Protein in Pigs. *Vaccine* **2017**, *35*, 4722–4729. [CrossRef]
14. Han, J.; Ma, H.; Cao, L.; Jing, J.; Xiao, P.; Sun, W.; Xie, C.; Wen, S.; Li, Y.; Tian, M.; et al. Immunogenicity of Recombinant Vaccinia Virus Vaccines Co-Expressing Gp3/Gp5 of European Prrsv and Cap Protein of Pcv2 in Pigs. *Appl. Microbiol. Biotechnol.* **2018**, *102*, 1145–1154. [CrossRef]
15. Cortey, M.; Pileri, E.; Sibila, M.; Pujols, J.; Balasch, M.; Plana, J.; Segalés, J. Genotypic Shift of Porcine Circovirus Type 2 from Pcv-2a to Pcv-2b in Spain from 1985 to 2008. *Vet. J.* **2011**, *187*, 363–368. [CrossRef]
16. Ku, X.; Chen, F.; Li, P.; Wang, Y.; Yu, X.; Fan, S.; Qian, P.; Wu, M.; He, Q. Identification and Genetic Characterization of Porcine Circovirus Type 3 in China. *Transbound. Emerg. Dis.* **2017**, *64*, 703–708. [CrossRef]
17. Kwon, T.; Yoo, S.J.; Park, C.-K.; Lyoo, Y.S. Prevalence of Novel Porcine Circovirus 3 in Korean Pig Populations. *Vet. Microbiol.* **2017**, *207*, 178–180. [CrossRef]
18. Molini, U.; Marruchella, G.; Matheus, F.; Hemberger, Y.M.; Chiwome, B.; Khaiseb, S.; Cattoli, G.; Franzo, G. Molecular Investigation of Porcine Circovirus Type 3 Infection in Pigs in Namibia. *Pathogens* **2021**, *10*, 585. [CrossRef]
19. Chen, N.; Xiao, Y.; Li, X.; Li, S.; Xie, N.; Yan, X.; Li, X.; Zhu, J. Development and Application of a Quadruplex Real-Time Pcr Assay for Differential Detection of Porcine Circoviruses (Pcv1 to Pcv4) in Jiangsu Province of China from 2016 to 2020. *Transbound. Emerg. Dis.* **2021**, *68*, 1615–1624. [CrossRef]
20. Ha, Z.; Yu, C.; Xie, C.; Wang, G.; Zhang, Y.; Hao, P.; Li, J.; Li, Z.; Li, Y.; Rong, F.; et al. Retrospective Surveillance of Porcine Circovirus 4 in Pigs in Inner Mongolia, China, from 2016 to 2018. *Arch. Virol.* **2021**, *166*, 1951–1959. [CrossRef]
21. Sun, W.; Du, Q.; Han, Z.; Bi, J.; Lan, T.; Wang, W.; Zheng, M. Detection and Genetic Characterization of Porcine Circovirus 4 (Pcv4) in Guangxi, China. *Gene* **2021**, *773*, 145384. [CrossRef]
22. Nguyen, V.-G.; Do, H.-Q.; Huynh, T.-M.-L.; Park, Y.-H.; Park, B.-K.; Chung, H.-C. Molecular-Based Detection, Genetic Characterization and Phylogenetic Analysis of Porcine Circovirus 4 from Korean Domestic Swine Farms. *Transbound. Emerg. Dis.* **2022**, *69*, 538–548. [CrossRef]
23. Piñeyro, P.E.; Kenney, S.P.; Giménez-Lirola, L.G.; Heffron, C.L.; Matzinger, S.R.; Opriessnig, T.; Meng, X.-J. Expression of Antigenic Epitopes of Porcine Reproductive and Respiratory Syndrome Virus (Prrsv) in a Modified Live-Attenuated Porcine Circovirus Type 2 (Pcv2) Vaccine Virus (Pcv1-2a) as a Potential Bivalent Vaccine Against Both Pcv2 and Prrsv. *Virus Res.* **2015**, *210*, 154–164. [CrossRef]
24. Chen, F.; Yang, X.; Pang, D.; Peng, Z.; Dong, M.; Liu, X.; Ouyang, H.; Ren, L. Expression, Purification and Antibody Preparation Using Different Constructs of Pcv2 Capsid Protein. *Int. J. Biol. Macromol.* **2014**, *67*, 289–294. [CrossRef]
25. Lou, Z.-Z.; Li, Z.-Y.; Wang, G.; Li, J.-Q.; Lan, X.; Li, X.-R.; Yin, X.-P.; Liu, J.-X.; Liu, S.-D. Prokaryotic Expression and Potential Application of the Truncated Pcv-2 Capsid Protein. *Virol. Sin.* **2010**, *25*, 86–97. [CrossRef]
26. Shin, M.-K.; Yoon, S.H.; Kim, M.H.; Lyoo, Y.S.; Suh, S.W.; Yoo, H.S. Assessing Pcv2 Antibodies in Field Pigs Vaccinated with Different Porcine Circovirus 2 Vaccines Using Two Commercial Elisa Systems. *J. Vet. Sci.* **2015**, *16*, 25–29. [CrossRef]
27. Wang, X.; Jiang, W.; Jiang, P.; Li, Y.; Feng, Z.; Xu, J. Construction and Immunogenicity of Recombinant Adenovirus Expressing the Capsid Protein of Porcine Circovirus 2 (Pcv2) in Mice. *Vaccine* **2006**, *24*, 3374–3380. [CrossRef]
28. Fablet, C.; Rose, N.; Bernard, C.; Messager, I.; Piel, Y.; Grasland, B. Estimation of the Diagnostic Performance of Two Elisas to Detect Pcv2 Antibodies in Pig Sera Using a Bayesian Method. *J. Virol. Methods* **2017**, *249*, 121–125. [CrossRef]
29. Mu, Y.; Jia, C.; Zheng, X.; Zhu, H.; Zhang, X.; Xu, H.; Liu, B.; Zhao, Q.; Zhou, E.-M. A Nanobody-Horseradish Peroxidase Fusion Protein-Based Competitive Elisa for Rapid Detection of Antibodies against Porcine Circovirus Type 2. *J. Nanobiotechnol.* **2021**, *19*, 34.
30. Jin, Q.; Yang, J.; Lu, Q.; Guo, J.; Deng, R.; Wang, Y.; Wang, S.; Wang, S.; Chen, W.; Zhi, Y.; et al. Development of an Immunochromatographic Strip for the Detection of Antibodies against Porcine Circovirus-2. *J. Vet. Diagn. Investig.* **2012**, *24*, 1151–1157. [CrossRef]

31. Li, R.; Tian, X.; Yu, Y.; Qiao, S.; Wang, Z.; Ma, J. Development of a Rapid Immunochromatographic Assay for Detection of Antibodies against Porcine Epidemic Diarrhea Virus. *Pol. J. Vet. Sci.* **2018**, *21*, 139–147. [CrossRef]
32. Liu, J.; Gao, R.; Shi, H.; Cong, G.; Chen, J.; Zhang, X.; Shi, D.; Cao, L.; Wang, X.; Zhang, J.; et al. Development of a Rapid Immunochromatographic Strip Test for the Detection of Porcine Epidemic Diarrhea Virus Specific Siga in Colostrum. *J. Virol. Methods* **2020**, *279*, 113855. [CrossRef]
33. Li, H.; Yang, J.; Bao, D.; Hou, J.; Zhi, Y.; Yang, Y.; Ji, P.; Zhou, E.; Qiao, S.; Zhang, G. Development of an Immunochromatographic Strip for Detection of Antibodies against Porcine Reproductive and Respiratory Syndrome Virus. *J. Vet. Sci.* **2017**, *18*, 307–316. [CrossRef]
34. Guo, D.-L.; Pan, Q.-W.; Li, K.-P.; Li, J.-Q.; Shen, H.-W.; Wang, X.-L.; Zhang, X.-Y.; Li, X.-S.; Fu, F.; Feng, L.; et al. Development and Clinical Evaluation of a New Gold-Immunochromatographic Assay for the Detection of Antibodies against Field Strains of Pseudorabies Virus. *J. Virol. Methods* **2015**, *222*, 164–169. [CrossRef]
35. Fu, H.-J.; Yuan, L.-P.; Shen, Y.-D.; Liu, Y.-X.; Liu, B.; Zhang, S.-W.; Xie, Z.-X.; Lei, H.-T.; Sun, Y.-M.; Xu, Z.-L. A Full-Automated Magnetic Particle-Based Chemiluminescence Immunoassay for Rapid Detection of Cortisol in Milk. *Anal. Chim. Acta* **2018**, *1035*, 129–135. [CrossRef]
36. Yang, Y.; Lv, C.; Fan, J.; Zhao, Y.; Jiang, L.; Sun, X.; Zhang, Q.; Jin, M. Development of a Chemiluminescence Immunoassay to Accurately Detect African Swine Fever Virus Antibodies in Serum. *J. Virol. Methods* **2021**, *298*, 114269. [CrossRef]
37. Liu, B.; Chen, P.; Song, C.; Li, N.; Tian, Y.; Dong, Y.; Zhang, Y.; Li, Q. Establishment of a Chemiluminescent Immunoassay(Clia) and Its Application. *Chin. J. Cell. Mol. Immunol.* **2018**, *34*, 315–319.
38. Zhao, H.; Lin, Q.; Huang, L.; Zhai, Y.; Liu, Y.; Deng, Y.; Su, E.; He, N. Ultrasensitive Chemiluminescence Immunoassay with Enhanced Precision for the Detection of Ctni Amplified by Acridinium Ester-Loaded Microspheres and Internally Calibrated by Magnetic Fluorescent Nanoparticles. *Nanoscale* **2021**, *13*, 3275–3284. [CrossRef]
39. Nicol, T.; Lefeuvre, C.; Serri, O.; Pivert, A.; Joubaud, F.; Dubée, V.; Kouatchet, A.; Ducancelle, A.; Lunel-Fabiani, F.; Le Guillou-Guillemette, H. Assessment of Sars-Cov-2 Serological Tests for the Diagnosis of Covid-19 through the Evaluation of Three Immunoassays: Two Automated Immunoassays (Euroimmun and Abbott) and One Rapid Lateral Flow Immunoassay (Ng Biotech). *J. Clin. Virol.* **2020**, *129*, 104511. [CrossRef]
40. Dichtl, K.; Zimmermann, J.; Koeppel, M.B.; Böhm, S.; Osterman, A. Evaluation of a Novel Clia Monotest Assay for the Detection of Anti-Hepatitis E Virus-Igg and Igm: A Retrospective Comparison with a Line Blot and an Elisa. *Pathogens* **2021**, *10*, 689. [CrossRef]
41. Ji, H.; Chang, L.; Zhao, J.; Zhang, L.; Jiang, X.; Guo, F.; Wang, L. Evaluation of Elisa and Clia for Treponema Pallidum Specific Antibody Detection in China: A Multicenter Study. *J. Microbiol. Methods* **2019**, *166*, 105742. [CrossRef]
42. Chen, D.; Zhang, Y.; Xu, Y.; Shen, T.; Cheng, G.; Huang, B.; Ruan, X.; Wang, C. Comparison of Chemiluminescence Immunoassay, Enzyme-Linked Immunosorbent Assay and Passive Agglutination for Diagnosis of Mycoplasma Pneumoniae Infection. *Ther. Clin. Risk Manag.* **2018**, *14*, 1091–1097. [CrossRef]
43. Zhang, M.; Li, Y.; Jing, H.; Wang, N.; Wu, S.; Wang, Q.; Lin, X. Development of Polyclonal-Antibody-Coated Immunomagnetic Beads for Separation and Detection of Koi Herpesvirus in Large-Volume Samples. *Arch. Virol.* **2020**, *165*, 973–976. [CrossRef]
44. Mattarozzi, M.; Giannetto, M.; Careri, M. Electrochemical Immunomagnetic Assay as Biosensing Strategy for Determination of Ovarian Cancer Antigen He4 in Human Serum. *Talanta* **2020**, *217*, 120991. [CrossRef]
45. He, N.; Wang, F.; Ma, C.; Li, C.; Zeng, X.; Deng, Y.; Zhang, L.; Li, Z. Chemiluminescence Analysis for Hbv-DNA Hybridization Detection with Magnetic Nanoparticles Based DNA Extraction from Positive Whole Blood Samples. *J. Biomed. Nanotechnol.* **2013**, *9*, 267–273. [CrossRef]
46. Wu, L.; Yin, W.; Tang, K.; Shao, K.; Li, Q.; Wang, P.; Zuo, Y.; Lei, X.; Lu, Z.; Han, H. Highly Sensitive Enzyme-Free Immunosorbent Assay for Porcine Circovirus Type 2 Antibody Using Au-Pt/Sio2 Nanocomposites as Labels. *Biosens. Bioelectron.* **2016**, *82*, 177–184. [CrossRef]
47. Song, F.; Zhou, Y.; Li, Y.S.; Meng, X.M.; Meng, X.Y.; Liu, J.Q.; Lu, S.Y.; Ren, H.L.; Hu, P.; Liu, Z.S.; et al. A Rapid Immunomagnetic Beads-Based Immunoassay for the Detection of β-Casein in Bovine Milk. *Food Chem.* **2014**, *158*, 445–448. [CrossRef]
48. Luo, L.; Chen, J.; Li, X.; Qiao, D.; Wang, Z.; Wu, X.; Du, Q.; Tong, D.; Huang, Y. Establishment of Method for Dual Simultaneous Detection of Pedv and Tgev by Combination of Magnetic Micro-Particles and Nanoparticles. *J. Infect. Chemother.* **2020**, *26*, 523–526. [CrossRef]
49. Yang, H.; Li, Z.; Jiang, Q.; Fan, J.; Zhou, B.; Guo, Y.; Lan, G.; Yang, X.; He, N.; Jiang, H. Ultrasensitive Detection and Subtyping of Porcine Endogenous Retrovirus Provirus Based on Magnetic Nanoparticles and Chemiluminescence. *J. Nanosci. Nanotechnol.* **2015**, *15*, 5597–5604. [CrossRef]
50. Xing, N.; Guan, X.; An, B.; Cui, B.; Wang, Z.; Wang, X.; Zhang, X.; Du, Q.; Zhao, X.; Huang, Y.; et al. Ultrasensitive Detection of Porcine Epidemic Diarrhea Virus from Fecal Samples Using Functionalized Nanoparticles. *PLoS ONE* **2016**, *11*, e0167325. [CrossRef]

51. Vrublevskaya, V.V.; Afanasyev, V.N.; Grinevich, A.A.; Skarga, Y.Y.; Gladyshev, P.P.; Ibragimova, S.A.; Krylsky, D.V.; Morenkov, O.S. Development of a Competitive Double Antibody Lateral Flow Assay for the Detection of Antibodies Specific to Glycoprotein B of Aujeszky's Disease Virus in Swine Sera. *J. Virol. Methods* **2017**, *240*, 54–62. [CrossRef] [PubMed]
52. Zhou, S.; Han, S.; Shi, J.; Wu, J.; Yuan, X.; Cong, X.; Xu, S.; Wu, X.; Li, J.; Wang, J. Loop-Mediated Isothermal Amplification for Detection of Porcine Circovirus Type 2. *Virol. J.* **2011**, *8*, 497. [CrossRef] [PubMed]
53. Zhang, M.-J.; Ren, W.-Z.; Sun, X.-J.; Liu, Y.; Liu, K.-W.; Ji, Z.-H.; Gao, W.; Yuan, B. Genechip Analysis of Resistant Mycobacterium Tuberculosis with Previously Treated Tuberculosis in Changchun. *BMC Infect. Dis.* **2018**, *18*, 234. [CrossRef] [PubMed]

*veterinary sciences*

MDPI

*Article*

# TLR Agonist Immunoadjuvants Provide Effective Protection Against PCV2 and PRV Infections in a Bivalent Subunit Vaccine for PCV2 and PRV

**Fulai Yu [1,2], Wei Xiang [2], Weiye Ou [3], Yang Li [2], Xinbiao Shu [2,*] and Xiaoliang Li [1,4,5,*]**

1 Department of Veterinary Medicine, College of Animal Sciences, Zhejiang University, Hangzhou 310058, China; yufulai@zju.edu.cn
2 Zhejiang Dovro Animal Health Products Co., Ltd., 298 Binhong West Road, Jinhua 321017, China; xiangwei1981213@126.com (W.X.); liyang@mebolo.com (Y.L.)
3 Zhejiang MEBOLO Biotechnology Co., Ltd., Binhong West Road, Jinhua 321017, China; ouweiye@mebolo.com
4 Institute of Preventive Veterinary Medicine, Zhejiang Provincial Key Laboratory of Preventive Veterinary Medicine, Zhejiang University, Hangzhou 310058, China
5 Zhejiang University-Xinchang Joint Innovation Centre (TianMu Laboratory), Gaochuang Hi-Tech Park, Xinchang 312500, China
* Correspondence: shuxinbiao@dovro.com (X.S.); xlli@zju.edu.cn (X.L.)

**Simple Summary:** Porcine circovirus type 2 (PCV2) is the causative agent of porcine circovirus-associated diseases (PCVADs) and immune suppression, which have been widespread in commercial farms in the swine industry worldwide. Pseudorabies virus (PRV) is the causative agent for pseudorabies, affecting swine of all ages and causing severe clinical signs, including neurological problems, fever, and respiratory distress in suckling and nursery piglets. The co-infection of pigs with PRV and PCV2, followed by secondary bacterial infections leading to porcine respiratory disease complex (PRDC) with more serious clinical symptoms and lesions, causes enormous economic losses in the pig farming industry. We developed a bivalent subunit vaccine for PCV2 and PRV with TLR agonist immunoadjuvants (FLICd), which induces high levels of PRV and PCV2 an-tibodies, providing effective protection against PCV2- and PRV-related disease until the pig fattening stage. Additionally, the bivalent subunit vaccines simplify the immunization programs by decreasing the frequency of vaccinations and reduce the likelihood of PR outbreaks caused by low antibody levels before the pigs are vaccinated by the attenuated pseudorabies vaccine.

**Abstract:** Diseases associated with porcine circovirus type 2 (PCV2) and pseudorabies virus (PRV) significantly affect the economy of pig farms, particularly when combined infections lead to bacterial co-infections. Antigens from the pseudorabies variant strain gB and gD proteins and PCV2 (genotyped) Cap protein were mixed with the pattern recognition receptor (PRR) agonist FLICd as adjuvants and formulated with a micro-hydrogel adjuvant into PCV2 and PRV bivalent subunit vaccines. Twenty pigs, aged 30–35 days, were divided into groups A (received bivalent subunit vaccine) and B (received bivalent subunit vaccines with recombinant FLICd adjuvant), as well as C (non-vaccinated challenge control) and D (blank control). Groups A and B showed no significant difference in average daily weight gain compared to the unvaccinated controls. Fourteen days post-second vaccination, groups A and B exhibited significantly higher levels of PRV and PCV2 antibodies than groups C and D. Group B showed significantly higher average titers of PRV-specific neutralizing antibodies than group A. Fourteen days post-second vaccination, a PRV (ZJM-1 strain) challenge test was conducted. The vaccinated group achieved 100% protection. Vaccination effectively reduced virus load post-challenge and shortened the PRV shedding period.


https://doi.org/10.3390/vetsci12010025

Vaccination with PCV2 and PRV bivalent subunit vaccines effectively prevents the onset of PCV2-related diseases and infections by wild pseudorabies strains.

**Keywords:** PCV2; PRV; bivalent subunit vaccine; pattern recognition receptors

## 1. Introduction

Porcine circovirus type 2 (PCV2) infections lead to subclinical infections and porcine circovirus-associated disease (PCAD) in pigs, which affect growth performance and cause significant economic losses to the pig industry [1]. Increasing evidence suggests that subclinical PCV2 infection results in lymphocyte depletion and the inhibition of dendritic cell activation, leading to immunosuppression [2]. Co-infection with other viruses results in severe porcine diseases. Pseudorabies (also known as Aujeszky's disease), caused by the pseudorabies virus (PRV), causes piglet mortality, sow abortion, and respiratory diseases in fattening pigs. Since 2011, new variant strains have caused widespread PRV outbreaks in China. Co-infection with PRV and PCV2 synergistically inhibits the activation of natural immune response pathways, causing more severe immunosuppression [3], which facilitates secondary infections with pathogens, such as *Actinobacillus pleuropneumoniae* [4,5], challenging the control of respiratory infection-related diseases in intensive pig farming and resulting in more severe economic losses [6].

Currently, vaccination against PCV2 and PRV effectively reduces economic losses, owing to PCAD-related diseases and PRV outbreaks in pig herds [7]. The prevention of both diseases can be achieved by vaccination with high levels of antibodies. Therefore, vaccines developed to achieve high titers and long-lasting antibodies in pigs are effective tools for controlling diseases at different production stages. Over the past two decades, the application of immunoadjuvants has increased rapidly, driven by an enhanced understanding of the mechanisms underlying adjuvant immunoenhancement. Toll-like receptors (TLRs) in PRRs mainly include extracellular TLR1, TLR2, TLR4, TLR5, TLR6, and TLR10, as well as intracellular TLR3, TLR7, TLR8, and TLR9 [8]. TLR5 agonists, such as flagellin proteins, activate bone marrow-derived dendritic cells, enhance CD4+ helper T cell-dependent B cell humoral immune responses, and produce high levels of humoral immune responses [9]. The NIH protection experiment of co-immunization with recombinant expressed Salmonella typhi flagellin protein adjuvant and freeze-dried inactivated rabies vaccine in mice showed a significant increase in vaccine potency from 3.75 IU/mL in the blank control group to 48.69 IU/mL in the adjuvant group [10]. Therefore, in this study, antigens from the pseudorabies variant strain gB and gD proteins and the PCV2 (genotype d) Cap protein were mixed with the pattern recognition receptor (PRR) agonist FLICd as adjuvants. These were then formulated with a micro-hydrogel adjuvant into two groups of PCV2 and PRV bivalent subunit vaccines, which were administered to pigs to evaluate their effectiveness.

## 2. Materials and Methods

The study design was approved by the Laboratory Animal Welfare and Ethics Committee of Zhejiang University (ID number: ZJU20230399).

### 2.1. PRV (ZJM-1 Strain) Virus and ST Cells

ST cells were cultured in MEM supplemented with 10% newborn calf serum (provided by Inner Mongolia Opcell Biotechnology Co., Ltd., Hohhot, China) at 37 °C and 5% $CO_2$. These cells, preserved by Zhejiang MEBOLO Biotechnology Co., Ltd., Jiaxing, China, were used to passage the PRV strain used in the challenge experiments and detect neutralizing antibodies in pig serum. The challenge virus was a PRV strain (ZJM-1) isolated in 2013. A sequence clustering analysis of gI/gE and gD indicated high homogeneity with the HeN1 (GenBank: KP098534.1) and TJ (GenBank: KJ789182.1) strains, classifying it as a genotype 2 variant. After plaque purification, the strain was cultured for six generations for use in the pig challenge. After being cultured for ten generations, this strain was used to challenge five piglets aged 50–60 days via intramuscular injection and nasal drip (1 mL, $10^6$ $TCID_{50}$/mL), resulting in 3/5 fatalities and 5/5 exhibiting disease. When used to challenge five fattening pigs aged 90–100 days, the mortality and morbidity rates were 1/5 and 5/5, respectively.

### 2.2. Preparation of the PCV2 and PRV Bivalent Subunit Vaccine

Three batches of PCV2 and PRV bivalent subunit vaccines were prepared according to the detailed formulation, including the specified amounts of PCV2 and PRV antigens, adjuvants per pig dose, and the volume percentage of adjuvants used in Table 1. Pseudorabies gB and gD proteins expressed in CHO cells were purified by Ni-TED affinity chromatography (the reference sequences of the designed PRV antigens were the TJ strain variants AIT55752 and AIT55803). The recombinant PCV2b Cap protein expressed in Hi5 insect cells infected with baculovirus was purified using cation exchange chromatography. The classical carrier adjuvant, microhydrogel adjuvant 7749, was acquired from SDA Bio Co., Ltd., Sanford, NC, USA. The immunomodulating adjuvant flagellin protein (FLICd) nucleic acid, designed based on the D1 active domain of Salmonella typhimurium flic flagellin [11,12], was cloned, expressed in BL21 (DE3) Escherichia coli, and purified by nickel column chromatography after ultrasonic disruption. The vaccine was formulated by mixing the adjuvant (7749) with an aqueous phase (a solution containing PRV gB and gD proteins, PCV2d Cap protein, and immunomodulating adjuvant) in a 1:9 volume ratio. Initially, the aqueous phase was mixed and stirred at 350 rpm/min in a beaker, followed by the addition of the adjuvant (7749) to the mixture, which was stirred for an additional 30 min to produce two distinct vaccine formulations for packaging.

**Table 1.** The formulation of the PCV2 and PRV bivalent subunit vaccine.

| Groups | Adjuvant (*v*/*v*) | Immunomodulating Adjuvant | Antigen | |
|---|---|---|---|---|
| | | | PRV | PCV2 |
| A | 7749 1/9 | Non | 60 μg gB 60 μg gD | 70 μg |
| B | 7749 1/9 | 100 μg FLICd | 60 μg gB 60 μg gD | 70 μg |

### 2.3. Schedule of Vaccination and Challenge

Piglets (30 to 35 days old), confirmed to be negative for nucleic acids of porcine reproductive and respiratory syndrome virus (PRRSV), African swine fever virus (ASFV), pseudorabies virus (PRV), and porcine circovirus type 2 (PCV2), were randomly allocated to four groups. Group A received the first bivalent subunit vaccine, group B received the second bivalent subunit vaccine, which included the recombinant FLICd adjuvant, group C was the non-vaccinated challenge control, and group D was the blank control group. The pigs in groups A, B, and C were challenged on day 35 post-vaccination with the PRV (ZJM-1) strain ($10^6$ $TCID_{50}$/mL) with a 1 mL nasal drip and 1 mL intramuscular injection in the neck. Serum samples were obtained prior to the primary vaccination (specifically on day 0) and on days 21, 35, and 50 post-vaccination. Subsequently, whole blood, throat swabs, and anal swabs were collected on days 3, 7, 11, and 15 post-challenge (detailed in Table 2). The timeline presentation was shown as below to provide better clarity.

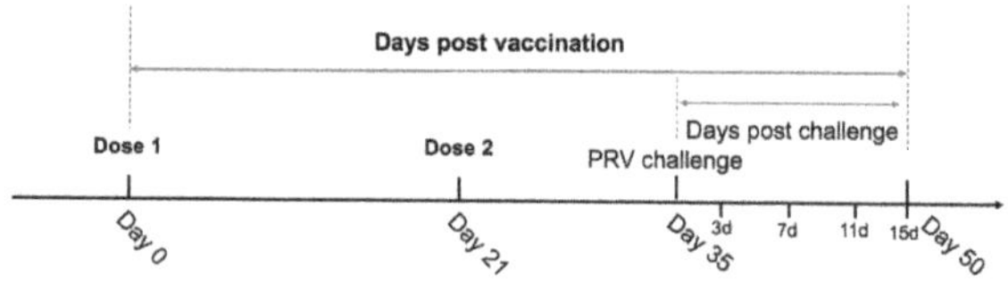


**Table 2.** Sampling schedule post-vaccination and post-challenge.

| Groups | Vaccine Groups | Vaccine Dosage (IM) | Serum Sample | Whole Blood, Throat Swabs, and Anal Swabs |
|---|---|---|---|---|
| | | | Days Post-Vaccination | Days Post-Challenge |
| A | 1 | 1 mL | 0d/21d/35d/50d | 3d/7d/11d/15d |
| B | 2 | 1 mL | 0d/21d/35d/50d | 3d/7d/11d/15d |
| C | none | None | 0d/21d/35d/50d | 3d/7d/11d/15d |
| D | none | None | 0d/21d/35d/50d | 3d/7d/11d/15d |

Note: "none" indicates that no action was performed.

### 2.4. Detection of Antibodies in Serum Samples

Serum samples were collected according to the study protocol and tested using commercial ELISA kits for PCV2-specific IgM and IgG antibodies (Ingenasa, Madrid, Spain) and PRV-specific gB and gI antibodies (kits were obtained from BioChek, Reeuwijk, the Netherlands). Testing was conducted according to the procedures outlined in the manufacturer's manual. In the PCV2-specific antibody ELISA test setup, the critical threshold for determining positivity was set at the average OD value multiplied by 0.4 and the average OD value of the IgG positive control multiplied by 0.3. Samples exhibiting IgM and IgG antibody OD values above these critical thresholds were classified as positive, suggesting either immunization or infection in pigs. For the PRV-specific gB antibody ELISA assessment, an S/P value exceeding 0.5 indicated a positive result, whereas an S/P value of $\leq$0.499 signified a negative outcome. In the PRV-specific gI antibody blocking ELISA evaluation, samples with S/N values > 0.55 were considered negative and those with S/N values of $\leq$0.55 were considered positive.

### 2.5. PRV Neutralization Assay

Pig serum samples were heat-inactivated at 56 °C for 30 min, then diluted in a 2-fold series and thoroughly mixed with the PRV (ZJM-1 strain, 6th generation) at a concentration of 100 $TCID_{50}$/well. The mixture was incubated at 37 °C for 1 h. The neutralized mixtures (100 uL/well) were then transferred into wells of a 96-well plate, where ST cells had been grown to a monolayer, and each dilution was repeated four times. The virus-positive control wells received an equal volume of PRV without serum, whereas the negative control wells received an equal volume of MEM containing 2% NBS. The plate was incubated at 37 °C in a 5% $CO_2$ atmosphere, and cytopathic effects (CPEs) were observed for 5 days. The half-protective dose (PD50) was calculated using the Reed–Muench method to determine neutralizing antibody titers. Additionally, the CPE of the virus-positive control well was calculated using the Reed–Muench method, and virus titers ranging between 100 and 200 $TCID_{50}$/0.1 mL were considered valid.

### 2.6. Clinical Observations Post-PRV Challenge

Groups A and B were vaccinated with the PCV2 and PRV bivalent subunit vaccine, and group C, the unvaccinated control, was challenged with PRV (ZJM-1 strain) (1 mL nasal drip and 1 mL intramuscular injection) on day 35 post-vaccination. Rectal temperatures were systematically recorded, and clinical signs in all pigs were observed during the post-challenge period. Weight was recorded before the challenge and necropsy to determine the average daily weight gain.

*2.7. Pathological Evaluation*

All pigs were euthanized for tissue collection from the brain, lung lobules, and inguinal lymph nodes on day 15 post-challenge. Samples were fixed in 4% paraformaldehyde for 24 h and processed using a series of dehydration, clearing, and paraffin embedding steps. Sections were cut to a thickness of 4 μm and stained with hematoxylin and eosin (HE), following the protocol provided by the reagent kits (Shenzhen Dakewe Biotech Co., Ltd., Shenzhen, China), and sections were examined under a microscope (100× magnification) after being sealed with neutral resin (Beijing Solarbio Science & Technology Co., Ltd., Beijing, China).

ImageView 1.1 software was employed to measure lung interstitial width at 100× magnification for quantitative analysis. Briefly, each group contains 5 pigs; after measuring 3 times for each pig, the mean value of the lung interstitial width for each pig and for each group was calculated.

*2.8. Quantification of PRV Load*

On days 3, 7, 11, and 15 post-challenge, blood, oral, and anal swab samples were collected from the pigs. DNA was extracted using the EasyPure® Blood Genomic DNA Kit (Beijing TransGen Biotech, Beijing, China) according to the manufacturer's instructions. The samples were then subjected to qPCR to amplify the PRV gE gene fragment, and the resulting fluorescence was quantified against a predefined standard curve. The primers used included PRV-F 5′-GCTGTACGTGCTCGTGAT-3′, PRV-R 5′-TCAGCTCCTTGATGACCGTGA-3′, and the probe 5′-HEX-CACAACGGCCACGTCGC CACCTG-BHQ1-3′. The reaction mixture was prepared in a total volume of 25 μL, incorporating various components such as 1 μL dNTPs(10 mM), 1 μL former and reverse primers (10 mM), respectively, 0.6 μL probe (10 mM), 1.5 μL Taq enzyme(5 U/μL), 2.5 μL 10 × buffer, 2 μL template DNA, and 15.4 μL water. Cycling conditions involved initial steps at 50 °C for 2 min and 95 °C for 2 min, followed by denaturation at 95 °C for 15 s, annealing/extension and fluorescence collection at 60 °C for 30 s over 40 cycles, and a final cooling step at 25 °C for 10 s. The Applied Biosystems StepOne real-time PCR system was used for amplification and quantification, with samples considered positive if the HEX channel Ct value was ≤40 and the amplification curve exhibited a typical S-shape; samples were deemed negative if the Ct value was >40. A standard curve was established using PRV (ZJM-1 strain) with a titer of $10^{8.0}$ $TCID_{50}$/mL, allowing for the conversion of Ct values into virus titers (half-tissue culture infective doses) based on this curve.

*2.9. Statistical Analysis*

The experimental data are presented as the mean ± standard deviation (SD) of replicates. Data were analyzed using JASP software (Team, 2020, JASP, Version 0.14.1) and Excel software using a one-way ANOVA (Levene's test for homogeneity of variances, Bonferroni method for multiple comparison tests). The level of significance was set at $p < 0.05$.

## 3. Results

*3.1. Safety Assessment of the PRV and PCV2 Bivalent Subunit Vaccine*

Rectal temperatures were recorded for the experimental pigs three days before the initial immunization and seven days following the first and second immunizations, with temperatures ranging between 38.5 and 39.5 °C. No significant difference was observed in rectal temperatures between the immunized groups A and B and the unimmunized control groups C and D. Similarly, no significant temperature variation was noted within the immunized groups before and after vaccination (Figure 1A). All pigs maintained a rectal temperature below 40.0 °C throughout the experiment, with no reduction in feed intake, noticeable swelling, or other adverse effects at the injection sites. On day 0 of primary vaccination and day 14 post-second booster, all pigs were weighed, and the average daily weight gain for groups A, B, C, and D was 0.43 ± 0.12, 0.39 ± 0.04, 0.40 ± 0.12, and 0.50 ± 0.04, respectively. No significant difference was observed in the average daily weight gain among the groups ($p > 0.05$) (Figure 1B), indicating the safety of PCV2 and PRV bivalent subunit vaccines.

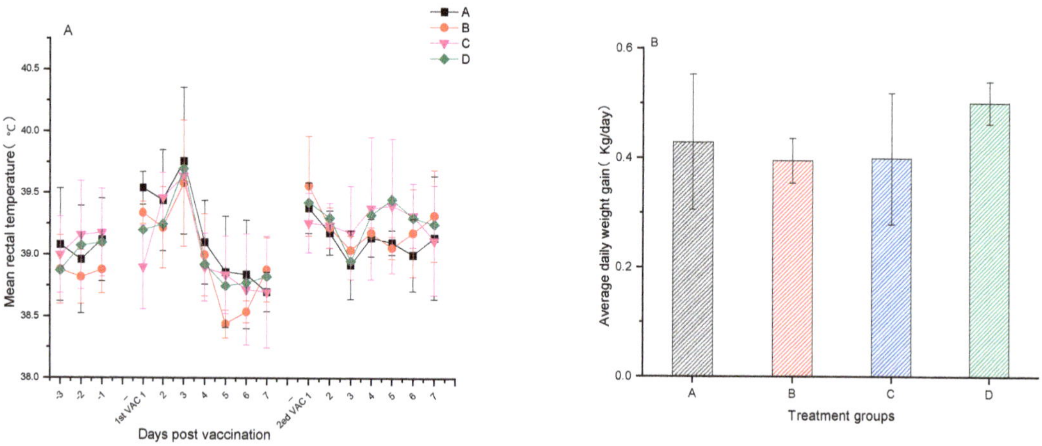


**Figure 1.** (**A**) Average rectal temperatures ± SD of experimental pigs in each group three days before vaccination and seven days after the primary and second booster vaccination. (**B**) Average daily weight gain ± SD of experimental pigs in each group from the primary vaccination to 14 days after the second booster vaccination.

### *3.2. Results of ELISA Antibody Tests Post-Immunization with the PCV2 and PRV Bivalent Subunit Vaccine in Experimental Pigs*

All 20 experimental pigs tested positive for maternal antibodies against PRV, with a 100% positivity rate for PRV gB-specific ELISA antibodies prior to immunization and the lowest S/P value of 1.5. Fourteen days following the booster vaccination, the average S/P values for PRV-specific gB protein ELISA antibodies in groups A and B were 1.07 ± 0.32 and 1.74 ± 0.83, respectively, showing no significant variance among the immunization groups ($p > 0.05$), all with a 5/5 positivity rate. The mean S/P values for gB protein ELISA antibodies in the unvaccinated control groups C and D were 0.42 ± 0.27 and 0.45 ± 0.78, with positivity rates of 1/5 and 2/5, respectively. Immunization with the bivalent subunit vaccine effectively generated PRV gB-specific ELISA antibodies in piglets carrying maternal antibodies (Figure 2A). Prior to the viral challenge, all groups were negative for PRV gI-specific ELISA antibodies, suggesting that the experimental pigs were not exposed to the wild-type PRV strain (Figure 2B).

Fifteen days post-challenge, the S/P values for PRV gB-specific ELISA antibodies in the vaccinated groups A and B and the unvaccinated control group C increased significantly to 2.57 ± 0.89, 3.65 ± 0.96, and 3.85 ± 1.07, respectively, with no significant difference in the average S/P values among the challenged groups ($p > 0.05$). The S/P value for the blank control group D remained at 0.20 ± 0.05, continuing to test negative in 5/5 cases. Seven days post-challenge, the S/N values for PRV gI-specific ELISA antibodies in all challenged groups were 0.28 ± 0.03, 0.31 ± 0.07, and 0.26 ± 0.08, respectively. The challenged control group C showed the lowest S/N value; however, no significant difference was observed compared to groups A and B ($p > 0.05$), with all challenged groups turning positive. For the blank control group D, the S/N values for PRV gI-specific ELISA antibody tests were all above 0.55, testing negative in 5/5 cases during the study period (Figure 2B), indicating no PRV infection.

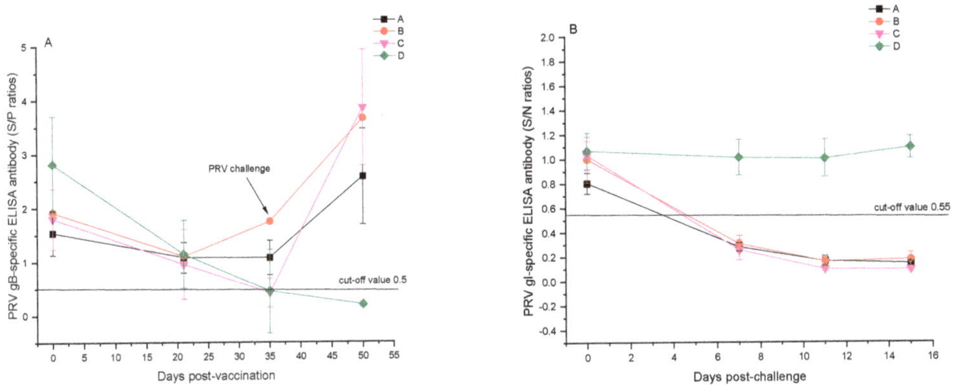


**Figure 2.** ELISA detection results for PRV-specific antibodies. (**A**) PRV gB-specific antibody ELISA detection results from pre-vaccination to 15 days post-challenge in experimental pigs from each group. (**B**) PRV gI-specific antibody ELISA detection results before and after challenge in experimental pigs from each group.

Before vaccination, the experimental pigs tested positive for PCV2-specific IgG antibodies and negative for IgM via the ELISA, indicating the presence of maternal antibodies and no infection with PCV2. Fourteen days following the booster vaccination, pigs in the PCV2 and PRV bivalent subunit vaccinated groups A and B exhibited PCV2-specific IgM antibody ELISA OD values of $1.3 \pm 0.5$ and $1.09 \pm 0.26$, respectively, all converting to positive (exceeding the standard threshold of 0.87). Group B had the highest mean OD value; however, no significant difference was observed in the mean OD values among the vaccinated groups ($p > 0.05$). Pigs in the non-vaccinated groups C and D had PCV2-specific IgM antibodies at $0.35 \pm 0.12$ and $0.30 \pm 0.06$, respectively, both remaining negative, indicating a robust immune response in groups A and B post-vaccination. Control groups C and D continued to test negative for IgM 29 days after the booster shot, demonstrating that the experimental pigs did not contract PCV2 during the study period (Figure 3A). Owing to elevated maternal antibody levels, pigs in group B had an average OD value of $1.64 \pm 0.54$ for PCV2-specific ELISA IgG antibodies 14 days post-booster, higher than those in groups A with $1.54 \pm 0.30$, with no significant differences between the groups ($p > 0.05$). The OD values for PCV2-specific IgG antibodies in control groups C and D were $0.79 \pm 0.35$ and $1.10 \pm 0.26$, respectively, with a notable distinction observed between the vaccinated groups and control group C ($p < 0.05$), whereas pigs in group D maintained high titers of maternal antibodies. Twenty-nine days following the second vaccination (50 days post-initial vaccination), the vaccinated groups A and B had average OD values for PCV2-specific IgG ELISA antibodies at $1.45 \pm 0.36$ and $1.57 \pm 0.43$, all testing positive in 5/5 cases, exceeding the unvaccinated control groups D and E with average OD values of $0.46 \pm 0.26$ and $0.87 \pm 0.28$, with positivity rates of 2/5 and 5/5, respectively ($p < 0.05$, Figure 3B). Vaccination with PCV2 and PRV bivalent subunit vaccines in piglets with high maternal antibodies resulted in the maintenance of elevated antibody levels in piglets aged 70–80 days.

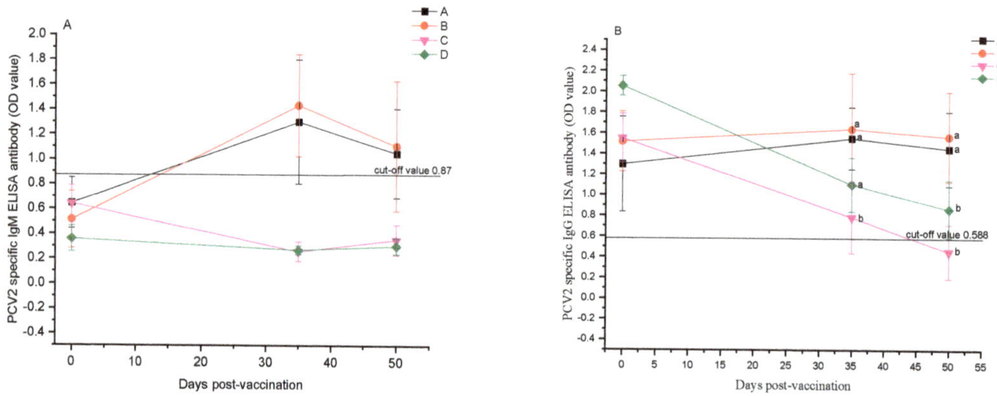


**Figure 3.** Outcomes of ELISA tests for PCV2-specific antibodies among various experimental pig groups. (**A**) Results of PCV2-specific IgM antibody ELISA tests conducted on different experimental pig groups prior to immunization and 14 and 29 days following the second vaccination. (**B**) Results of PCV2-specific IgG antibody ELISA tests conducted on different experimental pig groups prior to vaccination and 14 and 29 days following the second vaccination. The data were analyzed by one-way ANOVA, and alphabet letters (a and b) different between groups represent statistically significant difference.

### *3.3. Results of PRV-Neutralizing Antibody Tests*

Experimental pigs with maternal antibodies against PRV showed neutralizing antibody titers of 1:9.7 ± 3.9. Pigs in group B, vaccinated with the FLICd adjuvant-enhanced vaccine, exhibited PRV (ZJM-1 strain)-specific neutralizing antibody titers of 1:270.2 ± 102.8 14 days post-booster, significantly outperforming the adjuvant-free group A with titers of 1:108 ± 51.2 ($p < 0.05$), thus effectively boosting PRV-neutralizing antibody levels. In contrast, non-vaccinated control groups D and E tested negative for neutralizing antibodies (<1:4). PCV2 and PRV bivalent subunit vaccines successfully induced the production of PRV-specific neutralizing antibodies in vaccinated pigs. Fifteen days following the PRV exposure, the average titers of neutralizing antibodies in groups A and B increased to 257.2 ± 99.6 and 318.6 ± 42.3, respectively, whereas unvaccinated control group C was 1:15.4 ± 6.4 (Figure 4B). The observed trends in neutralizing antibody responses were similar to those observed in the PRV-specific gB antibody ELISA tests, although without a direct linear relationship between the actual figures. This indicates that neutralizing antibody testing in vaccine evaluations should not rely solely on the ELISA method.

### *3.4. PRV Challenge of Pigs*

Groups A and B comprised pigs vaccinated with the PCV2 and PRV bivalent subunit vaccines, and the unvaccinated group C was challenged with PRV (ZJM-1 strain). Rectal temperatures were meticulously recorded for 15 days post-challenge (Figure 5A). On day 3 post-challenge, all pigs in the unvaccinated control group C experienced body temperatures exceeding 41 °C, with temperatures above 40.5 °C persisting for 4 days. In contrast, in groups A and B, the temperatures in individual pigs exceeded 40 °C starting on day 3, though this elevation lasted for less than 48 h. Between days 2 and 6 post-challenge, the average rectal temperatures in groups A and B were significantly lower than those in the unvaccinated control group C ($p < 0.05$). Weigh-ins conducted before the challenge and 15 days afterward revealed that the average daily weight gains in the vaccinated groups A and B and the blank control group D were 1.10 ± 0.16, 1.09 ± 0.23, and 1.11 ± 0.15 kg, respectively, showing no significant variance among the groups. The unvaccinated challenge control group C had a notably lower average daily gain of 0.80 ± 0.23 kg, significantly lower than groups A, B, and D ($p < 0.05$, Figure 5). Vaccination effectively safeguards against PRV, mitigating the economic impacts associated with reduced daily weight gain owing to sickness.

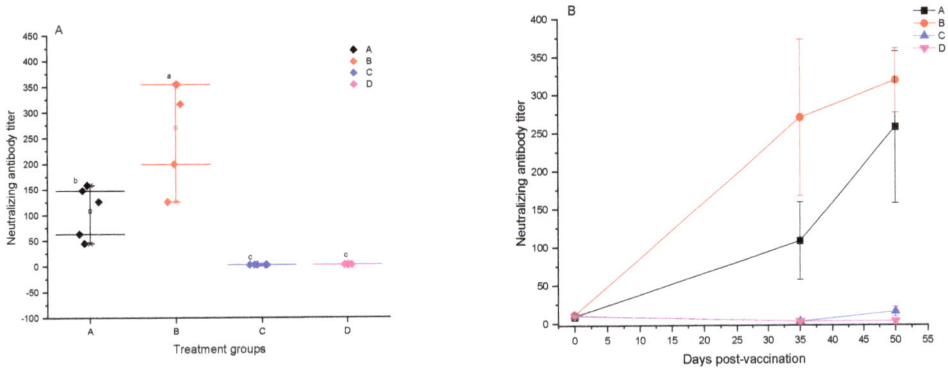


**Figure 4.** Results of testing for PRV-specific neutralizing antibodies across various experimental pig groups. (**A**) The average value ± SD of PRV-specific neutralizing antibodies measured 14 days following the second vaccination in different groups of experimental pigs. The data were analyzed by one-way ANOVA, and alphabet letters (a, b, and c) different between groups represent statistically significant difference. (**B**) The average value ± SD of PRV-specific neutralizing antibodies measured prior to vaccination, 14 days after the second vaccination, and 15 days post-challenge in different groups of experimental pigs.

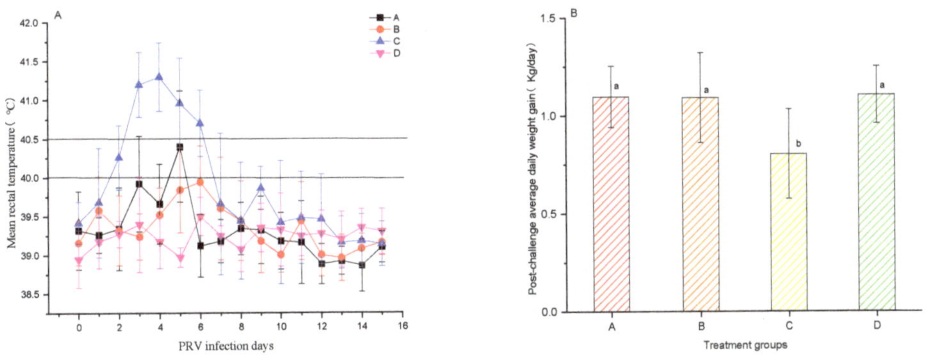


**Figure 5.** (**A**) Average rectal temperatures ± SD for various experimental pig groups at 15 days following exposure to the challenge agent. (**B**) Average daily weight gain ± SD for various experimental pig groups at 15 days post-challenge. The data were analyzed by one-way ANOVA, and alphabet letters (a and b) different between groups represent statistically significant difference.

### *3.5. Detection of Viral Load Post-PRV Challenge*

Following the challenge with PRV, the experimental pigs had their blood, throat swabs, and anal swabs collected as scheduled, and the viral load was determined by quantitative RT-PCR. All samples from the blank control group D tested negative across the board (5/5). On day 3 post-challenge, the whole blood samples from groups A, B, and C exhibited viral loads ($log10TCID_{50}$/mL) of 3.15 ± 0.70, 2.41 ± 0.79, and 4.00 ± 0.91, respectively, all testing positive, with group C showing a viral load higher than A and B ($p > 0.05$). By day 11 post-challenge, the viral loads ($log10TCID_{50}$/mL) in the blood samples of the vaccinated groups A and B were 0.85 ± 0.05 and 1.24 ± 0.25, respectively, significantly lower than the 2.53 ± 0.94 observed in group C ($p < 0.05$). Fifteen days post-challenge, the vaccinated groups A and B showed average viral loads (log10TCID$_{50}$/mL) in their blood samples of 0.31 and 0.21, respectively, with the challenged control group C recording an average viral load of 2.26 ± 0.54, demonstrating significantly lower levels in the vaccinated groups compared to the control ($p < 0.05$) (Figure 6A); the turning negative rates for the vaccinated groups A and B were 4/5 and 4/5, respectively, compared to 0/5 in the challenge control group C. The viral loads detected in throat swab samples showed a similar trend to those in blood samples, with significantly lower levels in vaccinated groups A and B than in group C on days 11 and 15 post-challenge ($p < 0.05$) (Figure 6B). By day 15 post-challenge, the ratio of throat swab samples turning negative was 2/5 and 4/5 for groups A and B, respectively, exceeding the 0/5 in group C. On day 3 post-challenge, the anal swab samples from groups A and B had average viral loads (log10TCID$_{50}$/mL) of 1.18 ± 0.35 and 1.35 ± 0.31, respectively, lower than group C's 3.09 ± 0.87 ($p > 0.05$). By day 11 post-challenge, the proportions of anal swab samples that turned negative were 4/5, 3/5, and 3/5 in groups A, B, and C, respectively. On day 15 post-challenge, all anal swab samples from the challenged pigs tested negative (Figure 6C). These findings demonstrate that immunization with the PCV2 and PRV bivalent subunit vaccines, in comparison to the non-vaccinated group C, effectively reduced the PRV load in the blood and excretion of pigs in groups A and B, thereby shortening the duration of viral shedding post-infection.

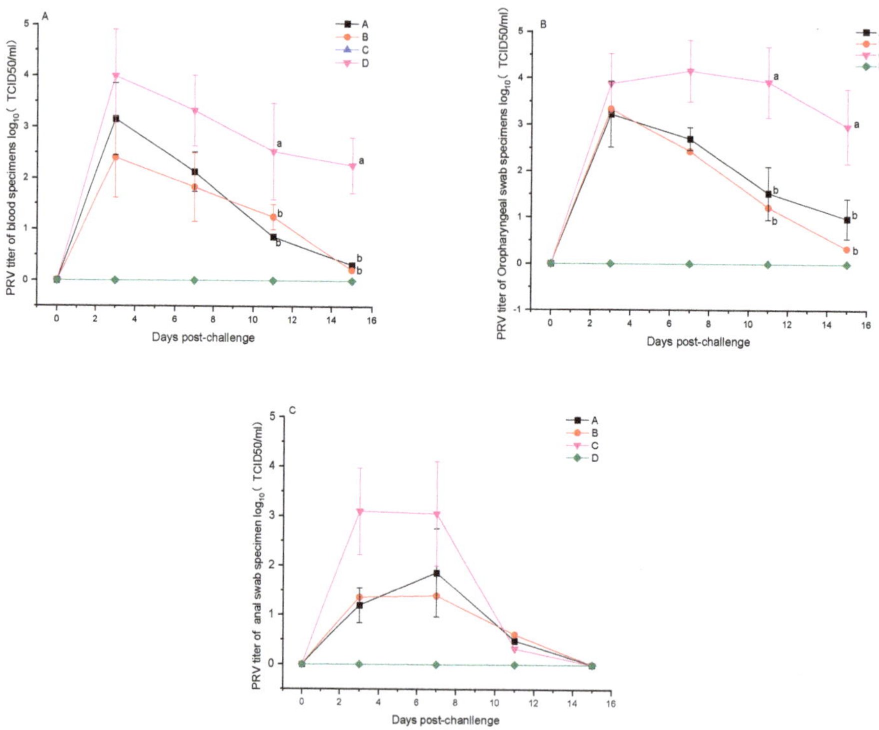


**Figure 6.** Viral load testing results for blood, throat swab, and anal swab samples from experimental pigs in groups A, B, C, and D before the PRV challenge and 15 days afterward for the five tests. (**A**) Average PRV viral loads ± SD in five whole blood samples from pigs for the four groups. (**B**) Average PRV viral loads ± SD in five throat swab samples from pigs for the four groups. (**C**) Average PRV viral loads ± SD in five anal swab samples from pigs for the four groups. For A and B, the data were analyzed by one-way ANOVA, and alphabet letters (a and b) different between groups represent statistically significant difference.

### 3.6. Post-PRV Challenge Clinical Signs and Pathological Examination

Fifteen days after the PRV challenge, clinical observations were made regarding temperature fluctuations and respiratory and neurological symptoms, including a compilation of disease incidence data for pigs in groups A, B, C, and D (Table 3). All pigs in the unvaccinated challenge control group C (5/5) exhibited disease symptoms, whereas none (0/5) in the vaccinated groups A and B showed any signs of illness. When increased to 70–80 days of age, pigs in group C, challenged with the genotype 2 variant strain of PRV (ZJM-1), developed signs indicative of interstitial pneumonia and exhibited high fever, poor mental condition, and, in some cases, tremors and neurological symptoms. These pigs also exhibited food intake decreases and impaired weight gain. However, pigs that were not affected by the disease exhibited a resolution of their clinical signs and regained their normal health. Conversely, pigs in the vaccinated groups A and B exhibited no significant clinical signs of pseudorabies infection.

**Table 3.** Summary of clinical signs and disease incidence of the pigs in different experimental pig groups for 15 days post-PRV challenge.

| Group | Number | Virus Titer ($TCID_{50}$/mL) | Challenge Dose (mL)/Challenge Route | Clinical Signs | | | Morbidity | Mortality |
|---|---|---|---|---|---|---|---|---|
| | | | | Recital Temperature (>40.5 °C) | Nervous Signs | Respiratory Signs | | |
| A | 5 | $10^{6.0}$ | 1/IM + 1/IN | 0/5 | 0/5 | 0/5 | 0/5 | 0/5 |
| B | 5 | $10^{6.0}$ | 1/IM + 1/IN | 0/5 | 0/5 | 0/5 | 0/5 | 0/5 |
| C | 5 | $10^{6.0}$ | 1/IM + 1/IN | 5/5 | 2/5 | 5/5 | 5/5 | 0/5 |
| D | 5 | None | None | 0/5 | 0/5 | 0/5 | 0/5 | 0/5 |

By day 15 post-challenge, all pigs in the challenge groups had their clinical signs resolved. A comprehensive pathological examination revealed no significant lesions in the lungs, brain, spleen, or inguinal lymph nodes of any of the experimental pigs. Tissues from the lungs, brain, and inguinal lymph nodes were fixed and processed for hematoxylin- and eosin-stained histological examination. The examination of brain and inguinal lymph node sections from all pigs in the challenge groups showed no significant signs of inflammatory cell infiltration or normal tissue structures. Lung sections from all pigs in the challenge groups exhibited varying degrees of alveolar interstitial widening compared with the blank control group D. Notably, the unvaccinated challenge group C exhibited more severe alveolar interstitial widening and more pronounced interstitial pneumonia than the vaccinated challenge groups A and B (HE staining at 100× magnification, Figure 7).

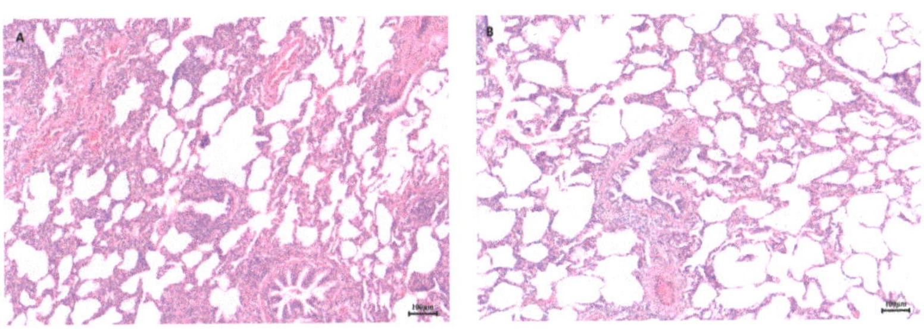


**Figure 7.** *Cont.*

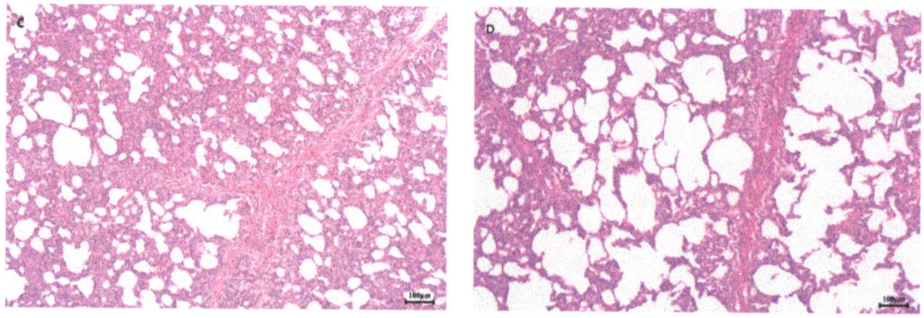

**Pulmonary interstitial thickness:**

| | Group | Nr of Pigs | pulmonary interstitial thickness(μm) |
|---|---|---|---|
| | A | 5 | 42.38±12.13[b] |
| Infected Pigs | B | 5 | 39.73±12.02[b] |
| | C | 5 | 70.20±16.89[a] |
| Uninfected Pigs | D | 5 | 34.51±6.986[b] |

**Figure 7.** Hematoxylin- and eosin-stained lung tissue sections from groups (**A–D**) of experimental pigs, 15 days post-PRV challenge. The lung sections are magnified by 100× and are represented as images (**A–D**) in the figure, representing groups (**A–D**). Compared to the blank control group (**D**), the lung tissues of the vaccinated groups (**A**,**B**) show a slight increase in the width of the alveolar interstitium. The most severe interstitial widening was observed in the unvaccinated group (**C**). The data were analyzed by one-way ANOVA, and alphabet letters (a and b) different between groups represent statistically significant difference.

## 4. Discussion

In the context of DNA viral infections, immune cells can recognize viral DNA through pattern recognition receptors, such as Toll-like receptor 9 (TLR9), cyclic GMP-AMP synthase (cGAS), and the DNA-dependent activator of IRFs (DAI). These receptors initiate a cascade of signaling pathways that result in the expression of type I interferons (IFNs) and pro-inflammatory cytokines, thereby influencing a specific immune response [13]. Viral proteins interact with host cell molecules; for example, PRV proteins like UL13, UL24, UL42 [14], and gE inhibit the NF-kB and cGAS-STING signaling pathways, suppress the innate immune response, and promote an environment conducive to viral replication within infected pig cells. Although no direct correlation exists between the strength of innate immune responses and the magnitude of antigen-specific immune responses, activating innate immunity to a certain threshold is essential for enhancing effective antigen-specific immune responses. Compared with the intricate immune responses triggered by viral infections, subunit antigens have a relatively limited capacity to activate innate immune responses. Immunomodulating adjuvants such as FLICd can activate dendritic cells, thereby enhancing specific B cell-mediated humoral immune responses [9,11]. Vaccination of groups A and B with PCV2 and PRV bivalent subunit vaccines demonstrated that pigs in group B, vaccinated with the FLICd adjuvant-enhanced vaccine, exhibited significantly higher mean neutralizing antibody titers against PRV (ZJM-1 strain) on day 14 post-second vaccination than those in group A ($p < 0.05$). Group B showed higher mean ELISA OD values for PCV2-specific IgG antibodies than group A, although the difference was not statistically significant ($p > 0.05$). This study indicates that FLICd can enhance antibody levels in vaccinated pigs and is a potential immunomodulatory adjuvant for subunit vaccines (Figure 8).

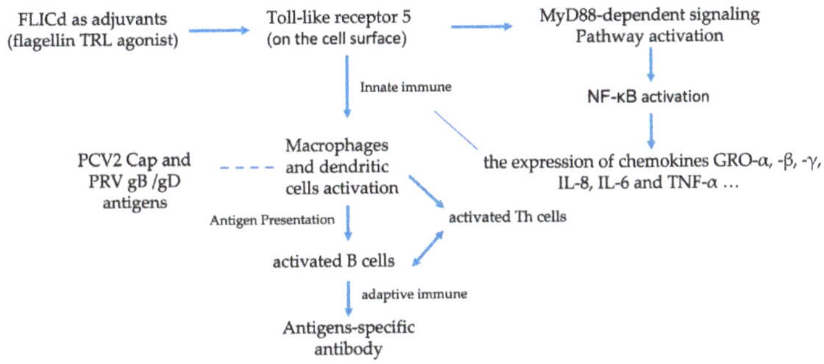


**Figure 8.** The recombinant flagellin TLR agonist FLICd binds to TLR5 on the surface of different cell types (macrophages, dendritic cells, lymphocytes, etc.). By activating the MyD88-dependent signaling pathway and the NF-κB signaling pathway, FLICd induces the expression of a series of cytokines, such as IL-6 and IL-8, and activates the innate immune cells. The activated macrophages and dendritic cells present vaccine PCV2 and PRV antigen proteins to B cells and helper somatic cells (CD4) Th cells, further activating effector B cells to express antibodies to PCV2- and PRV-specific antigens.

Rectal temperatures were recorded for seven days before primary vaccination and after vaccination with the PCV2 and PRV bivalent subunit vaccine, with no significant differences among groups A, B, C, and D ($p > 0.05$) and no cases of temperatures consistently exceeding 40 °C twice. After immunization, groups A and B exhibited no adverse reactions, such as the swelling of the injection site. Weights were measured before vaccination and on day 14 post-booster vaccination, and no significant difference was observed in the average daily weight gain between vaccinated groups A, B, C, and D ($p > 0.05$). The results showed that the PCV2 and PRV bivalent subunit vaccines were safe for pigs.

Eight genotypes of PCV2 strains [15], each with varying virulence levels, have been identified. Some strains cause high viremia and severe pathological damage [16]. Pigs were administered the PCV2 vaccine-induced neutralizing antibodies, which showed higher titers against the vaccine PCV2 strain genotype than that against other genotypes; however, no significant differences were observed, indicating cross-protection against different PCV2 genotypes [17,18]. A positive correlation was observed between the PCV2-specific IgG antibody levels, neutralizing antibody levels [19,20], and protection against PCV2 challenge [21]. Therefore, we did not perform a PCV2 challenge efficacy study. All pigs in groups A and B tested positive for PCV2-specific IgM antibodies on day 14 post-vaccination, and some pigs tested positive for IgM antibodies on day 29 post-second vaccination, indicating a robust, specific immune response to the PCV2 and PRV bivalent subunit vaccine. Although all piglets had high titer maternal antibodies, vaccinated groups A and B exhibited significantly higher PCV2-specific ELISA IgG antibodies than the unvaccinated control groups C and D ($p < 0.05$) on day 29 post-vaccination. All vaccinated piglets grown to 70–80 days of age had high titers of PCV2-specific antibodies that protected them against PCV2.

Reasonable concordance existed between PRV gB-specific antibodies detected by ELISA and PRV-neutralizing antibodies in vaccinated groups A and B, with similar trends but no linear relationship. Although there were similar average ELISA S/P values for gB-specific antibodies using ELISA kits across groups A, B, and C, the neutralizing antibody titer of group C was significantly lower than that of the vaccinated groups post-PRV challenge. These results suggest that ELISA cannot replace the evaluation of PRV-protective neutralizing antibodies. The positive rate of PRV gI-specific antibodies detected by ELISA on day 7 post-PRV challenge aligned with the rate of viral load in all samples, proving to be an effective method for evaluating wild virus infection. The challenge strain of PRV

(ZJM-1), a genotype 2 variant strain identified in 2013, demonstrated a mortality rate of 3/5 and a disease rate of 5/5 when used to challenge pigs aged 50–60 days using a similar method as in this experiment. All vaccinated groups showed 0/5 disease incidence during this challenge, and the challenged control group had 5/5 disease incidence; however, all challenged pigs survived. Studies indicate that as pigs age to 5 weeks, the viral load of PRV (Kaplan strain) infection significantly decreases in the central nervous system, such as the cerebellum and thalamus [22], as well as in peripheral nerves, such as the trigeminal nerve, which may explain the mild tremor neurological symptoms exhibited by only 2/5 pigs in this study. The experimental pigs (aged 70–80 days) exhibited clinical signs but no fatalities and respiratory signs recovered on day 15 post-challenge, possibly due to the absence of a secondary infection. All challenged pigs were euthanized 15 days post-challenge and showed no significant pathological changes in organs such as the brain, lungs, liver, kidneys, or inguinal lymph nodes. Lung pathology sections (HE staining) revealed slight interstitial widening in groups A and B, with the unvaccinated challenge control group C showing more pronounced widening, indicating that severe interstitial pneumonia occurred post-challenge, and these results showed that the vaccine provided effective protection. One week post-challenge, pigs in the challenge control group C experienced high fever, poor mental state, reduced food intake, and respiratory signs such as coughing and respiratory rapidity, likely contributing to reduced weight gain. On day 3 post-challenge, the viral load in blood samples from all challenge groups peaked and gradually decreased thereafter, with the positivity rates of blood samples in groups A and B on day 15 being 1/5 and 1/5, respectively, whereas the challenge control group C remained at 5/5 positive. Vaccination with the PCV2 and PRV subunit vaccines induced high-titer PRV-specific neutralizing antibodies, reducing the viral load of PRV (ZJM-1 strain) in the blood post-challenge infection, as well as reducing the viral load spread through the mouth and feces, shortening the shedding time, effectively preventing PRV disease, and reducing pig weight loss due to disease. The emergence of PRV variants increases the pathogenic virulence in pigs and has led to diseases in pigs vaccinated with the Bartha-k61 attenuated vaccine [23–25]. However, vaccination with the PRV Bartha strain-attenuated vaccine can provide 60% or complete protection against PRV variants [24,26,27], which corresponds to its lower specific neutralizing antibody titers against variant strains than classical strains. The neutralizing antibody titers produced by the variant strain-attenuated vaccine against the classical PRV strain (LA strain) were similar to those of the Bartha-K61 vaccine but significantly higher against PRV variant strains than against non-variant strains (Bartha-K61, HB2000, and SA215), effectively protecting against variant and classical strains in experimental pigs [26,28]. This indicates that PRV remains a single serotype and that improving the titer and affinity of neutralizing antibodies is crucial for PRV disease prevention and epidemic control. High-titer maternal antibody piglets vaccinated with the PCV2 and PRV bivalent subunit vaccines had a 100% positive rate for PCV2- and PRV-specific antibodies compared to the unvaccinated control group. This vaccination strategy effectively avoided maternal antibody interference and minimized the risk of PCV2 and PRV infection associated with delayed vaccination schedules. PCV2 and PRV bivalent subunit vaccines are safe, can effectively prevent PCV2 and PRV diseases, and can reduce the economic loss caused by respiratory diseases due to secondary mixed bacterial infections in high-density pig farming.

**Author Contributions:** Conceptualization, F.Y., X.L. and X.S.; methodology, F.Y. and W.X.; software, F.Y. and W.X.; validation, F.Y., W.X. and Y.L.; resources, W.O., X.S. and W.X.; data curation, F.Y. and W.X.; writing—original draft preparation, F.Y. and W.X.; writing—review and editing, X.L., F.Y. and W.X.; visualization, X.L., F.Y. and W.X., supervision, X.S., W.O. and X.L.; project administration, X.S.

and W.O.; funding acquisition, W.O. and X.S. All authors have read and agreed to the published version of the manuscript.

**Funding:** This work was financially supported by the High-level Talents Special Support Plan of Zhejiang Province (2021R52041), the Zhejiang Provincial Agricultural Major Technology Collaborative Promotion Program (2023ZDXT14), and the "Pioneer" and "Leading Goose" R&D Program of Zhejiang (No.2023C02023).

**Institutional Review Board Statement:** The animal study protocol was approved by the Ethics Committee of Laboratory Animal Welfare and Ethics Committee of Zhejiang University (ZJU20230399, 24 October 2023).

**Informed Consent Statement:** Not applicable.

**Data Availability Statement:** The raw data supporting the conclusions of this article will be made available by the authors upon request.

**Acknowledgments:** The authors would like to thank Jinhua Municipal Government for the policy support on this study.

**Conflicts of Interest:** Author Fulai Yu, Wei Xiang, Yang Li and Xinbiao Shu were employed by the company Zhejiang Dovro Animal Health Products Co., Ltd. Weiye Ou was employed by the company Zhejiang MEBOLO Biotechnology Co., Ltd. The remaining authors declare that the research was conducted in the absence of any commercial or financial relationships that could be construed as a potential conflict of interest.

## References

1. Alarcon, P.; Rushton, J.; Wieland, B. Cost of post-weaning multi-systemic wasting syndrome and porcine circovirus type-2 subclinical infection in England—An economic disease model. *Prev. Vet. Med.* **2013**, *110*, 88–102. [CrossRef]
2. Shi, R.; Hou, L.; Liu, J. Host immune response to infection with porcine circoviruses. *Anim. Dis.* **2021**, *1*, 23. [CrossRef]
3. An, T.Q.; Peng, J.M.; Tian, Z.J.; Zhao, H.Y.; Li, N.; Liu, Y.M.; Chen, J.Z.; Leng, C.L.; Sun, Y.; Chang, D.; et al. Pseudorabies virus variant in Bartha-K61–Vaccinated pigs, China, 2012. *Emerg. Infect. Dis.* **2013**, *19*, 1749–1755. [CrossRef] [PubMed]
4. Chen, S.; Li, X.; Zhang, X.; Niu, G.; Yang, L.; Ji, W.; Zhang, L.; Ren, L. PCV2 and PRV coinfection induces endoplasmic reticulum stress via PERK-eIF2α-ATF4-CHOP and IRE1-XBP1-EDEM pathways. *Int. J. Mol. Sci.* **2022**, *23*, 4479. [CrossRef] [PubMed]
5. Li, X.; Chen, S.; Zhang, L.; Niu, G.; Zhang, X.; Yang, L.; Ji, W.; Ren, L. Coinfection of porcine circovirus 2 and pseudorabies virus enhances immunosuppression and inflammation through NF-κB, JAK/STAT, MAPK, and NLRP3 pathways. *Int. J. Mol. Sci.* **2022**, *23*, 4469. [CrossRef] [PubMed]
6. Yu, X.; Sun, Q.; Ku, X.; He, D.; Li, Z.; Ghonaim, A.H.; Fan, S.; He, Q. The epidemiological investigation of co-infection of major respiratory bacteria with pseudorabies virus in intensive pig farms in China. *Vet. Med. Sci.* **2021**, *7*, 175–183. [CrossRef] [PubMed]
7. Alarcon, P.; Rushton, J.; Nathues, H.; Wieland, B. Economic efficiency analysis of different strategies to control post-weaning multi-systemic wasting syndrome and porcine circovirus type 2 subclinical infection in 3-weekly batch system farms. *Prev. Vet. Med.* **2013**, *110*, 103–118. [CrossRef]
8. Bowie, A.G.; Unterholzner, L. Viral evasion and subversion of pattern-recognition receptor signalling. *Nat. Rev. Immunol.* **2008**, *8*, 911–922. [CrossRef]
9. Mizel, S.B.; Bates, J.T. Flagellin as an adjuvant: Cellular mechanisms and potential. *J. Immunol.* **2010**, *185*, 5677–5682. [CrossRef]
10. Sokol, O.O.; Nikitin, N.A.; Evtushenko, E.A.; Karpova, O.V.; Matveeva, I.N.; Gryn, S.A.; Popova, V.M.; Ivanov, I.V.; Fedorov, Y.N.; Litenkova, I.Y. Protective activity of inactivated rabies vaccine using flagellin-based adjuvant. *Biochemistry* **2024**, *89*, 574–582. [CrossRef] [PubMed]
11. Smith, K.D.; Andersen-Nissen, E.; Hayashi, F.; Strobe, K.; Bergman, M.A.; Barrett, S.L.R.; Cookson, B.T.; Aderem, A. Toll-like receptor 5 recognizes a conserved site on flagellin required for protofilament formation and bacterial motility. *Nat. Immunol.* **2003**, *4*, 1247–1253. [CrossRef] [PubMed]
12. Yoon, S.I.; Kurnasov, O.; Natarajan, V.; Hong, M.; Gudkov, A.V.; Osterman, A.L.; Wilson, I.A. Structural basis of TLR5-flagellin recognition and signaling. *Science* **2012**, *335*, 859–864. [CrossRef]
13. Zahid, A.; Ismail, H.; Li, B.; Jin, T. Molecular and structural basis of DNA sensors in antiviral innate immunity. *Front. Immunol.* **2020**, *11*, 613039. [CrossRef] [PubMed]
14. Ye, G.; Liu, H.; Zhou, Q.; Liu, X.; Huang, L.; Weng, C. A tug of war: Pseudorabies virus and host antiviral innate immunity. *Viruses* **2022**, *14*, 547. [CrossRef] [PubMed]

15. Franzo, G.; Segalés, J. Porcine circovirus 2 (PCV-2) genotype update and proposal of a new genotyping methodology. *PLoS ONE* **2018**, *13*, e0208585. [CrossRef] [PubMed]
16. Cho, H.; Oh, T.; Suh, J.; Chae, C. Comparison of pathogenicity of 4 porcine circovirus type 2 (PCV2) genotypes (2a, 2b, 2d, and 2e) in experimentally infected pigs. *J. Vet. Med. Sci.* **2023**, *85*, 83–87. [CrossRef]
17. Opriessnig, T.; Xiao, C.T.; Halbur, P.G.; Gerber, P.F.; Matzinger, S.R.; Meng, X.J. A commercial porcine circovirus (PCV) type 2a-based vaccine reduces PCV2d viremia and shedding and prevents PCV2d transmission to naïve pigs under experimental conditions. *Vaccine* **2017**, *35*, 248–254. [CrossRef]
18. Yu, C.; Cao, M.; Wei, Y.; Liu, J.; Zhang, H.; Liu, C.; Feng, L.; Huang, L. Evaluation of cross-immunity among major porcine circovirus type 2 genotypes by infection with PCV2b and PCV2d circulating strains. *Vet. Microbiol.* **2023**, *283*, 109796. [CrossRef] [PubMed]
19. Pileri, E.; Cortey, M.; Rodríguez, F.; Sibila, M.; Fraile, L.; Segalés, J. Comparison of the immunoperoxidase monolayer assay and three commercial ELISAs for detection of antibodies against porcine circovirus type 2. *Vet. J.* **2014**, *201*, 429–432. [CrossRef] [PubMed]
20. Oliver-Ferrando, S.; Segalés, J.; López-Soria, S.; Callén, A.; Merdy, O.; Joisel, F.; Sibila, M. Evaluation of natural porcine circovirus type 2 (PCV2) subclinical infection and seroconversion dynamics in piglets vaccinated at different ages. *Vet. Res.* **2016**, *47*, 121. [CrossRef] [PubMed]
21. Kiss, I.; Szigeti, K.; Homonnay, Z.G.; Tamás, V.; Smits, H.; Krejci, R. Maternally derived antibody levels influence on vaccine protection against PCV2d challenge. *Animals* **2021**, *11*, 2231. [CrossRef] [PubMed]
22. Papageorgiou, K.; Grivas, I.; Chiotelli, M.; Theodoridis, A.; Panteris, E.; Papadopoulos, D.; Petridou, E.; Papaioannou, N.; Nauwynck, H.; Kritas, S.K. Age-dependent invasion of pseudorabies virus into porcine central nervous system via maxillary nerve. *Pathogens* **2022**, *11*, 157. [CrossRef] [PubMed]
23. Luo, Y.; Li, N.; Cong, X.; Wang, C.H.; Du, M.; Li, L.; Zhao, B.; Yuan, J.; Liu, D.D.; Li, S.; et al. Pathogenicity and genomic characterization of a pseudorabies virus variant isolated from Bartha-K61-vaccinated swine population in China. *Vet. Microbiol.* **2014**, *174*, 107–115. [CrossRef] [PubMed]
24. Tong, W.; Liu, F.; Zheng, H.; Liang, C.; Zhou, Y.J.; Jiang, Y.F.; Shan, T.L.; Gao, F.; Li, G.X.; Tong, G.Z. Emergence of a Pseudorabies virus variant with increased virulence to piglets. *Vet. Microbiol.* **2015**, *181*, 236–240. [CrossRef] [PubMed]
25. Bo, Z.; Miao, Y.; Xi, R.; Gao, X.; Miao, D.; Chen, H.; Jung, Y.S.; Qian, Y.; Dai, J. Emergence of a novel pathogenic recombinant virus from Bartha vaccine and variant pseudorabies virus in China. *Transbound. Emerg. Dis.* **2021**, *68*, 1454–1464. [CrossRef] [PubMed]
26. Jiang, C.; Ma, Z.; Bai, J.; Sun, Y.; Cao, M.; Wang, X.; Jiang, P.; Liu, X. Comparison of the protective efficacy between the candidate vaccine ZJ01R carrying gE/gI/TK deletion and three commercial vaccines against an emerging pseudorabies virus variant. *Vet. Microbiol.* **2023**, *276*, 109623. [CrossRef]
27. Zhou, J.; Li, S.; Wang, X.; Zou, M.; Gao, S. Bartha-k61 vaccine protects growing pigs against challenge with an emerging variant pseudorabies virus. *Vaccine* **2017**, *35*, 1161–1166. [CrossRef] [PubMed]
28. Sun, Y.; Zhao, L.; Fu, Z.F. Effective cross-protection of a lyophilized live gE/gI/TK-deleted pseudorabies virus (PRV) vaccine against classical and variant PRV challenges. *Vet. Microbiol.* **2022**, *267*, 109387. [CrossRef]

MDPI AG
Grosspeteranlage 5
4052 Basel
Switzerland
Tel.: +41 61 683 77 34

*Veterinary Sciences* Editorial Office
E-mail: vetsci@mdpi.com
www.mdpi.com/journal/vetsci

www.ingramcontent.com/pod-product-compliance
Lightning Source LLC
Chambersburg PA
CBHW040810250926
51850CB00007BA/1128